DOCUMENTS FROM
F. TAYLOR OSTRANDER

RESEARCH IN THE HISTORY OF ECONOMIC THOUGHT AND METHODOLOGY

Series Editors: Warren J. Samuels, Jeff E. Biddle and Ross B. Emmett

RESEARCH IN THE HISTORY OF ECONOMIC THOUGHT AND
METHODOLOGY   VOLUME 23-B

DOCUMENTS FROM F. TAYLOR OSTRANDER

EDITED BY

WARREN J. SAMUELS

*Department of Economics, Michigan State University,
East Lansing, MI 48824, USA*

2005

ELSEVIER
JAI

Amsterdam – Boston – Heidelberg – London – New York – Oxford
Paris – San Diego – San Francisco – Singapore – Sydney – Tokyo

ELSEVIER B.V.	ELSEVIER Inc.	**ELSEVIER Ltd**	ELSEVIER Ltd
Radarweg 29	525 B Street, Suite 1900	**The Boulevard, Langford**	84 Theobalds Road
P.O. Box 211	San Diego	**Lane, Kidlington**	London
1000 AE Amsterdam	CA 92101-4495	**Oxford OX5 1GB**	WC1X 8RR
The Netherlands	USA	**UK**	UK

First edition 2005

Library of Congress Cataloging in Publication Data
A catalog record is available from the Library of Congress.

British Library Cataloguing in Publication Data
A catalogue record is available from the British Library.

ISBN: 0-7623-1165-7
ISSN: 0743-4154 (Series)

♾ The paper used in this publication meets the requirements of ANSI/NISO Z39.48-1992 (Permanence of Paper). Printed in The Netherlands.

Working together to grow
libraries in developing countries

www.elsevier.com | www.bookaid.org | www.sabre.org

ELSEVIER   BOOK AID International   Sabre Foundation

CONTENTS

COURSES FROM MELCHIOR PALYI

LIST OF CONTRIBUTORS

Kirk D. Johnson   Dover, DE, USA

F. Taylor Ostrander   Williamstown, MA, USA

Warren J. Samuels   Department of Economics, Michigan State
                           University, MI, USA

COURSES FROM FRANK H. KNIGHT

NOTES AND OTHER MATERIALS FROM FRANK H. KNIGHT'S COURSE, ECONOMIC THEORY, ECONOMICS 301, UNIVERSITY OF CHICAGO, FALL 1933, INCLUDING F. TAYLOR OSTRANDER'S TERM PAPER "THE MEANING OF COST" PREPARED FOR FRANK H. KNIGHT'S COURSE IN ECONOMIC THEORY, ECONOMICS 301, UNIVERSITY OF CHICAGO, FALL 1933

Edited by Warren J. Samuels

The course in economic theory has been central to the graduate curriculum in the University of Chicago Economics Department – then and to the present day. F. Taylor Ostrander took the course from Frank Knight during the autumn quarter at Chicago in 1933.

Taught by Knight or Jacob Viner, the course was primarily Marshallian price theory modified or supplemented, as it were, by the ideas of other economists and

Documents from F. Taylor Ostrander
Research in the History of Economic Thought and Methodology, Volume 23-B, 3–86
Copyright © 2005 by Elsevier Ltd.
All rights of reproduction in any form reserved
ISSN: 0743-4154/doi:10.1016/S0743-4154(05)23101-2

of Knight and/or Viner. These others included early modern economists, especially Adam Smith and John Stuart Mill; turn-of-the century economists, including Eugen von Böhm-Bawerk, John E. Cairnes, J. B. Clark, F. W. Taussig and F. A. Walker; early twentieth century economists, such as B. M. Anderson, Thomas Nixon Carver, John Davenport, and Irving Fisher, plus several men now largely seen as more or less heterodox: Henry George, Walton H. Hamilton, and A. B. Wolfe, as well as others appearing on the Assignment list and in the notes. Recitation of these names tells the reader something of both the ideas that constituted "Economic Theory" in the early 1930s and the range of recognized contributors to the field.

The materials from the course include: (1) the printed list of assignments supplemented by names given orally by Knight and written down by Ostrander; (2) Ostrander's class notes, which run to 78 pages; (3) his term paper, "The Meaning of Cost"; (4) his notes on readings; and (5) his mid-term examination book. Only (1), (2) and (3) are published here. The materials have been minimally edited. When a modicum of doubt exists about an abbreviation, the completion is placed in square brackets, which are also used to signal editorial intrusions. In the assignment list, "Mim" undoubtedly is an abbreviation for "Mimeographed." Ostrander has, with help, proofread my transcription against his original hand-written notes.

Near the end of this Introduction, I write, "The lectures pale in comparison with his essays and articles, which are better thought out..." Ostrander says, in response, "Perhaps not! Perhaps it was F.T.O. who was more struck by his iconoclastic comments than by his good points on economic theory!" Of course, he is correct; he wrote the notes, I was not there. But I do not see iconoclasm interfering with the development of theory. Still, I do not want to make much of this, nor, for the reasons just given, can I.

Ostrander must have mastered economic theory as he combined his studies at Williams, Oxford and Chicago. Paul Samuelson, who, like Ostrander, worked at the War Production Board (W.P.A.) during the war, told me that the question of who was the better economist, Ostrander or James S. Earley was a topic of discussion (Notes from Earley's course at the University of Wisconsin during 1954–1955 on the history of economic thought and a biographical sketch appear in Archival Volume 21-B (2003). Notes from his courses on microeconomics and on monetary and macroeconomics will appear in future volumes).

The following are the editor's comments, musings, underscorings, efforts at historical perspective, and critique of what Knight is recorded by Ostrander to have said. The topics are primarily in the order in which they appear in the notes.

Knight certainly made his students think about economic theory, its foundations, its substantive content, and its nuances. Knight was prepared to criticize something about every element of economic theory but one, namely, the idea of the price system itself. In retrospect, the course was idiosyncratic and discursive, decidedly not the type of theory course one would have expected given the technical literature then extant (and subsequently made much of), but one surely congruent with what is known about Knight's technical and philosophical interests.

The economic theory that Ostrander mastered was pure Frank Knight. The first line of the course notes reads:

Getting rid of "original sin" – romanticism in economics.

And continues,

- Economics is mainly a problem of morals and philosophy.
- To get the proper, <u>objective</u> point of view.
- Change in name from <u>political economy</u> to economics was a most unfortunate one, especially in this day.
- There can not be a <u>science</u> of economics.
- Wertfrei (free of value judgement) is impossible.

There are enough tensions, if not conflicts, within these lines to require a major treatise to discuss fully. Knight explored many of them in dozens of articles and a book or two.

Knight's conception of economics does not exclude economics as either the pure logic of choice (constrained maximization) or Robbins's scarcity definition but it certainly includes 19th century liberal political ideology as its philosophic base. Price theory, he urges, must be understood against the backdrop of what he takes liberalism to mean.

Exceptionally candid and dramatic is the recorded statement, "All economics is <u>basically</u> political propaganda." Knight clearly sees economics as an emanation from a particular culture, whose simultaneous explanation and legitimation it pursues. Apparently also candid and dramatic, but unclear, is the statement "Most economic theory belongs in the realm of <u>social abnormal psychology</u>."

Knight is quoted saying, "Pioneering will be done if it pays a higher return than settling on old land." This may be compared with Adam Smith's discussion of how investment abroad may be limited by the investor's preference for security and certitude, presumably whatever the expected rate of profit, or return (The unintended consequence is an increase in domestic power – brought about by the invisible hand. Adam Smith, *The Wealth of Nations*, Book IV, Chapter II. Smith is unclear as to the nature of any tradeoff between return and security).

The language, "the instructor's belief that the whole of liberal education is encompassed in his course!," is pure Knight, even if it comes in an insertion by Ostrander.

Striking is the degree to which Knight juxtaposes his ideas and positions on issues to those he identifies being held by "radicals."

Also striking is Knight's emphasis on deliberative behavior, especially in light of Friedrich von Hayek's later emphasis on spontaneous order and the principle of unintended and unforeseen consequences, i.e. rationality versus the nondeliberative. The two are not incompatible, though the propaganda element in Hayek stresses the nondeliberative (More on this below).

More striking is Knight's cynical claim about the origin of democracy: "Democracy was, in origin, a device for state borrowing at a lower rate of interest – borrowing by an immortal society rather than by a mortal king."

More impressive is his remark, "Marx says history is the class war; spends the rest of 2400 pages belaboring people for not living up to, and acting as, their class, i.e. he merely wanted history to be a class war."

Knight discusses goods, wants, want satisfaction, and goods as services, in such a way as to suggest anticipation of the characteristics approach to demand. Yet in discussing "rational budgeting," he claims, "The classification of items is arbitrary. You can't draw up lists of commodities, the whole problem is an abstract one – even wheat's classification is arbitrary." It is curious, nonetheless, that he devoted so much time and effort to the distinctions between goods and services and between money and the goods acquired with money.

The seemingly unusual attention given to questions of capital and related topics reflects, in part, Knight's early preoccupation with and participation in the capital controversy of his day.

The notes have Knight raising the fundamental problem of reflexivity: "Knight wonders how any critical objective thinking is possible about society by a person in that society – lifting one's self by his bootstraps!"

Knight is recorded as having said, "Thus the payment for personal productive capacity is wages and is slightly (institutionally) different from payment for property productive capacity." Surely he does not mean that property, and thus rent and interest, is not a matter of institutions, only that the institutional arrangements are slightly different.

One of the remarkable passages in the notes is the following:

> – Persons: self-service is essentially outside the economic system, yet it can not be set off to any great degree – must be treated arbitrarily [In margin alongside: –Even if what is self-service to one person, is sold in the market by another – that market price can not be carried over to the self-services – are matters of intention].

It is remarkable because in 1991 Ronald Coase (of the University of Chicago Law School) was awarded the Nobel Prize for Economic Science, in part, for his 1937 article on the theory of the firm in which he argued that firms make decisions whether or not to self-supply and that such decisions are important to the formation and/or evolution of markets for those services. As for Knight's marginalia, Coase argued that market price was indeed one factor in the firm's decision. Later in the notes, the interjection "(what is a firm)" is recorded in a discussion of decreasing cost).

Coase and Knight have in common, *inter alia*, a dislike of formalism. In Ostrander's notes, Knight is recorded as saying, "There is too much curve-drawing in economics! Curves drawn for one industry with its limiting conditions – can not be extended to another." Coase disparages what he calls "blackboard economics."

Also remarkable is the extent to which Knight draws distinctions and makes points intuitively and *a priori* without any evident empirical or grounded theoretical basis. One example (another is on the immediately preceding line in the notes), is this: "Logically, in an economic world there is only one productive factor: capital. All these other – are mostly legal in source – categories are arbitrary, for the sake of practical talking." Two points: (1) Capital itself, as an economic category and claim upon income, is seen as "mostly legal" in character; and (2) Knight is clearly theorizing differently from Marshall. Knight's tools are sometimes tools for analyzing actual economies. They are also conceptual devices ("Logically") for constructing and manipulating hypothetical variables in a pure abstract a-institutional conceptual economy – itself a tool (One manifestation of this is the recorded statement, "The so-called 'production process' is a confusion of economic process with the observed physical process"). Knight denies meaningful "connection between this theoretical system and 'reality.' " Knight practiced – as will be seen below – elements of both types of economics.

Another remarkable statement is, "But, though there may be constant cost, there can never be decreasing cost."

Rather odd is the reported statement, "The whole theory of capital is in the capitalization formula." Also odd is his treatment of interest requiring perpetual capital.

Knight is reported to have said of institutional economics, "Institutional (largely a revolt for sake of revolt)." His basic position on institutionalism was different and more thoughtful, if still negative: the institutionalists dealt with important questions, but not in the right way (or came up with the wrong answers), i.e. less dismissive and less sarcastic. Nonetheless, his negative view shows when he said of "cumulated change" that it was "Veblen's phrase, didn't know what it meant."

Knight's position on institutional economics is ironic (see my Introduction to Ostrander's notes from Knight's course on institutional economics, forthcoming in Volume 23-B). The Chicago School opposed much institutionalist theory and policy, defending and rendering the market system an absolute, whereas the institutionalists were critical of the capitalist-dominated form of the market system and tried to demystify it as well. The institutionalists saw regulation as a mode of changing property rights, whereas the Chicago School saw it as redistributive and inefficient, and so on. The Chicago Economics Department during the pre-World War One and interwar periods had a number of institutionalists, including Veblen. Knight and other Chicagoans paid attention in their own way to institutionalist topics (see Malcolm Rutherford, "Chicago Economics and Institutionalism" (manuscript, 2002). For example, Knight refers here to John Maurice Clark's overhead cost.

Knight is also reported to have said both that "There is no <u>objective</u> measuring stick for <u>anything</u> in economics" and that "There is <u>no established</u> economic usage for anything in economics." Two comments: (1) Knight was in the unfortunate position of desiring closure and determinacy but believing that such was impossible in this world (and this was the only world there is); and (2) If these statements, especially the second, were widely believed among the leaders of the discipline, this might help explain the efforts, starting in the early 1930s with the formation of the Econometric Society and accelerating after World War Two, to tighten economic theory and practice. Driven in part by status emulation (being seen as "scientific") and the anxieties and insecurities of the Cold War, this meant mathematical formalism, an emphasis on pure a-institutional conceptualizations, and a research protocol that mandated unique determinate optimal equilibrium solutions, as well as efforts to establish the hegemony of such a new neoclassicism to replace the amorphous interwar combination of institutionalism and neoclassicism.

Knight's candor and overall position are indicated in the following statement:

> Social <u>dogmatism</u> – <u>unconscious</u> – is source of about all political conflict today. Is there too much or too little competition for <u>the system</u> to work?
> – This does not affect the truth of "the system."

When Knight is said to deny time preference – "Why should all society have a preference for the present? Why not a preference for the future?" – he is dismissing one explanation for the interest rate and one of the determinants of its height. "A preference for the future" would, of course, imply a propensity to save; Keynes's revolutionary ideas on saving were still in the future and, moreover did not appeal to Knight – but Adam Smith's "prudence" yielded the same result.

According to Knight, "– It is impossible to represent – without a third dimension – any <u>change</u> in the <u>wants</u> and demands, or any technical advance." Neoclassical economists knew that preferences were not given and could change (More than that, "In historical, cumulative change – wants, resources, ownership, technology, institutional setting – any change in any one of these brings about a complete change in the stationary circular system"). Still, they preferred to assume given preferences and not explore further – the diagrammatics and underlying mathematics were, presumably, too complex. Knight knew better – but neither emphasized nor promoted it. The stationary state (to which the assumption of given preferences especially pertains) was easier to work with. A family (say) of demand curves, due to changing wants/preferences, did not yield unique determinate results. This position is suggested, for example in, "The economic problem is the correct allocation of resources to wants" – "*the* correct allocation," as if ontologically only one exists – and in "Rational behavior is uniquely determined behavior, it is mechanical – could be predicted from known causes," thus equating the emerging neoclassical research protocol with rational behavior. Post-War economists developed the idea of a "core" of situation-specific optimal solutions but frequently, even typically, spoke of "the" optimal solution.

Knight says,

> At equilibrium no one would be allowed to improve his position without the consent of everyone else. Or, if he does improve his condition, it is theft. No one can improve his position as against the rest, without its being theft.

He may, or may not, be anticipating – and may or may not agree with – the argument (more complex than either need be or can be developed here) that in contractarian systems "what people consent to are individual transactions, rather than the order that spontaneously emerges from them. . . . the parties consent without really knowing what they are getting into" (David Schmidtz, "Justifying the State," *Ethics*, vol. 101 (October 1990), p. 100).

The continuation of Knight's discussion is also interesting. He is recorded as saying,

> Our very discussion of this subject assumes that we assume rationality; we must distinguish sales-talk, conspiracy, oratory, from Discussion – which, to be rational, can not be an attempt to get others to act so as to benefit the discusser. It must be impartial.

This is unrealistic and ungrounded. It is also in conflict with the later Chicago School' s argument (articulated first and/or most strongly by George J. Stigler) that individuals pursue their interests in all phases of life. When not a tautology, it must be subject to testing. And it may not be self-reflexive, i.e. it may fail to recognize that Knight had his own not impartial reasons for insisting on this conception of rationality and discussion – to limit collective (read: government) action.

A manifestation of the tendency for neoclassical economics to provide a defense of capital, capitalists, and capitalism (as well as interest and profit) is Knight's argument,

Some changes are the result of human acts, human calculations.

–Only changes in capital come under this head – human calculations will be reflected in some change in human wealth.

The problem, in brief, is that all economic actors make calculated decisions with consequences for their respective levels of wealth. The relevant literature tells two different stories, one in terms of a particular social class (with more or less open entry), the other in terms of a function performed by all economic actors.

Knight also desires to confine "history" to "only the non-calculated changes in technology or resources, or wants," i.e. "to every non-calculable, non-predictable change – for every predictable change is reflected in the capital market." Knight is implicitly using his famous and important distinction between risk and uncertainty to establish a model in which much if not all the stuff of history – political, social, class, and economic (broadly defined) – is excluded from history. The distinction is supplemented by the neoclassical reasoning that "every predictable change is reflected in the capital market" – even if information and power (income, other resources, organization) are asymmetric. This conservative model building is akin to later: (1) models of public choice without public choice or with public choice deemed sub-optimal; (2) rent seeking theory that would exclude most if not all efforts to change the law legally; and (3) externality theory that engendered the dichotomy of Pareto relevant, worked out through the market and therefore irrelevant for policy, and Pareto irrelevant, without substantive efforts to work out through the market and therefore also irrelevant for policy – all through the adept construction of definitions, also with the same asymmetries. The later models were produced by a brilliant and creative student of Knight's at Chicago, James M. Buchanan, who apparently learned his lessons well.

Knight argued that "stored up stocks . . . are a dead loss to their owners." This is a mystery. Involuntary inventories (from failure to sell) aside, voluntary inventories derive from efforts to be prepared for anticipated sales. One is reminded of Alfred Marshall's publisher lecturing, in letters, the foremost British (if not world) economist, on his misunderstanding of the organization of the book-publishing and –selling industry and its policies.

Throughout the notes one finds Knight discussing economic theory less as a set of tools and more as a definition of reality. In this, he was, increasingly, not alone; equilibrium, for example, is treated by Knight and numerous modern theorists as if it could be an observable phenomenon.

Knight's position on the use of "utility," however, is that of a tool, not truth. The notes record, "Knight does not defend utility as truth – it may be merely a rationalizing concept" and "We may doubt its validity, but must use its concepts." And again, "We can't rationalize choice without a concept of total utility – yet such a concept is unreal."

Knight seems to think, like many neoclassical economists, that economic actors should think like neoclassical economists. Thus, for example, Knight complains, "People seem to consider increment of utility without realizing the necessary revaluation of all previous units that is necessary." Knight was opposed to behaviorism and had to resort to the metaphysics and mechanics of utility theory. Thus he knows "We cannot <u>measure</u> (experimentally) marginal utility. We <u>assume</u> it from consumer's activity." He "does not think it necessary to pay too much attention" to "take account of mistakes or irrationalities in consumer's activity." But he does want to use utility theory; "If we are going to theorize, we might as well <u>do</u> it."

Knight's position on consumer's surplus is both correct and problematic. The notes read in part, "Consumer's surplus is not psychologically true, yet it seems to follow <u>logically</u> from utility theory." He is correct in distinguishing validity ("follow logically") from truth. He is problematic about what is true: it depends on what the consumer knows, understands, and values ("psychologically true"). His statement, "Deductive philosophy is the non-valuation philosophy," also applies to truth: Deductive validity is not the same as either truth or desirability.

Knight seems to have advanced a somewhat different rationale for maximization than, say, Paul Samuelson. Samuelson successfully promoted the logic of constrained maximization derived in part from mathematics and in part from the widely held general view that more goods is better than fewer goods. The notes have Knight arguing, "The concept of a maximum total is <u>vital</u> to the existence of economics (which <u>need</u> not exist)" and that the "Economy can not be rationalized except in terms of a maximum utility." He makes two other points. One likely derives from his pessimism: "The ascetic ideal can not be followed and keep the notion of economy – it results in universal suicide – which <u>may</u> be desirable." The other indicates his recognition of either the shaky grounds of the rationality assumption or the reason for its use (to finesse and/or otherwise evade problems): "The inconsistency here is no more than that implied in <u>any</u> attempt to find complete rationality in life." Quite incisive, however, and consistent with the broader analysis found in his writings, is the statement, "the <u>base</u> of psychological choice is shifted with every new choice."

Knight is incredibly dismissive of "the problem of marginal utility of money for rich and poor." He wants to dismiss redistribution of income and must dismiss any significance to the marginal utility of income for the rich being less than that for the poor, and therefore to dismiss A. C. Pigou's approach that would redistribute until the marginal utilities were equal. Pigou's approach raises serious problems and identifies what one needs to assume beyond the foregoing, but has more analytical if not putative policy substance than Knight gives the topic as a whole. Surely he cannot object to Pigou having the kind of problem that he himself minimizes in his own use (*supra*), which led to Marshall's rejection of the doctrine of maximum satisfaction itself. Be that as it may, Knight here argues,

"Total utility . . . is relative only to one person – not to comparison of total utilities of different people. There can be no <u>rational</u> basis for progressive taxation." The interpersonal comparability of utilities, or some equivalent assumption, is one of the assumptions of Pigouvian welfare economics; for Knight it falls short of rationality. The subsequent development of interdependent utility functions further complicates the matter, e.g. rendering inconclusive many of the usual formulations.

Knight says, "We reason about human nature on the assumption that it is reasonable." Recent work, in and outside of economics, suggests, first, that rationality (to which the idea of reasonableness leads) is more complex than the simplistic rationality assumption allowed, even within the economistic notion; and, second, that notions of rationalism exist beyond the economistic notion.

Knight seems to have anticipated the rational expectations theory of Robert Lucas and others: "The economic theory of human nature is that people act on the basis of correct and rational anticipations. Results do not enter in." This theory formed a basis of another generation of the Chicago School. Also, the phrase "The economic theory of human nature" affirms what is only an assumption – the claim of one school.

In what seems like an aside but is another aspect of Knight's conservative social theory, in a short section of the notes entitled "Language as a type of social phenomena," he follows Carl Menger and precedes Friedrich von Hayek in their theory of language. The theory, which is not wrong but can be treated too exclusively, holds that linguistic change is nondeliberative. Knight puts it thus: "Rational adaptation does not function in <u>linguistic change</u> – it can not be explained in terms of economics, mechanics, or aesthetics." Ostrander recorded, "This concept is what Knight thinks is <u>his</u> contribution to social sciences." Alas, among economists, Menger beat him to it, and by half a century.

One of the great lines in the notes is this: "People never get tired of making history." Knight uses the point to explain why the idea of historical equilibrium is "foolishness," for "historical changes . . . do not tend to an equilibrium." He does not explain the point itself, i.e. why people never tire to make history.

Knight next is recorded as saying, "<u>All economic changes are the result of price changes, and tend to obliterate the price changes that occasion them – but historical changes are not like this</u>, i.e. they do <u>not tend to an equilibrium</u>. Thus they can not be discussed in terms of cause and effect – but we can discuss only in those terms. Thus <u>most historical change</u> (as linguistics change) <u>is unintelligible</u>." The insightful point about economic changes obliterating the price changes that caused them aside, this statement is either wrong or the result of a narrow definition of economics. Many economists, especially Chicago School economists, hold that relative price, and changes in relative price, generates changes in behavior etc. But other economists, including the Chicagoan Ronald Coase, hold that changes in both prices and institutions produce changes in behavior. Gary Becker, Coase's Chicago colleague, emphasizes economic change as a function of price change. But Becker, an apostle of economics imperialism, insists that prices can take non-monetary forms and that prices therefore pertain to areas of life hitherto not deemed "economics." Apropos of (what Knight called) historical changes not tending to equilibrium, and the now-accepted point about the economy never actually reaching equilibrium (whatever *its* definition) aside, economists such as Vilfredo Pareto emphasized social and historical equilibrium, or the historical equilibrium mode of analysis, as an analytical tool, i.e. he developed a theory of general social and historical equilibrium.

Knight is comfortable with asking questions about fictional states – equilibrium and instantaneous production and consumption – as if these involved object facts; for example, "In an equilibrium position – all production and co nsumption are instantaneous – is anything done but maintenance? Yes."

When Knight is recorded as saying, "Both means and ends are <u>data</u>, are givens," he has the economic actor becoming a passive responder to stimuli; nowhere present is his later emphasis, *inter alia*, on the exploratory and the emergent aspects of behavior and of economic life in general. Of course, the present context is equilibrium. Quite fascinating is his subsequent statements, "Economist <u>assumes</u> people do not choose between ends – sets this off from the rest – but gets into all sorts of trouble this way" and "Market opportunities are a purely technical function," which seems to render the entrepreneur, or entrepreneurial function, impotent.

Perhaps no group of passages in these notes more clearly indicates Knight's preoccupation with the philosophy of economics than these:

> – There is <u>no</u> economic theory of international trade – what have political borders to do with it.
> – International trade is only one case of general economic theory. . . .
> – Economic norm is defined out of our civilization, is a norm which people in other civilizations may or may not follow.
> – If they don't follow this norm, if they are different from us, if they are <u>uneconomic</u>, then economic norm does not apply to them.
> – But it is still just as true as a norm and shows what they do <u>not</u> conform to.

The substance of the relevant philosophy is clear: The ubiquity and preeminence of price theory ("general economic theory") and the truth of the economic norm (constrained maximization). This is the Chicago version of neoclassical economics at its most fundamental level. That Knight treats the norm as a matter of Truth, rather than as normative matter, indicates the tendency of the later Chicago School towards an absolutist ontology. Furthermore, his disregard of personal distribution and narrow view of functional distribution are indicated by his recorded statement, "Cost of production <u>is</u> the distributive share." All three emphases – on price theory, on the economic norm, and on distribution understood as cost of production, especially the third – suggest that for Knight economic theory is undertaken from the perspective of business management. Such a view sounds radical but is entirely consistent with Adam Smith's and Thorstein Veblen's ideas on status emulation – here status identification with the dominant if not hegemonic class (so much so that economists who have identified with labor have been looked upon by the others as professionally unsafe).

When Knight is quoted or paraphrased as saying that "Production and consumption are <u>instantaneous</u> in the going system" and that the "Production process has no length," two possible understandings emerge. One (and *ceteris paribus* the issue of which is more apt) is that this is what Knight believes is involved in his *conceptual* model of the short run. Another is that he (also) ascribes these situations to *actual* economies; how else interpret the recorded admonition, "It is absurd to think of payments as interest for the use of money?" – when his point is that interest (and rent) are "payments for use of productive services." It is strange to make such a point in conceptual modeling – and Marshallian or Marshallian-like time periods are conceptual constructions.

One of Knight's brilliant expositions combines the economics of decreasing costs and the social psychology of policy making. He is recorded as saying that there is "[n]o equilibrium under decreasing costs" and that a "competitive stable price can not exist with them." The result is that equilibrium "[m]ust come from economies of organization; if available to an individual firm, it is a monopoly; so need limitations on size of firm – here it is really a question of social psychology." Echoes of Pareto's theory of general social equilibrium, including (social) psychology as an object of manipulation!

Knight remarks, for example, that "the total production . . . of the community" is "a difficult concept to define." In retrospect he therefore underscores the fact that assumptions, usually tacit, have necessarily been made in conventional terminology (as in national income analysis) that help define the economy (and therefore, in part, total production) for economists and others (see also end of immediately following discussion).

Knight's interpretation of substituting labor for leisure is that this leads to increasing cost and is due to "an element of irrational human behavior," although "subjective valuation enters in." The last clause helps protect him from calling behavior leading to increasing cost "irrational." But he then goes on to say that "Only tropical peoples are rational in this sense – i.e. pricing system doesn't work – only whiskey system does." The statement seems – and may be – invidious but modern Chicago theory, à la Gary Becker, would say that the latter is a price system too. As for Knight, his theory here of the backward-bending supply curve of labor is broached:

– Raise money income – will people work less and get same *total* income (money) as before?, or work more to get larger total money income, or some compromise in between.

and thus,

– Any supply curve finally comes down – at infinity an infinitesimal supply gets an infinitely large income.

This is illustrated by a diagram on which Supply of Labor is on the vertical axis and Wages is on the horizontal, with a curve commencing from the origin, rising, reaching a maximum and then, more slowly, falling. Knight next indicates his broader-than-usual view of preferences:

–What of work for sake of working?
–Importance of conventional standard of living.
  –Lower wages bring *more* work.
  –Raise wages, so far as people have no drive to live above the standard of living, less work.

On a different topic, when Knight says, "every time there is growth, the whole system is different and a new study of equilibrium is in order," the reader will notice the absence, common to many statements of neoclassical theory, of attention to a panoply of causes of disequilibrium or to the process of adjustment.

Oddly, Knight is indicated as saying that "Rent is the payment to capital and labor yielded by varying quantities of land" and that "Rents in theoretical sense are money costs to the entrepreneurs . . ." The latter is true enough, given that the identity of the "entrepreneurs" is not stated (they are farmers; one implication is that Knight possibly means by entrepreneurs not a definite group but a function). The former is misleading; rent is paid to the landowner. Further, his statement, "There is usually an element of price determined costs in every activity," relates to the conflict between rent as price determining and as price determined.

Another oddity (actually two) also comes in Knight's discussion of non-transferable resources. He is said to remark, "Fixed costs are not costs, but a bookkeeping entry. Are costs to the single person, but are not costs to the industry or to the product."

Knight's presentation of marginal productivity theory is a positivist version of John Bates Clark's blatantly normative one, and still falls short of the modern version, which is that factor return is a matter of equilibrium conditions and nothing more than that can be read into those returns. As it stands, Knight's is a vast tautology – useful to a point as an analytical tool but not the description of reality that it reads like. And as for the statement that "The only possible measure of the quality of land is its rent, i.e. what anyone will pay for it," this disregards physical-science analysis. Both that statement and another, "There can not be bookkeeping except in terms of marginal productivity," epitomize the absolutist quality of Knight's price theory, say, in comparison with Marshall' s version. It is also empirically wrong; bookkeeping exists without marginal productivity language.

Knight is remarkably candid in his view of price theory being economics. Economics is price theory, and only certain formulations of price theory – certain assumptions – qualify to be price theory and, therefore, economics. Here is Knight, in effect responding to Piero Sraffa's 1926 article by ruling his analysis out of economics:

> If competition is to yield the best organization, there must be diminishing returns in <u>both</u> cases. . . . If there is not decreasing returns in the <u>ratio</u> sense, there is not an economic situation.

The point could not be stated more baldly than this:

> Only possible economic theory is that of the price mechanism – nothing apologetic about this.

Knight's sensitivity to the subtleties of doctrinal history, for example, that certain terms can lead to undesirable doctrinal results, is reflected in his statement, "Increasing returns is a Cambridge phrase, is not good, all mixed up with wrong Ricardian rent doctrines."

Something of the same applies to Ostrander's report that "Knight doesn't believe there <u>are external economies</u> – hasn't ever seen one." Two points: First, external economies are associated with the economic theories of Marshall and Pigou, not exactly Knight's favorites. Second, the questions inevitably arise whether external economies and the like are to be understood as definitions of reality or tools for the analysis of reality and whether they pertain in either capacity to real economies and/or pure a-institutional conceptual ones – or, more likely, a combination of the two.

On the subject of definition, Knight seems unaware that different definitions of the same thing could be used for different purposes, that definitions are socially constructed and not derived from reality, and that definitions are tools. Knight seems to think that what makes sense to <u>him</u> is the definition of a word and therefore of reality and that discussion of definitions is sufficient, whereas the latter is ultimately a matter of which theory of an object should be ensconced in its definition (e.g. defining inflation as an increase in the supply of money, giving effect to the Quantity Theory, rather than the arguably theory-neutral definition, an increase – or sustained increase – in the price level).

Knight seems not to have been diffident in expressing his own opinions in forceful terms. We read such expressions as "foolish," "palpably false," "monstrous," "100% wrong," "nonsense" and so on. Perhaps more important is the sermonizing tone of numerous discussions.

Apropos thereof, I cannot refrain from quoting from Knight's review of Robbins. Knight wrote, "Nor is the pervasive sarcasm and some outright denunciation likely to make the 'case' more convincing." This is after he refers, on the same page, to "The people who talk 'mystical' nonsense about 'Harmony,' " and to how "The whole chapter of our intellectual history" comprising naturalism "is sad, depressing, a tissue of absurdities . . ." On the other hand, still on the same page, he defends Schmoller: "I must say that there is a vast amount of truth in historicism, and also that it affords a sorely need corrective to the naïve utilitarian individualism of the English Classical Economists" (Frank H. Knight, "Theory of Economic Policy and the History of Doctrine," *Ethics*, vol. 63 (July 1953), p. 280).

When Knight says, in part, "the ideal equilibrium position at one time, and the ideal equilibrium position at another time, are not unrelated, separate," he is recognizing path dependency at the highest theoretical level.

A textbook example of the embodiment of one theory or another in the definition of a word – and of the deliberative element in the evolution of language – is provided by the following:

> – Here arises the whole monstrous definition by Marxists of Capitalism – assuming the "Capitalist" to be in <u>power</u> over labor.
> – The capitalist is the man who rents his property to industry, he is in <u>exactly</u> the same position as labor.
> – Who controls whom? – the consumer controls by his control of expenditure.
> – In an imperfectly competitive system, the property owner can mortgage his property and wait – treating his capital as reserve and drawing on it.

A few lines later in the same discussion, Knight says,

> – A society which guarantees the freedom of an individual is setting up a limitation of freedom; the political authority refuses to enforce any <u>labor</u> contracts.

Two points: (1) When any Alpha and Beta are in the same sphere of action, government protecting one's interests or freedom negates the other's; (2) The refusal to enforce (for specific performance) is due, first, to an acceptance of the distinction between the worker and his or her services that Knight rejects and, second, to the constitutional prohibition against involuntary servitude – which is selectively interpreted. The distinction between the worker and his or her services also explains the limitation on one's freedom to sell oneself into slavery or indentured servitude – terms also selectively interpreted.

Further apropos of economics as apologetics, Knight insists, "By discussing the competitive system, we do not mean to uphold that system." But arguably more fundamental than apologetics is definition of reality. Normative statements constituting apologetics (or criticism) are clearly normative statements but defining reality in such a way as to finesse the normative issue – "That is the way it is, like it or not" – is a privileging linguistic move. When Knight says, "Relation of effort to return is the fundamental economic reality," he is avoiding the normative issue by defining reality that way – whereas a critic could say that his definition of reality is only a projection of a particular society's mindset.

Rather striking is the apparent aside that increases in the capitalized value of land are an "unearned increment": "Annual increase in capital value of land (unearned increment) is a capitalization of the increased future income."

Neoclassical constrained-maximization models inexorably require specification of that which is to be maximized. Different specifications are possible. Inasmuch as different interests may count, economic agents allocate resources to change the law (and the interpretation/application of moral rules) so that their interests will count (for more), i.e. others' interests will be economized. (For example, marginal-cost pricing requires choice from among multiple extant margins). This is recognized implicitly in Knight's statement, "Economy of effort is not different from any other economizing; effort is not the only thing to be economized."

Knight is recorded as saying,

> There is no such thing as long-run equilibrium – Marshall's Book 6 is nonsense; there is no ultimate saturation point in society.

One has to wonder in what sense this statement is made. For one thing, the economy is never in equilibrium, so the entire exercise is purely conceptual. For another, Marshall's set of time periods is purely conceptual, each period dealing with certain variables made variable or left unchanged. The language of the notes – which "sounds" right linguistically and which we must assume echoes Knight's – uses "is" as if Knight were referring to the actual economy. Knight also seems to confuse the Marshallian long run and the Ricardian stationary state.

At one point the notes record Knight saying, "The individual preference or choice scale becomes a social scale via the price system." At another point, "saving [is] a result of social mores." So Knight was seemingly willing to recognize the two-way causal relationship – though not together and very unlike conventional neoclassical practice (Knight also said, "The psychology of saving is chiefly a matter of getting ahead in life – a question of power" – this from an economist who later tried to exclude power from both his economics and his social theory). At another point Knight has labor supply as a function in part of standard of living and standard of living as a function in part of labor supply.

Knight says, "Demand curve is vertical: there is no limit to the speed at which capital can be invested; provided any investment is profitable; the demand of the capital market is indefinitely elastic." Some 70 years later it seems remarkable that no recorded attention is paid to liquidity preference, the expected rate of profit or to the opportunity cost of investment. Ensconced in the later Keynesian formula, investment is a function of the marginal efficiency of capital being greater enough than the interest rate – as obvious as that now seems. An infinitely elastic investment demand curve, indeed!

In a discussion of Marshallian wage theory the reader will find statements that anticipate and may even go beyond Coase's much later criticism of "blackboard economics." Part of what Ostrander recorded is, "Knight doesn't put any stock in any of this – it is a mere logical exercise," "We don't know anything about all this," and "Try to get some idea of realistic psychology." In the same discussion, Knight also said, "Criticism of utility theory after war – most of it worse than the theory itself."

Knight's lectures are a gloss – a series of commentaries – on the philosophy and methodology of economics and on the formulations and uses of price theory. They were not a systematic exposition of price theory as it then existed. The lectures were stimulating to certain students, perhaps more for tone, rhythm, and overall message than for brilliance of substantive presentation. The lectures pale in comparison with his essays and articles, which are better thought out; but the lectures provide entry into his mind precisely because they were relatively unrestrained and were neither edited nor expurgated. The lectures, and Knight himself, contributed to the evolution of the Chicago School.

The lectures have too much that is arguably wrong and/or, especially, idiosyncratic. Examples include, "Marshall's long- and short-run normal price analysis is foolish" and "Marshall's use of 'plant' is foolish." Other examples include his treatment of the relation of dictatorship to people's wants and his rejection of material interests as a basis of fighting.

It is not clear how Knight's contempt for religion would have left him in a world in which Christian fundamentalism is so much an ally of economic fundamentalism; his penchant for straight talk may well have gotten him into trouble of one kind or another. In such respects, Knight was to economic discourse what H. L. Mencken was to general public discourse. Interestingly, Knight believed that if some of what he wrote had a wider public audience, he might not have published it; he did not want to abet social and political radicalism, i.e. the unsafe. This, however did not bother Mencken, who has been called a "public crusader against conventional values" (Russell Baker, "Thus Spake Henry," *The New York Review of Books*, January 16, 2003, p. 11).

A final highlight of the notes is a marginal comment indicative of Knight's cynicism and his ability, even amenability, to criticize, however selectively, the world around him. The comment reads,

Ph.D. thesis on executive salaries – needs a cynical satirist, not a scientist to write – paid not to think, to sit and look, assure everyone that everything's all right. People with brains don't look or act that way. Are like movie contracts, advertising, pay high salary and fool people into believing he is great man.

"[N]eeds a cynical satirist," indeed. Frank Knight was right at hand.

(1) READING ASSIGNMENTS FOR FRANK H. KNIGHT'S COURSE, ECONOMIC THEORY, ECONOMICS 301, AUTUMN 1933

ASSIGNMENTS, 301

1. Law of Diminishing Utility
   J. S. Mill, Principles, Book Three, Ch. I
   Marshall, Book Three, Chs. III, VI
   Davenport, Value and Distribution, pp. 296–317
   Anderson, Social Value, Ch. V
   Downey, J. P. E., April, 1910
   (Mim. Outline, No. 3)
   Viner, J. P. E., Aug., 1925; Dec., 1925

2. Demand
   (Mim. Outline, Nos. 4, 9)
   Marshall, Bk. Three, Chs. III, IV

3. Supply
   (Mim. Outline, No. 5)
   Marshall, 6th ed., p. 142

4. Interaction of Demand and Supply, Temporary
   (Mim. Outline, No. 8)
   Marshall, Bk. Five, Chs. I, II
   Böhm-Bawerk, Positive Theory, Bk. Four, Ch. IV

5. Short-Run and Long-Run Equilibrium
   Carver, Distribution of Wealth, Ch. II
   Marshall, Bk. Five, Chs. III; IV, #5, 6; V; XII; App. H, pp. 805ff.
   Silberling, Am. Ec. Rev., Sept., 1924

6. Austrian Theory of Value
   (Mim. Outline, No. 11)
   Böhm-B., The Ultimate Standard of Value, Annals, Sept., 1894; or One Word
   More on Ult. S. Val., Ec. J., Dec., 1894

7. Joint Demand and Joint Cost
   Marshall, Bk. Five, Ch. VI

8. Monopoly Value

Marshall, Bk. Five, Ch. XIV
(Mim. Outline, No. 14)
Viner, J. P. E., 1925, pp. 107–111

9. Wages
J. S. Mill, Pol. Ec., Bk. Two, Ch. XI, #1; Bk. One, Ch. VI, #1
Henry George, Progress and Poverty, Bk. One, Chs, I, III, IV
F. A. Walker, Pol. Econ., Part Four, Chs. IV, V; Part Six, Sec. V
F. W. Taussig, Principles, Vol. II. Ch. LI
J. B. Clark, Distribution of Wealth, Chs. I, VII, VIII
W. H. Hamilton and S. May, Control of Wages, Chs. X, XII

10. Differences of Wages
A. Smith, W. of N., Bk. One, Ch. X
J. S. Mill, Pr. of Ec., Bk. Two, Ch. XIV
Cairnes, Pol. Econ., Part One, Ch. III, #4, 5
Taussig, Principles, Ch. XLVII.

11. Interest
J. B. Clark, Distribution of Wealth, Chs. IX, XIII
Böhm-B., Bk. Two, Chs. I-V; Bk. Five, Chs. I-IV; Bk. Six, Chs. I, II, IV, V, VI;
Bk. Seven, Chs. I, II
I. Fisher, Rate of Interest, Chs. II, III, (IV); Elementary Principles, Chs. XX,
XXI
A. B. Wolfe, Saver's Surplus and the Interest Rate, Q. J. E., Nov., 1920

12. Rent
Marshall, Bk. Five, Ch. X; Bk. Six, Ch. IX
Clark, Ch. XXIII

[The following names were written in by Ostrander]

Wicksteed –
L. von Mises – Archiv für Socialwissenschaft und Socialpolitik, 1930
O. F. Boucke – Critique of Econ.
Veblen – Place of Science in Civilization, "The Limitations of Marginal Utility,"
"Prof. Clark's Econ."
A. Smith – Book II, Chap. III [?] – Book I, Chap. VII
Marshall – Bks. I and II – Bk. V, Chap. V – Bk. 6, Chap. II 534, 532
Talcott Parsons – two articles
Cassal [sic] – Chap. 1 and 2
Jevons – pp.(50)

Spencer – Chap. on Unknowables
Janaism [?]
Viner – pamphlet on Cost Curves. Article on Cost in Soc. Sci. Encyclopedia
[Assignment:]
   Cost and Price – Paper, 2000 words (6–10 pages)
     [See Ostrander's term paper, "The Meaning of Cost," item (3) *infra*]

(2) F. TAYLOR OSTRANDER'S NOTES ON
FRANK H. KNIGHT'S COURSE, ECONOMIC THEORY,
ECONOMICS 301, AUTUMN 1933

Economic Theory (301) Knight

Philosophical Introduction – Problem of the Starting Point
  –Getting rid of "original sin" – romanticism in economics.
  –Economics is mainly a problem of morals and philosophy.
  –To get the proper, objective point of view.
  –Change in name from political economy to economics was a most unfortunate
    one, especially in this day.
  –There can not be a science of economics.
    –Wertfrei (free of value judgement) is impossible.
  –Mathematical economics – when economics is defined so closely as to become
    mathematical, it is no longer economics, and can not be political economy.
  –We tend to dispute propositions, not the truth.
  –This philosophical approach is true also of the natural sciences; the difference
    between economics and natural science is one of degree of enough importance
    to amount to one of kind.
  –Economics covers the whole gamut between the exhortive [sic] (stump-
    speaking) "come on, let's go" – "give 'em hell" – and the objective
    (mathematical economics), all a is b. ["fact and theory" connected with
    "objective" by an arrow]
    –A statement of fact is true, out of all relation to time and space – it is not
      factual.
    –Some statements establish a relation between facts. "In so far as," "if a, then
      b," caeteris paribus.
    –Asserting a hypothetical, functional relation.
 –Knight fights a pessimism that what is desirable in social sciences, is
    impossible.
    –People want magic, medicine, cure-alls, not thinking.

–Yet economics, without any attempt at social application, is form without content.

–1 – Rationalistic – theoretical – classical.

2 – <u>Institutional</u> (largely a revolt for sake of revolt).

    –a-Concrete, factual.

    –b-Sociological.

    –c-Policy – control.

–How find a starting point? that people will agree on.

  –Human behavior – in some aspects.

  –<u>Scientific approach</u> has definite limitations.

    –Even when directed towards nature, it explains some human relation; concepts of physics imply an <u>agent</u> – who think them; behaviorism lands in a paradox – the exponent is really another talking machine – or else different from everyone else!

    –When directed toward the <u>human</u> aspect of human beings, it is infinitely more limited.

      –It is "statements about" – sentences – utterances.

      –There is a <u>mutualism</u> (in natural sciences we assume a one-sided relation of nature and humans – we assume <u>it</u> is not studying <u>us</u>!) in study of human beings.

      –Manipulation and manipulator are not <u>social</u> terms. Social relations are not control relations; but are based on discussion (mutual).

      –Fundamental econnomic theory <u>implies democracy</u> – that one can publish the results. Not possible in dictatorship!

       –Has developed along with Western democracy.

    –Any one talking about <u>social control</u> and <u>meaning</u> it – hides his purpose. He does not write books about social control.

    –That is not the same as getting at a <u>consensus</u>.

    –All economics is <u>basically</u> political propaganda.

–Division between <u>fact</u> and <u>theory</u> is universal: exists in field of natural sciences, morality, <u>social sciences</u>.

  –It is intellectually insulting to be told that classical economics is at fault because it gives us no statistics. Instances do not have to occur to prove the correctness of any fact.

    –Arsenic is poison. Do we need instances?

  –We are <u>talking about economic behavior</u> – we do not have to tell <u>how much</u> economic behavior there has been.

–The prime source of confusion in economics is this mixture of science and propaganda.

–The drift is away from propaganda. (But towards what?) Socialists and communists claim there is no objective economics apart from propaganda.

–It is recent in the world's history that social problems could be discussed in the terms of political science – due to the development of the modern mind.

–Now, we have a political state in charge of economic life – in Middle Ages the two could not be separated.

–These two social categories are composed of the same people. Complicating the relation of controller and controllee.

–Salesmanship for one subject – as one of vested interest – is very vicious – the instructor's belief that the whole of liberal education is encompassed in his course!

–Human behavior – what is behavior? – All the muscular actions, or some of them? – Say, the deliberate, conscious behavior of man.

   –This will not suit the radicals who are not concerned with what enters behavior.

   –Knight says – if you do not limit behavior to deliberate b[ehavior] – you have no right to use the word economics without special definition and qualification.

      –The concept of discussion enters again; if we address ourselves to the subject matter of economics it must be deliberate.

–Delimiting the concept of behavior to arrive at the subject matter of economics.

   –General field of behavior – special field of economic behavior – special field of deliberate (sans habit/no reflex action) behavior – special field of problem-solving behavior.

   –Special field of problem-solving directed towards means and ends (economy, utilization of means, of resources).

   [Inserted between lines, likely intended to follow "technology" in next line: "(manipulation)"]

   –Special field excluding technology, i.e. allocating means between competing modes of use for a given end.

   [Single vertical line alongside notes commencing with "General field of behavior"]

   [In margin at top of page: (Defining economics by defining what it does not include)]

   –Special field of deliberation regarding use of means directed towards ends (excluding the consideration of ends).

      –Economics is – historically – hedonistic; "pleasure is the universal end" – but this tells us nothing about the end, or about pleasure. And we know that pleasure is one of several ends.

   –Special field of social behavior.

Price Theory Economics

–In real problems of social sciences, what strictly defined economics can offer is very slight.

    –Culture-anthropology, philosophy, psychology, political science, law, etc.

–But what little economics can say is extremely important.

    –The "economic moment" is tiny, but essential.

–Definitions: Marshall: "ordinary business of life." Ely: "social life." Fetter: "rational action."

    –What is Marx's definition of economics? – Under any valid definition of economics – 95% of Marxism is ruled out.

–The "economic moment" is one of allocation of resources among competing uses.

    –(1) Utility theory says: all economic behavior finally results in one particular quantitative, comparative end – want-satisfaction.

        –No qualitative distinction among those want-satisfactions

    – (2) If there were not diminishing effectiveness of those utilities, we would not have our world – there could be no choice [Double vertical lines alongside this line].

        –There could be no allocation, there could be no deferred spending.

        –This principle of diminishing utility is a part of the make-up of our minds – part of form of mind.

            –Is as near to a universal truth as we can arrive [at] in economics – or any other subject.

    –(3) The marginal concept – equalizing the utilities by allocation towards a maximum utility of the quantitatively limited resources.

    –(4) The nature of our subject is the content of some social order.

–Economic society is not aa society; each individual lives in a world of his own, having only technological relations to other people – each other individual is merely an economic opportunity.

–International trade does not exist as a part of economic theory [Single vertical line alongside this sentence].

–Society is an organization of individuals.

–Economics is a science of organization – we have been trying to define what it is an organization of.

    –An organization of individuals in possession of wants, resources, technology.

        –Each individual exists in a market, affording him technical opportunities or alternatives.

        –It is a "society" of individuals taken up with self-interest.

–How is it that <u>increased</u> organization brings about increased opportunity for using resources to satisfy wants? (Adam Smith's "amateur anthropologizing and philosophizing")

　　–Efficiency increased through specialization – in spite of the great cost of <u>exchange machinery</u>.

　　–Balance of <u>gains</u> in <u>technical efficiency</u> and <u>losses</u> through <u>restrictions on personality</u>.

　　　–Bringing in <u>value</u> – which <u>cannot</u> be avoided.

–Each person organizes with others for his own advantage.

–<u>Types</u> of economic constitution: (we can't help mentioning economy, even in plant life, or termite colonies)

　–"<u>Mechanical</u>" termite – extreme specialization of (economic) function.

　–<u>Caste</u> system – specialization of human (economic, etc.) functions.

　–<u>Anarchism</u> – voluntary caste system.

　–Autocratic – no separation of economic sphere and the political sphere, as in Egypt – caste imposed by one autocrat (<u>Within</u> a ruling group – as Fascism – there will be met all the problems they refused to meet in the open, free, system. – Easier <u>or</u> harder to solve).

　–Extreme organic theory of society – Max [sic] Spann.

　–Society is a mutual-aid society – Knight.

　–Man is a member of numerous societies – but not of <u>one</u> organism: society – French sociologists

　–(Socialistic) democratic – differ[enti]ation of political and economic spheres.

　–Democracy was, in origin, a device for state borrowing at aa lower rate of interest – borrowing by an immortal society rather than by a mortal king.

　–Exchange – a voluntary relation between individuals.

　　–How else separate political from economic other than by exchange.

　　　–<u>Political</u> implies compulsory action by a <u>group</u>.

　　　–<u>Economic</u> implies voluntary action by an individual.

　　–Political system is in control of economic system.

　　　–<u>Policy</u> is a constitutional question; is not an <u>act</u>.

　　a-Literal exchange system – handicraft – money is a convenience.

　　　–People <u>live</u> by exchanging products.

　　　–Disappeared long before it became dominant.

　　b-Enterprise – entrepreneur – people live by selling their services.

　　　1-Individual entrepreneur.

　　　2-Corporate

　　((Problem of <u>goods</u> against <u>services</u>.

–Knight says it is due to a confusion about this that economists have not got any further along.

–What are the "intelligent dimensions" of the flow? [Alongside, with arrow to "dimensions": –not too good a word]

 –Intensity x time.

 –Flow is essential.

[In left margin: Circular-flow diagram, with money moving counter-clockwise between Enterprise and Population, and consumption of goods and services moving clockwise. Below diagram:

–Two markets are really only one market. One phase cannot be discussed without bringing in all the rest of economy.

–Two sets of prices]

–Not wealth-J. S. Mill, but services is the essential economic unit-F. H. Knight. Wealth is a derivative.

–When we buy food we buy services, and we usually don't capitalize it.

–For durable goods, we must distinguish between the accumulated services and the good.

–Wealth is the summated, discounted value of services.

 –It is the intensity of service over some duration – discounted.

[Double vertical lines alongside preceding two sentences]

–In the circle [circular-flow diagram], "consumption goods" is bad [wrong] – services are exchanged against services.

 –"Services sold" are really personal property – one sells objects of his personality.

–What word means the opposite of both growing and contracting? – static is terrible – stationary not good.

–The individual in "population" is

 1. Leasing or loaning for dollars personal property

 2. Owner of productive capacity

 3. Consumer (of services)

 4. Member and participant in enterprise.

 –Mechanical analogies

 –Galileo – "constant force produces uniform acceleration" (exerted against inertia).

 –Velocity = distance per unit of time.

 –Acceleration = distance per second per second.

[Alongside in margin:

Constant force exerted against friction produces uniform speed.

All linear

–Elasticity – a mass is deformed – in proportion to the force applied.
  –Oscillations: arise in elastic system with inertia but no friction. – i.e.
  output varies inversely with price, but not in proportion.

Functional Analysis of Organization
  –Relation between organized units is essentially one of technique.
  –Mutual advantage between two economic units must exist – can not be
  discipline.

A. Stationary
  1. The price system is the value scale of the economic organization.
    –The individual preference or choice scale becomes a social scale via the
    price system.
  2. The price system is the guide or control of "production" (not ought to be).
    –a. Allocation of resources (individual and social) – social-wide system
    must tie together the allocation of individual resources – a social
    mechanism is the price system – a phase of society being organized.
    –b. Co-ordination (management, technique)
  3. The price system is the mechanism of distribution – at one time, or over time.
  –The main discussion within economics concerns lags in adjustment and
  distribution, and allocation and production.
    –Without such lags, if all adjustment were worked out instantly, there would
    be little to talk about in economics.

    [In margin at top of page:
      Marx says history is the class war; spends the rest of 2400 pages
      belaboring people for not living up to, and acting as, their class, i.e.
      he merely wanted history to be a class war]

B. Historical – longer run
  1–Price scale for consumption goods (embodiment of services) and services.
  2–Prices of productive services, allocate production between industries, and
  bring about efficiency within the industry.
  3–Money income and price being about distribution – price of services.

Do people want money or the goods it will buy?
  –They often want money – even when they want the goods it will bring, they
  don't really want those goods, they want something beyond, want-satisfaction,
  but when it comes to going further, why do they want want-satisfaction? –
  we are sunk, and leave the field of economics. Thus, for the economist, what
  people want, is the goods that produce want-satisfaction. The goods are taken
  as symbols of the wants; it is a game. People play the game for money, without
  wanting it, and also play the game for goods without wanting them.

–None of the commodities sold <u>correspond</u> to wants. <u>Nourishment</u> is not sold in the market – bread and meat is sold. They <u>provide</u> nourishment, but <u>are not</u> nourishment.

The Economic Moment
  –Each individual reacts only to services, not to other individuals as such.
  –Short run function – price system works out through speculative markets.
  –Historical, long run function.
  –One aspect of pricing is <u>interest</u>.
    –In stationary society it functions to allocate.
    –In historical society it functions as incentive to save.

Relation of <u>Wealth</u> and <u>Income</u>
  –The fallacy of wealth is fundamental.
    –Wealth is not what is produced and consumed.
    –<u>Services</u> are produced and consumed.
  –A sharp separation of <u>wealth-quantity</u> [Ostrander comments: "flow?"] and <u>services-quantity</u> = they are not equal. – one equalized only by <u>discounting</u>.
  –<u>Wealth</u> is a <u>service of some intensity</u> over <u>some period of time</u>. But the intensity is not at a constant rate (i.e. necessity of calendar).
    –Wealth must be discounted to give a magnitude; as such, it is only future income.
    – To understand what wealth is, one has to understand the <u>mathematical</u> theory of <u>compound interest</u>, compounded <u>instantly</u>, for an <u>indefinite period</u>. [Alongside in margin: "one year, six months, one day, one instant, zero] [Single vertical lines alongside preceding group of sentences]
    –But, services are not all possible of being capitalized into wealth – human services.
    –There never has been a society in which investments were made in human beings on a rational basis, i.e. never a real slave society – not even the theoretical capital value of human beings corresponds to cost of production.
    –The economic moment is very small and extremely important – as far from reality as its opponents have ever claimed – but not any more inessential for that.
      –Knowledge is about the consequences of acts, but the acts seldom happen – i.e. "arsenic is poison."
      –How do human relations work out when they are free (i.e. free exchange) – should have been the aim of economics – a critical, analytical study.
        –But economists have not done their job.
          –Knight has turned his attention to the <u>method of thinking about</u> this mechanism of free exchange.

–<u>Weber</u> and <u>Sombart</u> have thought about the sociological implications, the problem of <u>thinking about society</u>.

–Knight wonders how <u>any</u> critical objective thinking is possible <u>about</u> society by a person in that society – lifting one's self by his bootstraps!

Production

[Diagram similar to earlier one, now with Business Enterprise and People, with money flowing counterclockwise and productive services moving clockwise, and with Production defined as want-satisfaction services. Also: "<u>People</u> = those who <u>function</u> in the economic system, those who <u>spend money</u> – not dependents." Production as ultimate resources transformed into services and services into want-satisfaction. Productive services as personal, receiving wages, and property, receiving rent and interest]

–This should not be a linear concept, but should be <u>circular</u> and only another adaptation of the <u>circle of services</u>.

–Difference between interest and rent is only a formal one: – the difference between loaning productive equipment, i.e. leasing = rent, and loaning money, i.e. lending = interest.

  –Two forms of providing <u>the same</u> <u>service</u> [Double vertical lines alongside in margin].

–Difference between this service and wages is again only <u>institutional</u>.

  –Man can sell his personal productive capacity only by the laws. Our legal structure does not allow him to sell his personal productive capacity for an indefinite time, or even into slavery for a definite time.

  –Thus the payment for personal productive capacity is wages and is slightly (institutionally) different from payment for <u>property</u> productive capacity.

Final, direct services

–Rendered by <u>persons</u> and <u>things</u>.

–<u>Things</u> that render services may belong to the (a) beneficiary who receives services, or may belong to (b) someone else, in which case money changes hands (either rent or interest).

–<u>Persons</u>: <u>self-service</u> is <u>essentially</u> outside the economic system, yet it can not be <u>set off</u> to any great degree – must be treated arbitrarily [In margin alongside: –Even if what is self-service to one person, is sold in the market by another – that market price can not be carried over to the self-services – are matters of <u>intention</u>].

–Perishable goods, durable goods – usually used by consumer, but durable goods are not <u>always</u> used by owner.

–Indirect services.

–Consumers and producers goods.

–In between the selling of services, and the enjoying of services – lies the great complex of productive organization.

–No possibility of a general statement.

–The productive process has no ascertainable <u>number</u> of <u>stages</u>; it has no <u>time length</u> that can be set and measured.

–The use of <u>time</u> in <u>consumption</u> sets the existence of <u>intermediate</u> goods.

–The <u>physical stages</u> of production are largely accidental, and arbitrary. Can not be used in the <u>economic</u> analysis of "stages of production."

–We <u>can</u> trace the development of some physical <u>chunk</u> through physical units of time.

–The whole thing is a confusion of <u>physical</u> units with <u>economic</u> units.

–Economically, production and consumption are instantaneous.

–"The process of producing intermediate goods adds stages of production, adding to the final service," etc.

–It is just not true. Economically, this can not be dated or divided into units.

–<u>Capital accounting</u> is perpetual, not the physical units; society always has <u>maintained</u> and <u>advanced</u> its capital.

–Capital is like the Irishman's knife – it had had three new blades and two new handles, but it was still the same knife.

–Capital, when the physical unit is used up, is maintained by the reproduction of an <u>equivalent item</u>.

–The only way <u>we</u> figure interest and keep books – is on the assumption of the perpetual life of capital.

–We go bankrupt if we do not provide for <u>perpetual maintenance</u>.

–Maintenance can only be conceived of as perpetual, including replacement.

–The production of what is being consumed at every moment has gone on through all history. The consumption of what is being produced at any moment will extend indefinitely into the future.

–We should have to revise our interest theory if we were to take account of a possible society which saw its own doom and proceeded to liquidate.

–Even then it would not be a <u>very</u> different theory: substitution of minus for plus signs, etc.

<u>Labor</u>

–Classical economics tied up rent to wages, but <u>wages</u> in a <u>civilized</u> society have nothing to do with <u>subsistence</u> and little to do with product

–<u>wages</u>   are   <u>paid to keep the workers from working for somebody else – are competitive valuation</u>.

–are <u>mostly a matter of custom and habit</u>.

<u>Land</u>
  –Ricardo – as a general theory of economics – a tissue of incredible stupidity.
    –As an explanation of Napoleonic England, it is admirable.
  –[He was] Confused: Said agriculture was <u>an</u> industry.
    –Instead of a group of industries.
    –They are really stages in the production of final goods – not the same
      competition between subsidiary products or between parallel industries.
    –Used wheat – forgot all other products – considered it a final product.
    –Land was the home of the wheat industry.
  –Ricardo['s] situation confused again, by England's position in foreign trade.
  –"Original and indestructible powers of land."
    –"<u>Original</u>" has nothing to do with it – economics is not <u>history</u>.
    –"<u>Indestructible</u>" has a good deal to do with it, but nothing to do with <u>original</u>.
      –Goods may be indestructible, or not able to be used without destruction,
        or in between – <u>land is in between</u>.
    –Pioneering will be done if it pays a higher return than settling on old land.
    –Most economic theory belongs in the realm of <u>social abnormal psychology</u>.
  –Speculation – "risk-bearing is a <u>cause</u> that must be remunerative" – says the
    economist.
    –But anyone can see by the state of laws, that this is foolish; we have to threaten
      everything to keep everyone from spending all his money in speculation.
      –Yet the return (average) thereby becomes <u>very low</u>.
    –Industry is an activity producing a product.
      –Must distinguish between the final product and factors of production.
  –Any <u>thing</u> is a factor of production up to the moment of its disappearance in
    the act of giving final service. Lists of human wants, and lists of commodities
    and services will never coincide.
    –No <u>thing</u> is a final product; the service it renders may be a final product.
    –You buy the thing in order to buy the service.
    [Single vertical line in margin alongside preceding two lines]
    –In order to talk realistically about economics – for sake of policy, you
      must draw an arbitrary line, between factors of production and final
      products.
      –That line is best drawn where <u>interest</u> is no longer paid.
      –The final product is ordinarily never leased.
    –Logically, in an economic world there is only one productive factor: <u>capital</u>.
      All these other – and mostly legal in source – categories are arbitrary, for
      the sake of practical talking.

Capital
　　–All productive instrumentalities are really capital (including people).
　　–The gambling factor – nature causes our anticipations to be frustrated.

Nature of the Productive Process as a Whole
　　Stationary Economy – not subject to any "cumulative change" (Veblen's phrase, didn't know what it meant), i.e. accumulative and discumulative changes.
　　　　–Capital, wealth, property – are synonymous (persons may or may not be included).
　　　　–Logically there is no such thing as consumer's capital – although we use the term arbitrarily.
　　　　　–Only services are consumed.
　　　–Functional role of capital (property)
　　　　–Direct personal services use no capital.
　　　　–When capital enters into production (of services) it works and is worked upon.
　　　　–Property may grow in value (wine) without working or working upon.
　　　　–Distinction between materials and auxiliary aids.
　　　　　–Wheat is in the bread, the mill is not.
　　　　–The income is the given fact – the quantity of property is derived from that and the other given item, the rate of interest.
　　　　　–Capitalization is a formula.
　　　　　　–It is simple, if a given income, perpetual in dollars, is concerned. Compounded once a year.
　　　　　　–Otherwise it can not be figured arithmetically.
　　　　　　[Alongside in left margin: "–How often compounded? What of instantaneous compounding."]
　　　　　　–This is getting the summated worth of the present value of a dollar payable one year from now, plus the present value of a dollar payable two years from now, etc. $1/r$ (ratio)
　　　　　　　–Summating an infinite series of present worths.
　　　　　　–When the series is perpetual, it is easy to work out. But it is hard to work out if the series is limited (20 years).
　　　–The whole theory of capital is in the capitalization formula.
　　　–Abstinence – there is waiting in a sense, but no measurable length or quantity of it. The so-called "production process" is a confusion of economic process with the observed physical process.
　　　–The worker does not wait.
　　　–Does society wait? – Productive equipment may be used to provide future services, instead of present services. This seems like waiting. But, there is no

definite length of waiting; that future income is provided during every moment
of that future.

–Capital is perpetual – interest is the difference between gross and net income. Interest does not arrive until after the capital is made perpetual! [Double
vertical lines in margin alongside underlined sentence].

–Present wealth is converted into an income stream (Fisher: approximate).

–Investment

(a) Income may be figured for a limited life – summation of present values
of successive future incomes (discounted income finite period).

(b) Put in depreciation reserve, replacement fund, scrap value – perpetual
income.

–The two approaches yield the same formula.

–Why should a perpetual income yield an infinite present value? Böhm-Bawerk
was crazy.

– "The whole theory of interest has been a vested interest" – thing that makes
Knight most angry is Böhm-Bawerk's "Positive Theory of Capital" – only
thing worse is Fisher's criticism of it – creating a problem where there isn't
any.

–"and maintenance" should be tautological.

–No difference between man who makes the machine and man who repairs
the machine.

–The operation of society includes maintenance – the only way to keep books.

–If books are not kept to provide depreciation for perpetual income, one is
headed towards bankruptcy.

–If we were living in world headed for doom – society as a whole going
out of business – then income (present) could be increased for a space, by
stopping depreciation and replacement accounts – but what would be the rate of
interest.

–Postponement has a psychological and a physical side. Where any activity is
concerned, there is a physical alternative activity, and an alternative of valuation
[In margin alongside, double vertical lines].

–Economics assumes the effort to maximize return from resources [In margin
alongside, double vertical lines]

Interest – 365 daily rations of food – might be delivered on the last day of the year
– but, says the orthodox theory, there is a "reluctance to postpone" which results in
interest; Knight asks, why not deliver on the first day – or is there also a reluctance
to anticipate? – One doesn't want tomorrow's dinner to-day, or yesterday's, today.
But today's, today.

–The spacing in time of the rendering of services is a detail; it is the maximum
service rendered that is important.

–There is no time-spacing in consumption, it is continuous.

–Consumer's choice is the relative proportioning of items in a budget.
  –We carry a mechanical analogy over into a non-mechanical subject.
  –As "successive applications" in time.
  –Only in the investment of resources does time enter. Just as in disinvestment – time enters.
  –Society must either invest or disinvest, both cumulatively. It does no good to think of a competitive society which is not cumulatively investing.

Equilibrium = the amount produced equals the amount consumed.

  –When this is not true, the economics of equilibrium do not apply.
  –Capital can not be objectively defined except in terms of itself.
  –The quantity is a function of price (interest rate) which is a function of quantity.
  –How measure capital? There is no objective measuring stick for anything in economics. The denominator is always a valuation.
  –How define even any commodity, or measure it? I.e. there is no special difficulty with capital.
  –The thing priced is not a time comparison. Only in investment or disinvestment does time enter.
  – There is no established economic usage for anything in economics.
    –Service is either perpetual, stretching from now into infinity; or else is discounted from that infinity at an interest rate.
      –Or else, service extends over a set period and is incomparable to any other service extending over another set period of time.
  –Production includes the rendering of services, and the increase of the power to render service – connected by the interest rate.
  –Mill defines production – net additions to total wealth. There is only production when society is growing.
  –Capital must be defined to include a lot of intangibles – and this must be considered to understand production.
  –Knight – does not think there is any such thing as time-preference.
  –Why should all society have a preference for the present? Why not a preference for the future?
  –The essential thing is the sum of consumption income.
  –Knight suspects, even, a future preference. Finds this borne out through all poetry. The poet telling man to stop looking to the future.
  –In this circular system of economics – there arises the problem of the starting point.

[More elaborate version of earlier diagrams of circular flow: Clockwise flow of products (a. consumption and b. capital increase) and productive services, and counterclockwise flow of money. Entities are Enterprise and People (identified as "members of the <u>system</u> – heads of families"). Money flow from Enterprise to People represents cost of production to the former and private income to the latter. Money flow from People to Enterprise is cost of living to former and business receipts to the latter]

–It is best to think of this circle as a stationary economy – abstract capital expansion

  –It is impossible to represent – without a third dimension – any <u>change</u> in the <u>wants</u> and demands, or any technical advance.

  –There can not be discussion of a free enterprise system which is <u>not</u> founded on <u>growth</u>. Capital stagnation, as at present, is abnormal. (?)

–This equilibrium is never perfect – the adjustments around the circle are not perfect – dis-equilibrium.

–In stationary system – individual <u>wants</u> and <u>resources</u> are given and social <u>technology</u> is given.

(3)–In equilibrium – the activity of the circle is instantaneous. In any <u>frictionless</u> system, the velocity of money would be infinite, and <u>money</u> – other than a unit of account – would not exist. Money is an <u>oil</u> for a barter system.

(4)–In historical, cumulative change – wants, resources, ownership, technology, institutional setting – any change in any one of these brings about a complete change in the stationary circular system.

[The full 1, 2, 3, 4 enumeration is not in the notes. Knight may have used numbering on the blackboard. Problem also arises below]

–There is no more connection between this theoretical system and "reality" than there is between the "theory" that arsenic is poison, and the reality of people eating it or not. It is just as true as that other theoretical statement.

–Social <u>dogmatism</u> – <u>unconscious</u> – is source of about all political conflict today. Is there too much or too little competition for <u>the system</u> to work?

  –This does not affect the truth of "the system."

–<u>At equilibrium</u> no <u>one</u> would be allowed to improve his position without the consent of everyone else. Or, if he <u>does</u> improve his condition, it is theft. No one can improve his position as against the rest, without its being theft.

–Our very discussion of this subject assumes that we assume rationality; we must distinguish sales-talk, conspiracy, oratory, from Discussion – which, to be rational, <u>can not be</u> an attempt to get others to act so as to benefit the discusser. It <u>must</u> be impartial.

–<u>The</u> economic problem is the correct allocation of resources to wants.

(2)–<u>Seasonal</u>, and <u>short-term</u> system is of shorter time than the stationary system.

–Importance of speculation to distribution of <u>this</u> nature. <u>The ideal speculative market</u>.
 –Distributes the product over time and among people – as it would be distributed if every one bought his year's crop (of wheat etc.) on the first day of the year.
(1) –<u>Monetary</u> system – covers the imperfections of ideal speculative system.
 –These four periods are <u>factors</u> in the readjustment rather than a time-process; they proceed at varying rates.
 –Assume a change of consumption-expenditure:
   1.–<u>A momentary price</u>, highly speculative – price in this disequilibrium is a result of all the guesses in the market – a balance of speculative opinion.
    –If speculators <u>knew</u> the future course of prices.
   2.–A <u>short-time market price</u> would result.
    –i.e. that price that would exist if all the momentary factors worked themselves out.
    –Would equate the supply and demand over the period so as first to distribute the supply over the season.
    –It is <u>the</u> price which everyone in the momentary market is guessing at.
    –A price not affected by productive conditions – cost of production plays no role. A price which will distribute an <u>existing</u> supply over a certain time.

–<u>Stationary economy equilibrium</u>
 –A longer-run price brings about the shift of <u>production</u>. But this price starts at the same time as the other two:
   –cost of production = selling price.
   –value of productive resources = their contribution to value of product.
   –all resources are equally remunerated in all industries.
   –are three ways of saying the same thing [Referring to preceding three items, joined by brace in left margin].

–<u>Historical, cumulative changes</u>
 a. Some changes are the result of human acts, <u>human calculations</u>.
   – Only changes in capital come under this head – <u>human calculations</u> will be reflected in some change in <u>human wealth</u>.
   – Every productive act must yield either an *instantaneous* service, or store up future services in capital [Double vertical lines alongside this point in margin].
   – <u>Production</u> = consumption or capital-accumulation.
   – <u>Consumption</u> = capital dis-accumulation or production.

- The balance between those two sides of present and future (or present and past) is the rate of interest.
  b. Not call[ed] capital accumulation or dis-accumulation – historical change – but only the <u>non-calculated changes</u> in technology or resources, or wants. [double vertical lines in margin alongside this point].
    – Difficulty of defining the word "history."
    – Knight would like to confine the use of "historical" to every <u>non-calculable, non-predictable</u> change – for every predictable change is reflected in the capital market.
- Momentary, seasonal, etc. prices – only hold in the case where there are <u>accumulated stocks</u> – not to services consumed immediately.
    – The existence of stored up stocks is itself a speculation; they are a dead loss to their owners.
    – The essence of speculation is that the people doing it do not know they are, or at least how much they are [Double vertical lines alongside this point in margin].
    –With complete foreknowledge of demand situation, <u>inventory-cost</u> would be unnecessary.
    –In case of <u>physical</u> fluctuations of supply or of demand – there can be some inventory which is less or little a speculation.
    –In the momentary, speculative market [Diagram with amount on vertical axis and price on horizontal axis, downward sloping purchases line and upward sloping sales line, both highly inelastic] – the curves are very steep; on any day, the amount which would be bought slightly above the price is zero – the amount that would be sold – infinite.
    –Hook's law in physics – the strain is proportionate to the stress. The relation of pressure to displacement is linear – not perfectly linear in economics.

Limited Period –[Diagram with price on horizontal axis and with highly elastic consumption line, plus lines drawn perpendicular to each axis from arbitrary point on consumption line]. Most of a "demand curve" is pure dreaming. It is only in the vicinity of the normal demand that we can even guess anything – in such a short space we can not tell anything about slope.

  –Output is not a function of price in the case of (<u>constant</u> or) <u>decreasing</u> cost.
    –No curve can be drawn of supply under such conditions.
  [Diagram with cost on vertical axis and output on horizontal axis and with rising [positively inclined] curve. Under increasing cost – a usual curve can be drawn.
    –Cost affects price only as it affects the amount put on the market. What influences price is the amount put on the market.

[Two diagrams, bracketed, one with price on horizontal axis and downward sloping curve labeled rate of consumption, the other with output on horizontal axis, upward sloping curve labeled cost and downward sloping curve labeled selling price]. Are the same curve, rate of consumption a function of price; or price a function of output. Increasing cost can be put into the diagram.

   –But, though there <u>may</u> be constant cost, there can never be decreasing cost.

      –For the business man – all costs are money-costs.

         –Constant physical costs, produce increasing money costs; this is the limiting case.

Even in the <u>3rd</u> realm of <u>equilibrium</u>, there is shift of production from one industry to another – i.e. a short run, long run analysis of the stationary equilibrium. – <u>These</u> are Marshall's <u>short run, long run</u>.

   –But he did not distinguish his long run from <u>Knight's historical</u> – and grouped Knight's 1st, and 2nd together.

<u>Criticisms of Marshall</u> –:(1) Utter indefiniteness of his concepts of time-periods – what does he mean by long and short period? – He consciously avoided <u>long</u> period analysis.

   –"Normality is a condition of the system as a whole"; and (2) Marshall did not follow this precept – but analysed only a part of the system.

   –Distribution – cost of production <u>is</u> the distributive shares; two phrases for the same fact.

      –Division into three "productive factors" – land, labor, capital – which classical economists chose because they represented the three main social classes of the time (poorly!).

      –<u>Social</u> share in national income – all nonsense.

   –Distribution is money payments.

   –Price is income.

(2)[sic] Marshall did not tie the system together, did not discuss distribution at all in [his] Book 5, Chap. 5 but treated it in another book (VI) – <u>and</u>, (3) he treated price theory from the point of view of "stationary equilibrium" (marginal shepherds, meteoric stones – rough treatment of distribution). But <u>distribution</u>, in the next book, is discussed in terms of "<u>historical movement</u>."

   –i.e. giving the impression that distribution comes about only in the ultimate historical equilibrium (historical movement discussed in terms of an ultimate historical stability) – and discusses only wages and interest.

(4) And, he does not ever define what he means by his "<u>long</u> run period."

(5) Nor does he even tie together population growth and capital.

Knight's (3) <u>Stationary economy</u> is the same as Clark's <u>static</u> state.

–Normal price, normal <u>cost</u> (<u>distribution</u>).

    (4) <u>Secular economy</u> = <u>cumulative, irreversible changes</u>. "Historical" is not a
    good word. What we mean is <u>development</u> as against mechanical cause –
    <u>changes</u> in <u>resources, wants</u>, technology.

    (3) Has two aspects: <u>statics</u>, the formulation of equilibrium; <u>dynamics</u>, the theory
    of the movement of the system towards equilibrium.

    (4) Has two aspects: "<u>mechanical</u>" – investment and disinvestment; "<u>historical</u>,"
    in narrow sense.

–Under (3) – Marshall distinguished two aspects – short run normal, and long
run, but not adequately – and did not distinguish long run from secular.

    –Does he mean <u>growth</u> of total resources – or mere transference of resources
    from one use to another [In margin alongside: "p. 535"].

–Curves can never be drawn the way economists always draw them – you must
always start with a <u>point</u> – and study the way things will happen as the point
moves in <u>each</u> direction – and which of the four <u>periods</u> being studied must be
explicitly stated.

[Diagram with output on vertical axis and price on horizontal axis. One vertical
and three upward sloping lines intersecting horizontal line at common point]

    –<u>How</u> does industry <u>adjust itself to changes</u> – with reference to the <u>time-length</u>
    of said changes. The longer the time, the more flexible the shift is.

    –Given <u>unlimited</u> time for adjustment, cost will not rise appreciably, i.e. the
    curve will be "vertical."

    –Need of three-dimensional curve.

    –At least half of investment in an industry – in the historical sense – is irrevocable
    – except in case of an industry that is built for a specified period of time – as
    <u>World's Fair</u> – then investment comes out.

        –<u>Equipment</u> as such may often be transferred – though usual case is for
        transference by <u>replacement</u> in a new field. If equipment (even specialized)
        is short-lived, transference by replacement is rapid.

          –Unskilled labor = unspecialized labor.

      –You <u>have</u> to know whether the shift in output represents a shift <u>away</u> from
      another competing field – or a <u>new</u> increase or decrease of the system.

      –Marshall's long- and short-run normal price analysis is foolish.

      –Money expense is cost of distributive shares [Diagram with Products a, b,
      c, and Productive Services a, b, c; each of the former connected by lines to
      each of the latter].

      –What happens when there comes a shift in demand – is that some resources
      get a greater or lesser quasi-rent while adjustment goes on.

      –Process of readjustment is a problem in economic dynamics.

–What lies behind all this problem of readjustment is the rate of interest.

–Equilibrium = indefinite time for adjustment to work out, with historical change abstracted.

–Marshall's use of "plant" is foolish. The problems limiting adjustment are different for every case.

    –There is too much curve-drawing in economics! Curves drawn for one industry with its limiting conditions – can not be extended to another.

Marginal Utility

Curves showing utility

[Diagram with total consumption on horizontal axis, inverted U-shaped curve, horizontal at the top, originating from origin point and labeled "A" and downward-sloping curve labeled "B."]

–A = curve of total utility rises but at a decreasing rate – until it becomes horizontal, satisfaction complete – satiation, utility falls off.

B = curve of marginal utility falls off – each increment of total utility becomes less. When B crosses the zero line, A becomes horizontal.

–These curves are related by a mechanical rule [and] can be derived from each other by differential and integral calculus – these curves are calculus.

    [In margin, diagram with x good on horizontal axis; downward-sloping curve du/dx; and "he can not call this curve du/dx"]

–Jevons did not draw curve A!, the curve of summation, but only the curve of derivative.

    –Knight uses incremental for what is usually called marginal (a word used because economists did not know their a.b.c's). Jevons used final degree of utility.

    –Curve A = total utility of total supply: Curve B = incremental utility of successive increments.

    [In bottom margin: "1642 – Galileo died – Newton born (Leibnitz born 1632)"]

–These curves are basically in terms of force and motion (what do we mean by force and change in economics?)

    –A basic problem – how far the analogy holds.

–All of the theoretical mechanics is deduction.

    –After one bit of experimental observation (notion of quantity – calculable quantity – is one thing we don't pick up by knocking about in our childhood).

    –But arithmetic – as a science, or as an education – was impossible until the introduction of Arabic notation – Leonardo Pisano 1204 (year of capture of Constantinople by Venetians). – Not introduced generally into Italy till 15th century, England after the American Revolution, 18th century.

–What is the action of <u>force</u> against <u>mass</u> – it is always mixed up with friction. Aristotle: heavy bodies fall faster than light. Galileo dropped iron and wood balls from Leaning Tower of Pisa – rolled balls down inclined plane, with water clock.

–Found that <u>force</u> exerted on a <u>mass</u> – in <u>absence off friction</u> – results in <u>acceleration</u> [Double vertical line in margin alongside preceding sentence; also formula, $l = Kt^2$] – this was a <u>basic</u> discovery (It is only when <u>all</u> force is exerted to overcome <u>friction</u>, that velocity is constant).

  –How find <u>velocity</u> at any point? Motion must be continuous, yet if it is always increasing – <u>we</u> can not evaluate it. We have to separate the force used to keep motion constant, from the force used to increase motion [Double vertical line in margin alongside preceding sentence].

–But we consider a <u>constant force</u> as one that gives a <u>constant rate of change</u>.

–Study the <u>limit</u> – as $\Delta l/\Delta t$ approaches 0, when Δl and Δt are 0, 0/0 means nothing, but dl/dt = the last determinate integer (fraction)

dl/dt= slope of tangent to the curve at that point.

–If $l = Kt^2$, $dl/dt = 2kt$.

–The same relation holds for <u>any functional</u> relation, as utility: total changes and rate of change du/dX.

$Y = f(x)$ – it goes y distance in x time, thus $y + \Delta y$ distance in $y + \Delta x$ time

$y = Kx^2$, $y = 16x^2$ – we seek the distance covered in Δx and time.

$x + \Delta x$ $y + \Delta y = 16(x + \Delta x)^2$

Δx time $\Delta y = 16(x + \Delta x)^2 - 16x^2$

$\qquad\qquad\qquad = 16x^2 + 32\Delta x \cdot x + 16x^2 - 16x^2$

$\qquad\qquad\qquad = 32x \cdot 4x + 16\Delta x2$

$\qquad\qquad\qquad \Delta y/\Delta x = 32x + 16\Delta x$

$dy/dx = 32x$

$y^l = 32x$

As Δx approaches the limit multiply <u>a</u> by the coefficient and reduce the coefficient by 1.

  –Plot this – <u>a straight line</u>.

  –Velocity <u>is</u> a linear function of time.

$S = 1/2(gt^2)$ <u>Acceleration</u> (y) = 32.16 feet per second per second

  –Distance gone in first second – 0 to x, divided by 2 gives velocity.

  –Velocity = one-half sum of average and final velocity.

  –Average velocity in any interval = $1/2(0 + \text{final velocity}) = 1/2(gt)$

  –Distance gone = velocity x time = $1/2(gt^2)$

–Displacement of center of force sets up <u>oscillations</u> whose <u>period of time</u> is independent of the amplitude of the oscillation.

–Marginal utility is the <u>rate of increase</u> of total utility.

Total utility is the area under.

–Integral calculus gets <u>Area</u>.

–Utility is the name for the <u>n</u>th increment of good.

–In order to discuss choice – some <u>absolute quality</u> and some <u>relative quantity</u> are necessary.

–Utility did not come into economics till 100 years after Adam Smith.

  –Knight does not defend utility as <u>truth</u> – it may be merely a rationalizing concept.

  –We may doubt its validity, but must use its concepts.

–Quantity of <u>free satisfaction</u> called consumer's surplus.

  –In reality – you just <u>don't get</u> this.

  –This question throws us into the philosophical problem of utility.

  –Utility problem must relate to <u>some</u> given, fixed condition of choice (Cf. Marshall's answer to Nicholson).

  –Under such a given condition, no free satisfaction exists.

–Consumer's surplus is not psychologically true, yet it seems to follow <u>logically</u> from utility theory.

–We can't rationalize choice without a concept of total utility – yet such a concept is unreal [Double vertical lines in margin alongside this point].

–People seem to consider <u>increment</u> of utility without realizing the necessary revaluation of all previous units that is necessary.

  –Pragmatic philosophy is the <u>revaluation</u> philosophy.

  –(The poets preach that we should accept free goods – not bother <u>about</u> anything. The only real way to satisfy wants is to annihilate them. But civilization can not exist on this plane).

  –Deductive philosophy is the non-revaluation philosophy.

  –People buy units with respect to their <u>specific significance</u>, not considering re-valuation.

  –The concept of a maximum total is <u>vital</u> to the existence of economics (which <u>need</u> not exist). The ascetic ideal can not be followed and keep the notion of economy – it results in universal suicide – (which <u>may</u> be desirable).

  –Economy can not be rationalized except in terms of a maximum utility.

  [Double vertical line alongside in margin]

  [In margin, unlabelled diagram, with two curves ascending from origin, one with continuing ascent, the other turning down and reaching horizontal axis – at which point commences line to vertical axis with slope of c.30 degrees]

  –The inconsistency here is no more than that implied in <u>any</u> attempt to find complete rationality in life.

–Von Wieser – free goods have no utility; economy of consciousness, utility present only where is some impulse to action.

[In margin alongside preceding and following sentences: "du/dx, and Δx necessary to relate du/dx to some scale."]

–Valuations depend on comparisons – would not exist without comparisons.

　–Whole of distribution theory arises from the complementarity of productive units.

　　–It is just as true that consumption theory arises from complementarity of goods. $U = f(x)$ must be changed to $U = f(x_1 x_2 x_3 \ldots x_n)$

　　[In margin alongside preceding two points: "Without complementarity there would be only two truisms – no theory or need of it."]

　　–i.e. Jevons should have written $\partial u / \partial x_1$.

　　　–Partial derivative; partial change of utility due to change in one factor x_1, all others remaining the same

　　　–Utility would not arise if $U = f(x)$ only.

　　　–If we had the data, we could not solve the mathematics of the formula. Yet we must use the concept of the formula.

　　–i.e. the base of psychological choice is shifted with every new choice.

　　–There is some basis of comparison – and we know nothing more – nor need anything more to discuss economics.

　　[Diagram with rising total utility curve and, below it, falling marginal utility curve, both starting on vertical axis; re: former, "is this cut off from any base – or absolute numerical value," and re: latter, "is still perfectly definite."]

–We cannot measure (experimentally) marginal utility.

　–We assume it from consumer's activity.

　–How are we to take account of mistakes or irrationalities in consumer's activity. Knight does not think it necessary to pay too much attention to that. If we are going to theorize, we might as well do it.

　–There is not any connection between total utility and the problem of marginal utility of money for rich and poor. Jevons was perfectly clear on this, but Austrians and Marshallians talked nonsense about it for a half century.

　　–Total utility is a quantity that arises out of one man's comparisons between alternative ways of spending money – is relative only to one person – not to comparison of total utilities of different people. There can be no rational basis for progressive taxation, only common sense.

–The utility curve must be drawn as it would be if the consumer knew everything, made no errors.

–The <u>demand</u> curve is a curve of utility for <u>one</u> individual of <u>one</u> commodity <u>against money outlays</u>. Any point on that curve is the marginal utility for <u>all</u> purchasers – rich or poor.

[In margin and at bottom of page: Diagram, axes unlabelled, with one downward sloping curve and arbitrary point P on it: "x_1 (one commodity, <u>one</u> individual)." Second diagram, with total amount purchased on vertical axis and price on horizontal, and with one downward sloping curve; diagram as a whole labelled "Market D_1 curve – the <u>total amount</u> is the total bought by <u>all</u> the individuals (with their <u>individual</u> utility curves) at any price. – If there is a certain price, this curve shows how much <u>will be taken</u>." Third diagram, with price on vertical axis and quantity on horizontal axis, and with unlabelled downward sloping curve: "If there is a certain quantity, competition will set a certain price"].

[Next page: Diagram 1, with quantity on vertical axis and price on horizontal axis, with downward sloping curve labeled demand: "–but <u>not</u> allowing for long-run changes." Diagram II, vertical axis unlabelled and horizontal axis labelled Demand, with line bisecting quadrant from 0 point: "–Changes in this demand are <u>long-time</u>.

a-Assumes instantaneous adjustment of demand to price in [Diagram] I.

–<u>The demand curve abstracts from the process of adjustment</u>.

b-But demand is itself a function of price.

[Three-dimensional diagram, with vertical plane rising perpendicular from edge of horizontal plane; common side is time. Vertical plane has quantity on other side; horizontal plane has price on other side. Several downward sloping curves]

–The theory of advertising depends on a deception. It <u>ought</u> not to work if people were <u>on</u> to it. But they are so contrarily ignorant that it works anyway.

–We reason about human nature on the assumption that it is reasonable. We can't stop to rationalize the extent to which people are reasonable – every time we make a statement.

–The <u>economic</u> theory of human nature is that people act on the basis of <u>correct</u> and rational anticipations. Results do not enter in.

–<u>Language as a type of social phenomena</u>.

–This concept is what Knight thinks is <u>his</u> contribution to social sciences.

–Rational adaptation does not function in <u>linguistic change</u> – it can not be explained in terms of economics, mechanics, or aesthetics.

–Economic theory of short-period does not assume long-run rationality, but considers only short-run anticipations; it can study definite <u>miscalculations</u> of the short period – most important part of business cycle theory is this tendency to capitalize anticipations. Taylor's theory, consecutively corrected anticipations – studied by calculus. [In margin alongside "most important... capitalize anticipations," two vertical lines].

   –People think the future will be as it is now, but also think the future will change at the rate of present change, and that it will change as present changes [Single vertical line alongside in margin].

[Two diagrams. Diagram I with quantity on vertical axis and price on horizontal axis, Diagram II with the reverse; both with downward sloping curves]

–Producer's attitude may be either, i.e. may assume either price or quantity fixed.

–Under competitive conditions neither is fixed.

–From the consumer's point of view, it is always [Diagram] I, price fixed, quantity varies with demand.

Elasticity – of unity: if demand for a commodity is such that a change in price leads to a reciprocal change in quantity:

$-p \times q =$ constant (i.e. total receipts from total sales)

[Diagram III, entirely unlabelled, with curve of unitary elasticity and set of rectangles formed from lines drawn perpendicular to axes from points on curve]

–Every rectangle in [Diagram] III must be equal in area.

–Reciprocal: means equal and in opposite direction.

$$p \times q = C$$
$$100 \times 100 = 10,000$$

1% increase $101 \times 99.009 = 10,000$

–"Changes in Conditions of Demand" – include tastes, incomes, prices of other goods (complementary, competitive).

   –Taste = choice between two commodities can be represented by a curve [Diagram with y on vertical axis and x on horizontal axis, curve apparently of unitary elasticity and upward-sloping line from origin, labeled "taste" above the line and "income" below the line, intersecting curve] and introduce income change brings a third dimension. But to bring in all goods means $n + 1$ dimensions, times number of these systems, one for each person.

   –Taste = an individual's utility-surface (Edgeworth).

Data of an economic system = Tastes (utility surfaces)

                              = Resources, Technology

Change any one of the elements in Taste, in a given state in which adjustments have been made, and it may take an infinitely various number of lengths of time to work out. – there is no equilibrium except the stationary.

–Shift of production might take longer time or <u>might take shorter</u> time to work out to new equilibrium than consumer's demand (after change in taste). – the latter alternative, Knight has never seen stated.

–The <u>givens</u> in any economic system are individuals each having tastes and resources, and a social technology.

–<u>No process in nature can be discussed except in terms of a tendency towards equilibrium</u> [Double vertical lines in margin alongside].

–In stationary state, <u>historical changes must be stationary</u>; no individual can be attempting to accumulate or disaccumulate resources or tastes.

–Prices of commodity-services, prices of productive services, prices of goods.

–Quantity of final commodities produced; quantities of final commodities distributed to each individual – one and the same [In margin alongside, double vertical lines]

–Historical equilibrium is foolishness. People never get tired of making history. It was in this sphere that neo-classic economists made their greatest mistake.

–Even with <u>all data unchanged</u>, with conditions of demand <u>unchanged</u>; the <u>demand curve</u> can change – due to working out of <u>former</u> shifts in tastes.

–<u>All economic changes are the result of price changes, and tend to obliterate the price changes that occasion them</u> [Double vertical line alongside].

  –But <u>historical changes are not like this</u>, i.e. they do <u>not tend to an equilibrium</u>. Thus they can not be discussed in terms of cause and effect – but we can <u>discuss</u> only in those terms. Thus <u>most historical change</u> (as linguistics change) <u>is unintelligible</u> [Vertical line alongside last sentence].

–<u>Elasticity</u>: (Cf. Early Notes in Marshall's Mathematical Appendix): $xy^n = c$

	Price	Quantity
$p \times q = c$	100	100
	$100 \times 101/100$	$100 \times 100/101$

$E = 1$ when: price (p_1) is multiplied by any coefficient (γ) the quantity (q_1) will be multiplied by the reciprocal of the coefficient; i.e. by $1/\gamma$

$E = 1 = pq$ is a constant
$p_1/p_2 = q_2/q_1$ –price <u>change</u> in above case is 101/100
                                     –quantity change is 100/101

When $E = 2$, price change results in quantity change of $1/\gamma$ multiplied <u>twice</u> (not $1/2\ \gamma$, but $1/\gamma$ twice, i.e. $(1/\gamma)^2$

Price	Quantity
100	100
$100 \times 90/100$	$100 \times (100/90)^2$

$xy^n = c$ is the equation for all demand curves of constant elasticity.

$E = 1$, when $pq = c$, $\gamma p \times 8/\gamma = pq$

$E = 2$, $(\gamma p)^2 \times q/\gamma^2 = p^2 \times q = $ a constant

$xy^n = c|xy^2 = c$

n in Marshall's equation is E in Knight's.

$E = (dq/q)/(dp/p) = (dq/dp)(p/q)$ –Instead of $100 \times 1/100$, a 1% change, we use $100 \times dq/100$, an infinitesimal change.

[Diagram with quantity on vertical axis and price on horizontal axis, and negatively sloped line running from t′ on vertical axis to T on horizontal axis; point P on line t′T at point bisected by line from origin; lines drawn from P perpendicular to axes]

$dq/dp = $ slope of line at point P.

Elasticity $= $ slope of the demand curve <u>times</u> quotient of p and q.
or, $(dt/dT)(P/q)$

–Cournot saw this and defined elasticity in 1838.

a-Price must be put on the base line, as it is the independent variable. Marshall put quantity on the base line, but defined elasticity in conventional terms with price the independent variable.

b-It <u>is possible</u> to put quantity on the base line, if it is treated as the independent variable.

  –In both cases some time must elapse for working out – it is not an instantaneous process. Except that the time necessary for equilibrium is <u>much</u> longer in case of (b), (when quantity is set and price is dependent).

  –<u>Agriculture</u> has a highly inelastic demand, and also a highly inelastic supply – so that lower price brings no diminution of supply (small increase of output brings lower price).

    –In fact, a lower price brings a <u>larger</u> output – an increased elasticity.

    –A case when perfect competition defeats itself.

<u>Production</u> – we are most interested in production and distribution – though will discuss price and costs.

  –<u>Mises</u>: naïve belief in utility theory; it provides us with a <u>rationale</u> for consumer behavior; classical theory provides us with a rationale for producer behavior. We do think in terms of more and less – that is utility theory.

  –<u>Rational</u> behavior is uniquely determined behavior, it is mechanical – could be predicted from known causes.

  –Change from a utility curve to demand curve.

–Many authors (like Carver and Mises – who see everything in terms of black and white, clarity – but miss all the fundamental problems) go direct from the utility curve to the demand curve, without more ado.

–The demand curve is a curve of <u>relative price</u>.

  –A curve of comparison of commodities in terms of a set amount of income (We are <u>not</u> talking about the utility of income as a whole). [Alongside in margin, three diagrams: one with utility on vertical axis and curve starting at origin and rising; another with quantity on horizontal axis and negatively sloped line; third with quantity on vertical axis and price on horizontal axis, and negatively inclined demand curve, possibly intended to be of unitary elasticity]

  –It is merely another illustration of the concept of allocation.

  –Marginal utility curve is a curve of the <u>utility</u> of one commodity.

  –Demand curve is a curve of the monetary <u>demand</u> for that one commodity, but in terms of the relation of the commodity to demand for all other commodities.

–Money income is almost <u>more</u> an <u>end</u> than goods and services. We do <u>want</u> money income.

–From standpoint of public policy, we do have to treat goods and services as ends.

  –<u>Utilitarianism</u> – (peculiarly Anglo-Saxon concept) – value the act because of its result.

    –This is foolishness as a philosophy of life. But public policy must accept this.

    –It treats <u>freedom of action</u> as the best means for achieving maximum return (as Adam Smith).

    –But we really treat freedom of action as an end in itself.

Problem facing democratic government: to enforce fair rules of the game people <u>want</u> to play. Government can not decree the <u>ends</u> which people will want.

  –Dictatorship merely crystallizes those wants, understands <u>them better</u> than the people. It does not try to <u>mold</u> wants. No social science could tell, even afterwards, what the effect of a dictatorship on the <u>wants</u> of the people had been.

  –Society is a <u>game</u>, more than anything else.

  –Nothing more lunatic than the economic interpretation of social process – that man fights for <u>material</u> ends.

  –Opinions and tastes are closely connected with interests, opinions <u>are</u> interests, but the causal relation goes both ways.

  –Human nature seems to be competitive – even the attempt to be non-competitive would be competitive.

–Practically all the values of 20 years ago are going to be reversed in <u>our</u> generation.

–dq/dp is the slope of the curve, it is not the elasticity [Diagram alongside with quantity on vertical axis and price on horizontal (and on the next two diagrams), with curve resembling curve of unitary elasticity, labeled dq/dp]
   –Elasticity is not arithmetic increments, but is <u>ratio</u> increments.
–$\Delta p_1/p_1$ –grows from infinity to zero as the base p_1 increases.
$\Delta q_1/q_1$ –grows from infinity to zero as base line $q1$ increases.
   [Diagram with negatively inclined straight line connecting axes]
–<u>Elasticity</u> of a curve at any point, is the slope of the curve at that point divided by the slope of the line drawn from that point to zero. Marshall.
   $= -(dq/dp)(P/q) \; ((-dq/dp)/(q/p))$
   [Diagram the same as preceding diagram, but with line from point on negatively inclined line to origin]

–The process of demand is one of rational <u>budgeting</u>.
   –The classification of items is arbitrary. You can't draw up <u>lists</u> of commodities, the whole problem is an abstract one – even wheat's classification is arbitrary.
<u>Equilibrium</u> – defined in stationary terms, abstracted from historical changes.
   –Adam Smith – market price, normal price.
   –Stationary equilibrium breaks down into stages:
      a/–rate of production, and speed with which production can be adjusted.
      b/–the effect of accumulation.
   –Relation between means and ends is perfectly symmetrical in economic theory.
      –Both means and ends are <u>data</u>, are givens.
      –We are only interested in certain relations.
   –Robbins – his point is just not so – except as he contradicts himself, later in the book.
   –Economic theory deals with
         1–<u>means</u> (instruments, services of goods and persons)
         2–<u>ends</u>
         3–process
            –allocation
            –technique, not economic
      –There are certain functional forms of procedure which are involved no matter what the social organization.
      –We <u>must</u> distinguish between <u>means</u> and <u>processes</u>.
      –Economist <u>assumes</u> people do not choose between ends – sets this one off from the rest – but gets into all sorts of trouble this way.

–We are concerned with a <u>means-end</u> relation.

–Brings up the whole problem of how far scientific method <u>can</u> be applied at all to any study of man.

–We assume man to be a maximiser.

–Market opportunities are a purely technical function.

–There is <u>no</u> economic theory of international trade – what have political borders to do with it.

–International trade is only one case of general economic theory.

–i.e. Meade's cases, subsequent on a change in demand, are only in fact a study of effects of a change of demand, on the economic system [Single vertical line in margin alongside this point].

–Economic norm is defined out of our civilization; is a norm which people in other civilizations may or may not follow.

–If they don't follow this norm, if they are different from us, if they are <u>uneconomic</u>, then economic norm does not apply to them.

–But it is still just as true as a norm and shows what they do <u>not</u> conform to.

<u>Equilibrium</u> – what are the factors one would have to know to predict perfectly the price.

–Under what conditions would price change or not change? *

–A certain quantity of a perishable commodity in the market.

–Even then <u>time</u> makes a difference. The past and the future are definitely there, making a difference in demand for the commodity to-day.

–How much <u>storage</u> is possible[;] how much is there?

–Where there is an appreciable <u>time-interval</u>.

–Speculation – people look forward to future conditions of supply and demand.

–Abstracting the existence of stocks we get normal short-run theory.

–The assumption behind all normal short-run theories <u>is</u> that the <u>rate of flow</u> of production onto the market is the same as the <u>rate</u> of flow of consumption out of the market.

–Cost of production <u>is</u> the distributive share.

–Production and consumption are <u>instantaneous</u> in the going system.

–Production process has no length.

–Output is not a function of price except under conditions of increasing cost.

–Supply curve can not be drawn except on assumption that there is an adjustment of cost and selling price.

–Which assumes the productivity theory of distribution – no loss-no profit.

–Think of payments in production as payments for the use of productive services.

–It is absurd to think of payments as interest for the use of money.

–And that interest and rent are different; they are both payments for the use of productive services.

–Forget interest as payments on money.

–The answer to Böhm-Bawerk is entirely a matter of capital accounting.

–Maintenance of capital must be included in production costs in equilibrium.

–If there is a change, in historical sense (4) – then the time it takes the production process to work out that change, must be treated as time taken to change, and no more.

–Hayek, etc. – are really thinking of the time it takes to get capital out of its fixed state and liquid again.

–This item of time-length has very little to do with anything we are interested in.

–A catalogue of all the parts of the capital of any enterprise, and their time-length of their life – will tell nothing of the time it takes to change that enterprise to another field.

–You must consider the enterprise as a whole, in accounting terms.

–Product is the excess of total production over maintenance.

–Product above maintenance may be consumed, or not. If it is not consumed, it is saved, it is growth.

–But equilibrium analysis assumes no growth.

–Labor units introduce a new factor of difficulty and must be abstracted.

–No accounting can register the increase of production capacity of labor due to investment.

–You don't produce except as you produce above maintenance.

–Even if you plan for liquidation at a definite date (World's Fair), you must get enough income before that date to cover the loss of value of those capital items which are non-transferable to any other use at that date.

Supply curve is a cost curve assuming production in equilibrium, and perfectly worked out, and all payments are for productive services – property or personal; property divided into rent and interest. But interest not considered as payment for money.

–Start with a system with no production costs.

– Smith's beaver-deer example.

– Suppose beaver hunters and deer hunters are two separate castes, and can not do the other's work

– Suppose beaver grows on two units of land, deer on one unit of land – but land is transferrable.

– Takes two days to kill a beaver, one day to kill a deer.

    – To make any sense of this argument – you must <u>assume</u> a <u>demand</u> curve
    for one commodity expressed in terms of the other.
      – Such that the <u>demand</u> <u>clears</u> <u>the market</u>.
      – Equilibrium demand makes an equal utility ratio between the two.
      – The amount of one <u>considered</u> equal to some [or same] amount of the
      other [Single vertical line alongside].
      – i.e. utility is essential to the argument.
–If we assume, instead of man-days, <u>land-years</u>, then we can get somewhere – then
the subjective side does not at once enter in.
–We should never make a ratio of two things which are not perfectly homogeneous.
    –If 2 deer = 1 beaver, then does 1 beaver = twice as much as 1 deer.
    –If so, we violate the principle of utility.
    –If the two are equal, we can express a ratio – but not otherwise.
–Suppose beaver men and deer men can not do each other's work.
a-Do not desire each other's product – no exchange.
    –We can not use pleasure-pain concepts – which depend on establishing a zero-
    point, which we can not do.

    [Diagram in margin: straight line (no axes), x marking midpoint, also labeled 0,
with minus sign to left and plus sign to right of 0]

    –We can not make a demand curve for <u>one</u> good divorced from any consideration
    of any other good.
    –The demand curve <u>is</u> a comparison curve.
–Purchasing power is either the amount of deer or beaver brought to the market,
or the amount of skill for getting deer or beaver each man has.
[Two diagrams, axes unlabelled, one with negatively inclined curve labeled Utility
and positively inclined curve labeled disutility, the second with negatively inclined
curve labeled Utility of Commodity X and positively inclined curve labeled
Competing Utility of Other Commodities. Statement, "Knight does not like this
curve – we can not measure it," related to first diagram by arrow]
    –More utility of the Commodity X, means that it will be gained at the expense
    of other utilities from other goods.
    –This can be called <u>cost</u>.
      –Demand and supply, utility and cost – are all the same curves – essentially
      – and are all a question of <u>alternative allocation</u>.
b-Beaver men – deer men – costs. If they can exchange beaver for deer – then
producing beaver becomes a means of producing deer.
    –There is no difference between <u>land</u> and <u>labor</u>.
      –There is a leisure use for both.

–Personal powers are property; or property is an attribute of personality.

–No distinction on ethical or causal grounds.

–Both are inherited, result of environment, earned.

–No distinction on logical grounds.

–Some distinction on legal grounds, much difference on many other grounds.

    – But do they cause a difference for economic theory?

–Main distinction for the economist is that labor can not be capitalized, can not be sold outright, can only be leased.

    –History has never shown a real slavery system.

–To apply economic theory to a communist system, we consider that the State makes the consumer's choices, and the state makes the producer's choices.

    –The State is in same position as the person who makes both consumer's and producer's choices, in our system.

        –As any case of self-service – e.g. shaving at home – it could be priced – so that we give a man 0.15 [cents] for his income, every time he shaves himself, but subtract 0.15 for expenditure everything he spends it.

        –In communist system every thing would be like this – but there would be the same problem of how much to include in the pricing (arbitrary) – and how much to leave out.

–Deer man – beaver man, etc.

    –Twofold alternative in use of product after it is gotten. – use it himself, or use it to get the other good.

    –Can be represented:

[Two diagrams, axes unlabelled. One with negatively inclined curve labeled Utility of X, and positively inclined curve labeled Competing Utility of Y. "Or" in between the two diagrams. The other only with a negatively inclined curve labeled Utility of X in terms of Y]

–Because one beaver exchanges for two deer in the market, yields no equality of costs – for the two costs are in different individuals' heads.

[Three pairs of immediately foregoing diagrams, entirely without labeling, under the three titles, Deer Man, Beaver Man, and Market]

–Introduce an entrepreneur who hires both men – costs will be equal to money prices in both cases, but money prices are not caused by costs – are only derived from prices.

–For more specialized goods [Alongside, diagram with Quantity of Deer Hunting Offered on vertical axis and Wage on horizontal axis. Negatively inclined curve

"?a" intersected at same point by almost horizontal, i.e. slightly negatively inclined curve "?b" and by positively inclined curve "?c"].
  –If you assume the two things – commodity and leisure – as independent – curve a – but they are not independent.
–It is possible to get joint shares of joint cost – in fact, a competitive market does do this.
–(Add partial derivative, and working both ways – to Marshall's statement on p. 393).
–In the perfectly competitive system there are no entrepreneurs.
  –If we assume resources to be drawn into the production system by the possibility of return – of course the return to each share is the value of each share's contribution to the total.

  [Diagram with Bedeutung (tr. "major meaning": "resources?") on vertical axis and Allocation of Activity to Production of Commodity X on the horizontal axis, and one unlabelled negatively inclined curve. Alongside: i.e. as more is allocated in this direction [along the horizontal axis] it means less in respect to other resources]

Transferability, mobility, fluidity vs. specialization to a particular use.
  –Joint activity: heterogeneity between different elements, or a subjective valuation – make real cost measurement impossible – on money valuation.
  –One homogeneous transferable resource – then relative objective cost = price.
  –Subjective value comes out in intensity.
  –Homogeneity necessary for measurement.
  –Producer can work at only one occupation. But consumer buys a composite collection of all products.
  –There is no comparative relation between two commodities if they are not consumed by the same person.

1–Stationary economy – problem of expansion of an industry [No 2, 3 present].
  a-Transfer of resources – if these resources are homogeneous.
  (1)–Costs are increasing in dollars.
  (2)–Physical alternative commodity costs are constant.
  (3)–Resource cost – constant – if character of resources is fixed.
  –If 2 or 3 increase – it is due to non-homogeneity – a limit to the product must be paid for.
  –Alternative product cost is price determining.
  –If resources are fixed, resource cost is price determining.
b-Resource terms lose all relevance once the character of resources changes with changes in the scale of industry.

–With expansion of one industry usual changes are:

(1) Changes in proportions.

(2) Changes in kind

Under either of above:

(1) From competing industries

(2) From outside, extra marketing uses

(3) Intensity in use

All result in ascending cost curve.

But quantity of resource is meaningless.

c-Thus alternative commodity costs – is the only real basis to use with changing composition of cost.

–Hence, yesterday's point – costs not even measured as utilities (by same person) except a few marginal souls.

–For consumption: great mass on the margin.

–For production: very few, even on the margin.

–For shiftable resources – principle holds, as subjective evaluations are non-relevant and no reservoir of non-market uses, and cost = values.

–Resistances to transfer will increase short-time cost curve more rapidly.

d- No equilibrium under decreasing costs. Even under stationary conditions, decreasing costs are really historical changes.

–A competitive stable price can not exist with them.

–Must come from economies of organization; if available to an individual firm, it is a monopoly; so need limitations on size of firm – here it is really a question of social psychology.

Reaffirms: –(1) Increased productive power comes only from other uses, or from leisure activity. (2) But growth, differentially, is really the chief mode of transfer – allow more rigid specialization.

Increasing alternative product costs: change in kinds of productive resources – new supply less adapted, progressively less efficient.

Understanding of the what and why of variations in cost with output.

–No profit, selling price = cost, distribution problem accurately solved. Think of costs as homogeneous, split them up into distributive shares later, if different productive services enter into production.

–Abstract from growth; consider growth of an industry, i.e.

(1) at the expense of another – for we consider that the total production (a difficult concept to define) of the community remains the same.

–Although a very large part of the fluidity of productive services among industries comes as a direct result of growth. We have had such rapid growth, that we have found it possible to develop <u>rigid</u> specialization much more than would be the case if the <u>total</u> of productive capacity were to be stable.

    –Quasi-rents should be accompanied by quasi-wages.

  (2) or growth at expense of leisure.

–Cost is the amount of <u>other possible</u> products that will have to be given up to produce a certain one, i.e. <u>increasing alternative-product cost</u>.

1. <u>Productive agencies transferred are less adapted to their new use than to their old</u>.

2. <u>Productive agencies applied more intensively to a fixed agency</u> – (land fixed, labor moved: <u>increasing output per unit of land, decreasing product per unit of labor – this is law of diminishing returns.</u>

    –What is the product per unit of productive resources? We can not tell, unless we have some common denominator for land and labor.

    –We throw out old concept of <u>real</u> cost.

3. Drawing in extra services of productive capacity from leisure.

    –Not sure that this leads to <u>increasing</u> cost for an element of irrational human behavior and subjective valuation enters in.

    –Only tropical peoples are rational in this sense – i.e. pricing system doesn't work – only whiskey system does.

    –Raise money income – will people work less and get same <u>total</u> income (money) as before?, or work more to get larger total money income, or some compromise in between.

–Any supply curve finally comes down – at infinity an infinitesimal supply gets an infinitely large income.

[Diagram on which Supply of Labor is on the vertical axis and Wages is on the horizontal, with a curve commencing from the origin, rising, reaching a maximum and then, more slowly, falling]

–What of work for sake of working?

–Importance of conventional standard of living.

    –Lower wages bring <u>more</u> work.

    –Raise wages, so far as people have no desire to live above the standard [of living], <u>less</u> work.

    –In general, Knight follows Austrian theory – necessity of considering transferability.

    –A major fault of English economics is its concept of variability of supply – which brings in <u>growth</u>.

    –It is necessary to fix the totals of the production agencies – then supply curves show alternative product cost.

–Labor and productive capacities may be used to bring services that pass through the market, or services that do <u>not</u> pass through the market (Knight calls this <u>leisure</u>).

–There could be no laws of mechanics if the elements were not fixed and invariable.

–Economics can not deal, in terms of theory, with any more than this concept of fixed elements. But it does have to deal with growth, too, i.e. every time there is growth, the whole system is different and a new study of equilibrium is in order.

–High pressure work management is an historical change.

Short run-long run analysis

–Capital agencies are fluid in the long run. Even human beings are fluid in long enough run – so far as capacities are acquired.

–Practically all of Book 6 (Marshall) deals with historical changes.

[In top margin: Ph.D. thesis on <u>executive salaries</u> – needs a cynical satirist, not a scientist to write – paid not to think, to sit and look, assure everyone that everything's all right. People with brains don't look or act that way. Are like movie contracts, advertising, pay high salary and fool people into believing he is great man]

–Tie up short-run, long-run analysis with <u>quasi-rent</u> and <u>quasi-wages</u>.

–Specialized capital and labor take it in the neck during a change. But productive services will be paid equal to cost at all times.

–Shor<u>te</u>r-run: (1) Cost and selling price equated, but at a different place from ultimate equilibrium. (2) Productive services will get a different share than in ultimate equilibrium.

–Quasi-rent and short run-long run analysis both belong in a stationary economy.

–Different processes happen at different rates.

–Friction and inertia are inevitable sources of confusion.

–Inertia without friction gives perpetual oscillation after any movement.

–Friction ends oscillation, which can not overcome viscosity. Inertia is swamped. Movement shifts down to zero, stays there.

–<u>Economic system</u> has quite level viscosity and <u>some</u> inertia.

–Sometimes it has a condition of unstable equilibrium (instead of pendulum condition a cane balanced on a point – elastic top).

–Resource cost to any entrepreneur is the cost of getting it away from all other uses.

–But money cost, due to friction in the market, does not always correspond to resource cost.

–Thus prime and supplementary costs enter the problem.

–<u>Physical</u> obstacles to immediate adjustment – bringing in quasi-rent (only another aspect of <u>overhead cost</u>).

–<u>Monetary</u> obstacles to immediate adjustment (inertia is a figure of speech).

–<u>Intellectual</u> obstacles – mental lag of comprehension.

–<u>Legal</u> obstacles – fixed contracts.

–"Perishableness" of <u>labor</u> is foolish – everything is perishable in some degree.

–Entrepreneurs' unit-cost at a <u>moment</u>, and what it would be at equilibrium – what are the factors that enter into the difference between these two. [Single vertical line alongside]

    –Speed of accumulation of replacement fund is determined by the anticipated return.

    –There isn't, in fact, much hope of getting physical units <u>out</u> – they are <u>sunk costs</u>.

        –Legal costs, planning cost, marketing cost, human adjustment costs.

        –Can <u>not</u> be gotten out after once being invested.

    –The physical units are sunk, monetary costs are planned to include replacement fund on these irretrievable units.

    –Actually, replacement is largely a function of <u>growth</u> in the community – speed of growth gives possible speed of replacement.

        –We can never tell even whether there has been growth. Cf. Marshall, p. 431 – how could he build an economic system without separating growth to the total and shift within the total of investment?

–Ricardo never talks about <u>land</u> as bearing rent. Rent is the payment to capital and labor yielded by varying quantities of land.

    –<u>Rents</u> in theoretical sense <u>are money costs to the entrepreneurs</u>, although are not alternative commodity costs.

    –There is usually an element of price determined costs in every activity.

    –Emphasizing relativity (to price) of specialization (suitability) of various resources.

    –Two modes: (1) Direct changes of use. (2) Indirectly, through depreciation fund.

    –In short run all agencies are specialized. In long run all agencies are transferable.

    –"Value producing capacity" is the thing transferred – measured by this value; constituency is a technical problem.

    –There is no ideal index number – no constants in nature – no indestructibility of matter in economics.

    –Summary:

1. In general, competition of products for services (resources) makes costs price-determining.
2. If <u>non-transferable</u>, payment for services is differential – a pure Ricardian rent.
   –Here, value of specialized resources depends upon competition of entrepreneurs.
   –Reversal of usual relationship of resources to services.
   – Fixed costs are not costs, but a bookkeeping entry.
   –Are costs to the single person, but are not costs to the industry or to the product.
–Classical economics in error by identifying natural resources and non-transferable resources.
–Problem: Do marginal payments exhaust the product? Euler's theorem proves they do.
   –If y is a first order homogeneous function of a, b, c . . .,
      $y = dy/da \cdot a + dy/db \cdot bt \ldots$
   –So, if double variables, double function.
   –So, if all products are paid increment ($dy/da \cdot a$) these payments, summated, will equal the product.
   –Hence, the residual is equal to the marginal product in these differentials of true rent.
   –Hobson and Clark in 1890s both saw that we can consider any agent as fixed and dose it by another.
   –No meaning to separate product other than this incremental effect, i.e. partial differentiation.
   –Robbins was crazy when he said that the decrease of one factor involves (1) decrease of product of that increment, (2) loss, due to lesser productivity, of other factors, i.e. inferior organization of industry.
   –Thus, $p = f(a, b, c, \ldots)$

A. A. Young (in Ely's Outline): Residual product of land is axiomatically its marginal product (Same true of all factors).

Residual, Ricardian Theory of Rent

[Diagram with horizontal line intersected from above by negatively inclined line, with rectangles formed by drawing perpendicular lines to each axis from series of points on sloping line. Rent on Land is sum of areas above horizontal line and Return to Labor and Capital is sum of areas below horizontal axis]

[In margin alongside following initial notes: What makes a piece of land better than another? What is <u>good</u> land?]

Marginal product – the amount the total will be reduced by the withdrawal of the last unit of cultivation.

1. To say that a piece of land is worth its difference over the marginal land in product – is to say that its value is equal to the difference between nothing and something.
2. The land with highest net product where the same amounts of labor and capital are applied.
    - –But the <u>same</u> amount of labor and capital will never be applied to two pieces of land.
    - –The amount that <u>will</u> be applied depends on the product – you apply <u>that</u> amount of labor and capital to the two lands that will make the first increment of output the same.
    - –What one <u>does</u>, is get a maximum total by making the final increments to production in the two cases the same.
    - –As must happen when free mobility of labor and capital are assumed. Free competition <u>will</u> bring this about.
    - –The residual theory of rent <u>is</u> the marginal theory of productivity.
    - –The residual paid to any <u>one</u> factor of production after the other factors have been paid their marginal productivity is <u>its</u> marginal productivity [single vertival line in mrgin alongside this point].
    - –It is merely the inherited approach to the problem to use <u>land</u> as a residual.
    - –It is true that you don't move land about in space, physically, though it was naïve to extend this into economic theory.
    - –The only possible measure of the <u>quality</u> of land is its rent, i.e what anyone will pay for it.
    - –There is no sense in the theory of rent.
    - –The residual falling to any small increment of land is its marginal productivity – the best possible application of labor and capital being assumed.
    - –Product is same function of the factors used.

p $=$ f(a, b, c, . . .) (<u>not</u> plus signs)
    - –To allocate resources so that product per laborer was everywhere the same would bring an enormous loss of efficiency, or else would bring war between laborers in high productivity and those in low productivity industries.
        - –There can not be bookkeeping except in terms of marginal productivity. Could we accept a bookkeeping that <u>showed</u> such inefficiency?

$p = (a + b + c + d \ldots)$ is the simplest function.

But there is no sense in it for economics, for there would be no complementarity between the factors.

$p = (a^r + b^s + c^t + d^u \ldots)$ if $r + s + t + u = 1$

There are only two factors worth distinguishing, labor and property (they are the same productive service); if we are to use any more factors, we may as well extend them to infinity [Double vertical lines in margin alongside this point].

$p = a^r \times b^s$

$p = a^{1/h} \times b^{h-1/h}$

$p = 1^- \times c^{1/4}$

(Seems to be the division between labor and capital in the U.S.)

[Diagram pertaining to the foregoing, alongside, axes unlabelled, with Curve of Diminishing Returns commencing at origin]

[Diagram pertaining to the following, alongside, with Total (Product) Yield on vertical axis and Variable Factor (Fixed Situation) on horizontal axis. Two parallel positively inclined lines, the leftmost commencing at origin]

–Application of a variable factor – a linear relation.

–Product starts at zero, factor starts at same positive integer.

Diminishing returns may mean that as you increase the variable factor in any ratio, the ratio of increase of the product will be less.

Diminishing returns may mean that as you increase the factor by equal increments (arithmetically), you get decreasing increments of total product.

–Depending on shape of curve: straight line, equal increments = equal increments of yield bending upward, equal increments of factor = increasing increments of yield, bending forward, equal increments of factor = decreasing increment of yield.

In the case of ratios, it is a matter of direction of the curve, whether the tangent passes through the X axis, Y axis or origin.

If the variable factor, considered in the ratio curve, yields increasing returns, the fixed factor is being wasted.

	Land	Labor	Product
(1)	100	100	100
(2)	100	200	300
(3)	50	100	150

–If 2 is true, it is also true that a larger product could have been obtained with a smaller amount of land, as 3.

If competition is to yield the best organization, there must be diminishing returns in <u>both</u> cases.

[Diagram with curve resembling a bell curve, first section labeled I, second, II and third, III – as described just below]
–The best way to draw the curve – diminishing returns in both cases, after a period of increasing returns.
I-returns increase more than proportionately
II-returns increase less than proportionately
III-returns decrease

–I stage with one factor treated as variable, is the identical curve as III with the other factor treated as variable. (Cf. F. M. Taylor – Principles)
   –In one case you increase the numerator of a fraction by equal steps, in the other you increase the denominator of a fraction by equal steps.
–IInd stage, is the only part of the curve that makes sense anyhow – and is the simplest part.
$p = a(1/r), b((r - 1)/r)$
Take $b = 10,000$, then $p = a(1/r) (10,000)1/r$
$$p = 100\sqrt{a}$$

[Diagram with p on vertical axis, a on horizontal axis, with curve with only <u>supra</u> stages I and II, unlabelled. Alongside vertical axis, ascending numbers: 1.00, 1.41 [?], 1.79]

–If there is not decreasing returns in the <u>ratio</u> sense, there is not an economic situation.
–Rent has a bearing on the difference between cost of production and price.
   –The alternative-product cost at the margin (final increment) determines the price.
   –More and more transfers of non-homogeneous factors brings less efficiency of their use.
   –Proportion of more mobile factors to less mobile ones brings less efficiency – decreasing return.
–Decreasing cost leads straight to monopoly.
   –Unless there are presumed exceedingly complex technical structures.
   –The realities of competition between firms determines the size of the firm.
   –It is possible to say something about the technical unit in decreasing cost.
   –But practically impossible to say anything about the size of the firm (what is a firm) and its relation to efficiency.

–Competition must have brought each firm to its size of maximum efficiency before we start to discuss decreasing cost (Increasing returns is a Cambridge phrase, is not good, all mixed up with wrong Ricardian rent doctrines).

–Distinction between <u>plant, firm</u>, and <u>industry</u> is essential to the problem. – Cambridge school has not done this.

–Knight doesn't believe there <u>are external economies</u> – hasn't ever seen one.

–If there are external economies available to the industry, but not available to the firm – then there can be decreasing cost.

   –Marshall's illustrations are palpably false.

–External economies of one industry are internal economies of another.

–A true external economy would have to be such for the <u>whole process</u> of producing a product – getting into the intricacies of the whole production system.

–Decreasing cost is not theoretically impossible, but when we have defined the conditions in which it can exist we are so far from reality that we can never prove its existence.

–Much of the problem is a <u>moral</u> one, of the bureaucratic abilities of men.

–Any organization can – and naturally does – build up a much larger temporary size than it can permanently run.

–Barbarians could conquer Rome – but could not <u>run</u> the Roman Empire.

–There is a good deal of elementary geometry in this problem of the size of best technical efficiency.

   –Why are birds not as large as elephants.

      –Why are men not <u>twice</u> as large – volume cubed, muscles and bones squared.

         –We are not infinitely small only because of the limitations on the division of matter.

      –Arctic animals are all heavy, big – more space for radiation of heat.

      –Many mechanical measures are not possible.

         –Horsepower X necessary to drive ship Y. What horsepower necessary to drive ship ΔY (twice as big)? – cannot be calculated.

      –Size of human being is indivisible, fairly.

         –Width of ditch for 6″ or 24″ pipe is the same – to get a man into it.

         –i.e. many limitations on the efficiency of pure <u>size</u>.

            –"Snowball grows of its own momentum" – but has anyone ever seen a 100 foot snowball? – It falls apart of its own weight before then.

         –Tremendous importance of the size of firm in an industry.

            –If it went down all the way, there would be only one firm in the industry.

–If it went up all the way, there would be zero number of firms in the industry.

–The usual cost curve for the firm.

[In margin alongside the last three points, three diagrams: a downward sloping, an upward sloping, and a U-shaped cost curves – otherwise unlabelled]

–Marshall's representative firm: Not worth much attention, too useless.
–Marshall's predilection for biological analogy.
  –But even then, he didn't take account of the factor of uncertainty of the future.
  –The firms could not <u>know</u> of their life cycle or planning would be very different.
  –But the economist is supposed to know it.
  –Either known or unknown, the life cycle is unimportant.
    –If we can talk about it at all, the life cycle of a firm changes its rate too often.
  –The analogy of the forest is very bad – the economic forest is not the same all the time, but changes in every respect.
  –Only in exceptional cases can a firm make provision for its own end. [Single vertical line alongside]

–Relation of size of firm to market area (in geographic sense).
  –Are transport costs included in cost of production?
  –Cost and price to be F.O.B. as delivered.
  –Perfect competition impossible if cost and price on Delivered [basis].
    –Two firms of different efficiency, perfect competition, a hyperbole if [unfinished sentence] (Cf. Economic Journal – Joan Robinson)
  –Most of recent discussion of monopoly is palpably false.
  –Suppose equilibrium (this assumes the creation of a new economic system overnight), the ideal equilibrium position at one time, and the ideal equilibrium position at another time, <u>are</u> not unrelated, separate; if our second equilibrium grows out of a change in the first, it will be <u>very</u> different from the other concept of economic equilibrium.
  –The concept of commodity is an <u>area</u> concept – production is expanded by expanding <u>area</u> of sale – this is the same as <u>encroaching</u> on other markets in <u>any</u> other way – advertising, etc.
  –What <u>is</u> the concept of commodity – a, b, c, d?
    –Or a <u>continuum</u> – abcd <u>arbitrary</u> points.
    –Possible to go from one to another commodity without going through any empty space.
      –As <u>color</u> – there are no colors, except by arbitrary definition – an indefinite number of colors, an indefinite number between any two points.

–A firm is a region in commodity space (Advertising is a part of the saleable quality of a commodity – if the purchaser pays for it – but is the same as if he thinks the product is <u>better</u>).

–Thus the size of the firm is important.

–We can not start from a position of theoretical tabula rosa in discussion of monopoly, but must start from some actual point.

$P = f(a_2\ b_2\ c_2\ d_2\ e_2\ \ldots)$ –Remuneration of any productive agency can be called a wage or a rent.

  –At equilibrium there is no profit.

  –Theory of profit is a discussion of why imputation theory doesn't work well enough, so that profit and loss is left over.

  –Every interest contract [i.e. contract providing for interest payments] could have a rent contract substituted for it.

  –All distributive payments we have assumed to be payments for things, not money.

  –Property items, like labor items, make the <u>owner</u> take care of <u>replacement and upkeep, maintenance</u>.

    –Hirer of labor leaves that up to the worker.

    –Renter of property leaves that up to the owner.

    –If a man owns a farm, but has a mortgage on it to its full value, does he own the farm or not?

    –Here arises the whole monstrous definition by Marxists of Capitalism – assuming the "Capitalist" to be in <u>power</u> over labor.

    –The capitalist is the man who rents his property to industry, he is in <u>exactly</u> the same position as labor.

    –Who controls whom? – the consumer controls, by his control of expenditure.

    –In an imperfectly competitive system, the property owner can mortgage his property and wait – treating his capital as reserve and drawing on it.

      –The laborer – not being able to mortgage or sell is productive services (involuntary labor, etc.) due to our legal system [Alongside in margin: Inalienable rights] – is in a worse position – in a state of <u>pressure</u> to sell his <u>services</u>.

      –A society which guarantees the freedom of an individual is setting up a limitation of freedom; the political authority refuses to enforce any <u>labor</u> contracts.

–Only possible economic theory is that of <u>the price mechanism</u> – nothing apologetic about this.

–Universal instinct of humans, when anything is wrong, to find a witch and burn him. Preachers were looking for a "bad man" to knife him – preferably in the back.

–The existing <u>power-endowments</u> are taken as given data – no ethical implication.

–Are no different from property-endowments.

–When liberty ceases to be a value, discussion ceases to be a possibility. (No discussion, in objective sense as a basis of action, is possible).

–The intellectual processes involved in <u>discussion</u> are <u>entirely</u> different from the intellectual processes involved in scientific analysis and discovery.

    –Failure to realize this will probably bring the downfall of Western civilization.

    –The scientific method is manipulative; used in society it will set everybody at everyone else's throat.

    –Rules for manipulating subject matter are <u>very</u> different from rules setting up the rules of the game – the latter is immensely difficult.

–By discussing the competitive system, we do not mean to uphold that system.

    –Put any cure for social evils into the hands of mankind, and it becomes impossible to criticize that cure from an objective point of view, without being accused of upholding the social evils.

–Economics defined as the science of wealth, has been the greatest confusing element.

–The basic economic unit has a <u>time-aspect</u>.

    –There is no such thing as light unless it is moving at 1,860,000 feet per second.

    –There is no magnitude: light: only a flow.

    –There is nothing but a flow, but nothing flows.

[Note at top of page: But we <u>do</u> increase capital cumulatively – does it become a larger share than consumed services, or do consumed services increase anyway? FTO]

    –We probably can't conceive of light except as flowing through ether – must have <u>such</u> a space concept.

    –Gallons of water exist whether they flow or not.

    –But light doesn't exist if it doesn't flow.

–"Relation of effort to return is the fundamental economic reality."

    –Why limit economy to labor, it is as applicable to inert resources.

    –Classical economist said <u>original purchase price</u> was cost of labor.

        –Really it is the other goods that have to be given up that is cost.

–Mill defined production as anything that adds to total social wealth.
  –If nothing but this were production, there would be no production.
  –Really, this is not production – it is an increase of the possibility of consumption in the future.
  –This doesn't satisfy any wants. [Arrow indicating that this should come before "–If nothing . . ."]
–To count the value of the capital created in a year, and then to count the income created by that capital in later years is to count the same thing twice.
–Production is just what does not add to total social wealth – Mill 100% wrong. [Alongside in margin: "In sense of 3, not in sense of 4a"]
  –Even maintenance is not production.
–Labor is betwixt and between – is more outside than within the concept of production.
–Shall we count the maintenance and replacement of labor in the same way as capital?
  –The difference is entirely legal, institutional.
–If entrepreneurs hired all property as well as all labor – they would have no capital account (for maintenance and replacement would be up to the owners).
  –He has no capital account as far as labor is concerned, now. The services are hired from the owner of them, and he is left to take care of maintenance and replacement.
–These property and labor owners come into the economic system only as they are hired, only as they are sold. [Alongside in margin: Volkswissenschaft]
  –The replacement and maintenance is left to individuals in their private decisions – privatwirtschaft.
  –The economic system is not concerned with what these private decisions are – governments may be.
–There is no property right to labor – law will not enforce any. Any laborer can lay down his tools and quit at any time.
  –There is some actual infringement of this.
  –Labor force (or consumers force) can not be formed anew every day.
–The capital created in any year is the accumulated value of its product.
–Nothing is produced until equipment is first maintained.
a-Rent – assumes the productive agency is kept in maintenance by its owner – no capital account is kept.
    –The individual owner may keep a capital account.
  –In an equilibrium position – all production and consumption are instantaneous – is anything done but maintenance? Yes.
–Wages is the same as rent. The individual owner is responsible for maintenance account – though it is practically impossible to keep that account.

–Laborer knows his time-length of service – not perpetual. FTO

b-<u>Interest</u> – is a% rate.

–If a reproducible item of capital's return doubles, its value rises for a period – quasi-rent – as far as the <u>individual</u> goes, price determines marginal utility.

–As far as any individual case goes – capital value of any item is determined – in the short run – by capitalization.

–But in the longer run – the capital value does not double, when the short-run return doubles, but the return does not remain at that high point. So many new capital items are made that the return is brought back to where it is in line with the rate of interest.

–Historical cost has nothing to do with capital value.

–Rate of interest <u>is</u> the residue, after maintenance has been provided, divided by the capital value.

–Whether instruments are durable or not, their value is calculated by return times going rate of interest.

  –Demand curve for the capital item, rate of construction and its time – must be known in short-run.

–The interest rate itself enters into every item entering into the calculation for interest – this doesn't matter – <u>one</u> unknown in the formula.

–Annual increase in capital value of land (unearned increment) is a capitalization of the increased future income.

–Waiting is a bad word.

<u>Abstinence</u> is the right word – one <u>does</u> abstain – because the individual investment is never liquidated – he [the individual] liquidates by selling out – but he has the capital in money form, then.

–If you know that you can take advantage of an investment opportunity, by sinking $1000, and get – above perpetual maintenance – $50 return per year for good – the rate of interest is 5%.

  –This is all there is to the rate of interest theory.

  –Demand for capital is infinitely elastic.

– Rate of interest is ratio of money sum of return and money value invested, capitalized at rate of interest.

  – Capital instruments are produced at a cost.

  – In the long run their capitalized value must equal that cost – in any situation where capital instruments are being continuously produced.

  – We are abstracting from profit – as long as capitalized value equals cost of production, there <u>is</u> no profit. Profit is discrepancy therefore.

  – Capital instrument is <u>any</u> material good that yields a rental and is saleable.

  – Cost of production of a capital instrument is the cost of the productive services put into it.

- "Present goods" – live so short they are not counted as capital goods; we do not distinguish their final service from any other service.
- Capital value is found by a discounting process.
  - Wealth is a discounted present value of future services.
    - If life of a good is 15 minutes, it can have a discounted value as wealth, but we do not consider it so, rather we considerate to yield its services instantaneously, i.e. no rental possible, only instantaneous sale possible.
- Cost of a capital good is the value discounted back to the present.
- When its production is finished, its rendering of service begins.
  - An abstraction, there is no such line, we should consider when the cost of production exceeds the rendering of services as this line
- Cost of production is accumulated up to that line, the rental return or future income stream is discounted back to that line.
  - It doesn't matter where the line is, the sum of both sides will be the same.
  - Discounting is done (1) by setting aside for perpetual maintenance, or (2) setting aside for definite period maintenance – by a discounting formula. [In margin alongside: "Mathematically [(1) and (2) are] the same, but (1) is less simple."]
- Any investment is the diversion of productive agencies from producing present to producing future services.
  - The point where this dispersion takes place is the point where cumulation of cost of production of the investment begins.
  - (1) above – perpetual maintenance provided for, the return of any one year is divided by the cost of production of investment – to get rate of interest.
  - This is the correct, and common sense view of interest.

What of supply and demand curves for capital?
  - Rent is an economic pons asimorum. Why have best economic minds thought that rent implied economy of effort? Economy of effort is not different from any other economizing; effort is not the only thing to be economized.
  - Theoretically there is a total productive capacity, but it does not all enter into the pricing system at any time. Cf. Edgeworth vs. Böhm-Bawerk.

[Two diagrams: (1) With price (interest rate) on vertical axis and quantity of capital on horizontal axis; no curve, but label "cost curve." (2) Axes reversed; also no curve, but label "supply curve."]
– All theorizing about productive equilibrium depends on the assumption that the cost curve is the supply curve [Double vertical lines in margin alongside].

–Any cost curve for a commodity is expressed in terms of time, quantity is quantity per unit of time [Diagram with cost on vertical axis and quantity (wheat) on horizontal axis, with unlabelled demand and supply curves crossed]. [Double vertical lines in margin alongside].

–Marshall uses it in this sense about half the time.

–3/4 of textbooks are wrong at this point.

–In case of capital, the current production is added to the previous production. This is the only difference between supply and demand curves for capital and for other goods.

–The supply and demand curves will have to show the price at which new units of capital will enter the market, taking account of the previous accumulation, which changes every time a new point is reached.

–It can be worked out in terms of a quantity.

  –Taussig's wrongness is most pedagogically picturesque.

[Usual demand and supply diagram, with axes unlabelled]

  –Demand curve is one over long times, in terms of quantity.

  –Supply curve is for a given time – in terms of flow.

–Marshall draws the demand curve in terms of quantity – fixing the rate of interest, the demand curve will show how much capital society will consume over an indefinite period of time.

  –An equilibrium rate, at which, at some future time, no further additions will be made to capital.

  –Assumes a stationary demand curve while capital is accumulating – but while it accumulates, the demand curve must change, and the supply must [do the] same.

  –The supply curve especially (and demand curve too) is drawn out of the imagination.

  –What happens is that the demand quantity curve moves ahead almost as surely as capital accumulation moves ahead.

  –We do not move along these curves, but, as capital accumulates, the curves move.

  –There is no such thing as long-run equilibrium – Marshall's Book 6 is nonsense; there is no ultimate saturation point in society.

–The only theory of interest is a short-run one.

  –In terms of an equilibrium of rates of flow – production and consumption taking place at the same flow.

  –Any supply curves in interest theory are bad.

  –Rate at which people will save is a function of the rate of interest – supply curve.

  –Only assumption is that people invest at the best opportunities.

–The rate of interest in any transaction is the best rate that could be obtained
in any other transaction.

–Rate at which capital market will consume savings is a function of the rate of
interest – demand curve.

[Diagram with Rate of Absorption on vertical axis and Rate of Saving, of Flow,
on horizontal axis; with horizontal line intersecting vertical line, both unlabelled]

　　–We do not know what shape the supply curve takes　　[Referring　　to　　the
　　diagram].

　　　　–Probably is about constant.

　　　　–Saving a result of social mores.

　　　　–The psychology of saving is chiefly a matter of getting ahead in life – a
　　　　question of power.

　　　　–As French desire for a fixed money competence.

　　　　–The rich man has a margin of saving as much as the poor man.

　　–Demand curve is vertical: there is no limit to the speed at which capital can
　　be invested; provided any investment is profitable; the demand of the capital
　　market is indefinitely elastic.

　　　　–In short run, since it takes time to make plans, a sudden new increase in
　　　　rate of flow may cause a glut.

　　　　　　–Hayek right in thinking the business man always counts on the same
　　　　　　supply of savings forthcoming as he has been used to.

　　　　–Thus a decline in demand for capital – as Fisher draws the curve.

　　–If commercial banks increase credit, there is "locking," especially if there is
　　no unemployment.

　　–Business men always have plans.

　　　　–There is a certain rate at which the market is set to absorb capital.

　　　　–If saving comes into the market at a changed rate, suddenly/unexpectedly,
　　　　there is a glut, or leaves the market – low rate of interest.

　　　　–Suddenly, there is a squeeze – "distress bidding" – high rate of interest.

　　　　–Thus, for the short run – until plans are changed – the demand curve for
　　　　savings slopes to the right, downward.

　　　　–But given time to make plans, the demand curve is perfectly elastic.

　　　　–With constant cost, cost determines price – no matter what demand is,
　　　　price is cost.

　　　　　　–With increasing cost – it is impossible to say whether supply or demand
　　　　　　determine price.

　　　　　　–Constant cost means infinitely elastic supply.

　　　　　　–Which is the demand side, which is the supply side in the determination
　　　　　　of price?

–What is really bought and sold in the capital market, is not the services of capital, but-interest bearing securities.

–It is a matter of constant cost and infinitely elastic demand.

  –There is no variation in the cost of a perpetual income.

[Diagram with Rate of Saving on vertical axis and Quantity of Investment on horizontal axis, with line, infinitely elastic over much of its range and then slightly turning down]

–No matter how much is invested, the rate does not change much.

–If the demand curve remains the same – and if it were unity, we would have to double the total of saving (500,000,000,000) to reduce interest by one-half, then double 1,000,000,000,000 to reduce interest rate by another half.

Wages in Marshall, Book VI

–No distinction of long and short run and historical.

–No distinction of <u>labor capacity</u> from labor actually given.

–Cf. Economic Journal, December 1930, Harrod.

[Diagram, with Price of Income on vertical axis and Money Income Earned on horizontal axis; downward sloping curve, from point P on which to point Q on horizontal axis runs a straight line; neither labeled]

–Worker is in position of getting money income by giving up leisure.

–A curve drawn so that the work done is always the same, irrespective of wage.

–P, point at low income, at which worker works as hard as he can.

–Q, point at which, if you raise income enough, work stops altogether.

  [Diagram, no content except for Wage on horizontal axis]

    –<u>Wage</u> is reciprocal of price of income.

    –Axes reversed.

–Knight doesn't put any stock in any of this – it is a mere logical exercise.

  –We don't know <u>anything</u> about all this.

  –We don't know how hard a person <u>can</u> work.

    –Couldn't know till he was dead.

–Try to get some idea of realistic psychology.

–Criticism of utility theory after war – most of it worse than the theory itself.

  –Part of human psychology to throw over anything whose consequences they don't like.

  –We dispose of competitive system today, without trying to find out what was wrong with it to bring its consequences – kill dog instead of finding malady.

–English chain worker works three times as hard when wages raised three times. Mexican would work one-third as much when wages raised three times.

Malthusian theory (better stated before him)
  –Raise wages, population increases – long run equilibrium theory.
  –Smith and Ricardo have, not a subsistence theory, but a standard of living
    theory – are perfectly certain in this.
    –Standard of living does set a long-run wage level if there is a line, below
      which, in standard of living terms, no one marries or does anything but try
      to get above.
  –Ricardo's "bargaining power theory of wages" is based on assumption that
    there is no competition in labor market.
    –Employer would pay any particular worker additional amounts of wages as
      long as these made him work more.
    –This logic underlies Marx and Lassalle's theory of wages.
    –Employer pays what it is worth to him to pay.
  –Marx's dialectical, metaphysical theory, labor cost is basis of exchange.
    [Diagram, unlabelled, with horizontal line starting at vertical axis just above
horizontal axis, then becoming vertical]
    –Below standard of living line, labor supply decreases; above that line, labor
      supply increases indefinitely.
    –Long run equilibrium has meaning in wage theory.
    –The lower limit does operate, over a moderate period of time.
    –Except that standard of living would not remain the same, it is itself a function
      of labor supply and wage conditions.
    –Mill saw this.

Profit theory – in different category from other theory.
  –In U.S. especially, profit is usually defined – as it should be – as the residual
    left over after normal payments have been made to distributive shares.
  –Profit is a result of failure of imputation theory.
    –Are profits and losses balanced out – so that distributive payments do exhaust
      the product in the social sense.
  –If entrepreneurs foresaw events correctly, there would be no profits and losses
    – every productive resource would be put to its best use, paid its due share.
  –If competitive market had the same foresight, no one would be able to pay
    anything but this share.

–Profit theory must open subject of uncertainty.
  a – technical uncertainty – small in some industries, large in others (agriculture).
  b – business calculation uncertainty.
  –In mathematical probability there is a law of probability on shape of frequency
    curve.

–But this does not hold in relation, even to equal probability[,] in connection with <u>human</u> affairs.

–Life insurance, statistics are midway between.

–In case of business calculation, there is no <u>law</u> of distribution of probabilities. Problem of estimate brings up problem of knowledge – how do people make estimates?

    –It is a matter of estimate of immense complexity.

    –The accuracy of estimate is itself an estimate.

      –No chance of <u>equal</u> probability.

–What is the average of success in business estimation?

    –Any hirer of productive resources bids those resources away from other bidders.

    –Only if business estimation is <u>pessimistic</u> does profit appear uniformly positive.

    –You can't tell whether an industry was "profitable" until it ultimately liquidates – most industries don't ultimately liquidate.

    –A study of those that do is poor sample.

–Any person who goes into business is an <u>optimist</u>.

    –"Pure enterprise" would hire <u>all</u> its productive resources.

      –[But] It does not exist. Would people who use their own productive res ources make lower return than those who rent.

(3) F. TAYLOR OSTRANDER'S TERM PAPER, "THE MEANING OF COST," PREPARED FOR FRANK H. KNIGHT'S COURSE IN ECONOMIC THEORY, ECONOMICS 301, UNIVERSITY OF CHICAGO, AUTUMN 1933

Introduction

The paper published below was prepared by Taylor Ostrander for Frank Knight's course, Economic Theory, Economics 301, during the Fall 1933 quarter.

The paper is obviously the work of a bright graduate student who learned his professor's approach to the topic very well – and equally well was becoming socialized into the ways of the economics discipline. The paper closely reflects Knight's procedure and tone, as well as substantive treatment of the material.

Knight instructed his students on some of the limitations of his approach to economic theory. Other limitations were not brought up and examined, for example, that a theory of the meaning of cost and a theory of cost (given some meaning) are not the same thing; that a theory and a model are not the same thing;

that the theory of cost that Ostrander produces is derivative of his assumptions, i.e. of the variables he has included and how he has structured them; that the theory is limited by other meanings of cost and other theories, including the variables he has excluded; and so on – the meaning and significance of several of which were not more fully appreciated until later in the 20th century.

Among the technical issues one might raise, given the theory, are: the apparent change in the numbers of entrepreneur-producers and of consumers, from one to plural, i.e. whether his case changes from bilateral relations to competition; and his negative view that an individual can compare utility and disutility. Is the former instance a matter of "relax[ing] the assumptions, bring[ing] in qualifications and establish[ing] the theory in some closer relation to the actualities of economic organization?" Ostrander's statement in the latter instance is worth specific notice:

> even when the comparison of utilities and disutilities is made in the mind of the same person, while we can admit with only slight qualification that that person might be capable of comparison, even of some commensuration of different utilities, or of different disutilities, it is a much greater assumption to make the claim that that individual has any basis on which to base comparisons between utilities and disutilities.

This position seems to negate the calculatory process at the heart of neoclassical economics. Thorstein Veblen, in his essay on "Why Is Economics Not an Evolutionary Science?", satirized and repudiated the conception of man as "a lightning calculator of pleasures and pains"; Ostrander's paper, and what he learned from Knight, seem to deny any basis for the neoclassical version. Or, is there confusion between a methodological assumption and a definition of reality? And if so, is it due to Knight's pessimistic and dismissive-critical attitude? Elsewhere in the paper, Ostrander writes of "a margin of indifference between market and non-market uses of productive power." Surely the calculatory process is operative here. And surely it is operative on the margin of transferability – "margin of indifference" – discussed near the end of the paper in connection with dropping his third assumption, in which the matter of "costless transferability" also enters, so important to Ronald Coase's theory of the first, a few years later.

As with the other Ostrander documents (principally class notes, of course), editing has been kept to a minimum. Typographical and punctuation corrections have been made. Some awkward phrasing is noted (so the reader will not impute misprinting). Minor restructuring has been done. With Ostrander's permission, Knight's handwritten comments and corrections are placed in the text within brackets []. The editor's interventions are within braces {}.

A coda is in order. Early in 1934, Ostrander had been asked to teach at Williams College during 1934–1935. The invitation was made by Walter B. Smith, chairman of the Department of Economics, who had been Ostrander's professor at Williams.

During the negotiations over the offer, Smith suggested that Ostrander consider submitting a paper for the David A. Wells Fund Prize Competition at Williams College. Ostrander sent Smith this paper on cost theory and another paper on the Elizabethan era building industry, written for John U. Nef's course in economic history. Ostrander submitted "The Causes of Changes in the Supply of Building Materials, and the Effects of Such Changes on the Forms and Styles of Buildings, Studied in Relation to the Elizabethan Period" for the 1934 competition. Neither his paper nor that submitted by someone else won the Prize of $500 plus printing. In an early handwritten note to Smith, Ostrander wrote,

> I have spoken to Knight about this. He has worked about two years, in his way, and has just finished a 50 page article on the subject "Cost and Utility." He is very decent about my using this paper on Cost for a prize essay, but would like not to have it printed before his comes out. I don't feel I should like to have it printed at all – unless it were more original.

Knight's apparently published the aforementioned material in "Bemerkungen über Nutzen und Kosten" in *Zeitschrift für Nationalökonomie*, vol. 6, nos. 1 and 3, 1935, pp. 28–52, 315–336, and *Notes on Utility and Cost*, Chicago, IL: University of Chicago Press, 1936.

THE MEANING OF COST

F. Taylor Ostrander

The problem is to answer the questions: what is the meaning of cost; what factors determine cost; and the relations of cost to price. The method will be to study an arbitrary case, based on rigid assumptions, from this to generalize and state a theory of cost, and then relax the assumptions, bring in qualifications and establish the theory in some closer relation to the actualities of economic organization.

I

The arbitrary case is best described by stating the assumptions on which it rests.

(1) Assume one single, homogeneous productive resource, fixed in its total quantity. This may be expended in pecuniary or non-pecuniary use, any expansion of the market use being at the expense of the non-market use.

(2) Assume that the resource may be used in either of two uses, but that it is wholly used in these two uses, any expansion of the market use of the resource by the one outlet being at the expense of the market use of the resource in the other

outlet. The two uses result in the production of two commodities, which we shall call, Alpha and Beta.

(3) Assume that one entrepreneur produces both commodities, but that he keeps separate books for each.

(4) Assume that the productive service is feely transferred between those two uses, without the introduction of any new elements due to transfer; in other words, assume that cost is constant whatever relative quantities of the two commodities are produced.

(5) Assume that the demand ["money expenditures"] for these commodities is of a fixed total extent, but that it may vary in the proportions in which it is distributed between the two commodities; demand being always expressed as demand for Beta commodities in terms of Alpha commodities, or as demand for the Alpha use of productive resources in terms of the demand for the Beta use. Demand must clear the market.

(6) Assume that there is only one consumer of both commodities.

(7) Assume that both consumers and producers are guided by the principle of maximizing the return to expenditure; that the economic problem of proper allocation of both expenditures toward the end of greatest return is an axiom; i.e. assume "intelligent behavior in a market situation." {Knight seems to have questioned the use of singular producer and consumer in items 3 and 6 and plural ones here in item 7; see also 11 and passim.}

(8) Assume that the consumer's demand is affected by diminishing usefulness of successive increments of commodities.

(9) Assume an economy where production, distribution, and consumption are instantaneous; all changes in the unit of account and all lags in adjustment being abstracted from. That is to say, we assume that the economy is in equilibrium, the amounts of each commodity which produces equal unit-utility to the consumer, being the same as the amount of each commodity which the producers choose to supply.

(10) Assume all changes in the "Historical" sense to be abstracted from.

(11) Assume that the distributive process works exhaustively, total return being distributed to the productive services according to their incremental contribution to the physical output; the imputation of a product share to a particular service is carried out by competitive bidding among entrepreneurs.

It is hardly necessary to carry out the description of the arbitrary case any further; these assumptions have defined it and described it at the same time. We shall consider the case under two different positions of demand for Alpha in terms of Beta.

a-Assume the variable ratio of demand to be temporarily fixed at a ratio of 2 to 1; that is to say, two units of Alpha commodity have equal unit-utility to the consumer with one unit of Beta commodity. By definition, then, this is the ratio in which these commodities will exchange in the market, and the ration in which they will be produced. As there is only one productive resource, one unit of it used in the production of Alpha will produce two units of product, while one unit of it used in the production of Beta will produce one unit of product. Cost per unit of resource being, of course, constant, the production of two units of Alpha is equivalent to, or is, the production of one unit of Beta.

These two commodities have become the objects of two sets of choices. The consumer chooses between them in the expenditure of his income, and the producer chooses between them in his decision as to how much productive resource to use and in what direction to use it. In the equilibrium that we have assumed, the amounts of each commodity that create equal unit-utility to the consumer are the amounts that the producer chooses to supply. Prices, then, must correspond to producer's choices and to consumer's choices, and cost may be said to be the sole determinant of price, for the only thing that determines whether the amounts of consumer's demands will be met by the amounts of the producer's supply is the cost of producing these relative amounts of Alpha and Beta. In other words, since by definition there can be no lag or increasing cost in the transfer of resources, demand is bound to be met by an instantaneous equivalent supply (The only possible limitation on supply is also described as scarcity, utility being assumed, and in the case scarcity has, by definition, equal influence in each line of production).

b- Let us assume this variable ratio of demand to shift to a new figure $2\frac{1}{2}$ to $\frac{1}{2}$; two and a half units of Alpha being desired for every $\frac{1}{2}$ unit of Beta. Producers of Beta will instantaneously yield $\frac{1}{2}$ unit of productive resource to the producers of Alpha, and, as there is no additional cost per unit of output due to this shift, production of the commodities will be in the ration $2\frac{1}{2}$ to $\frac{1}{2}$, ["At what price(s)?," "Inelasticity of demand?" and "Used jointly in fixed proportions?"] and they will exchange in the market at a market price of the same ratio. A new equilibrium is established and cost is again the sole determinant of price.

From these cases we can draw two principles: (1) When all the assumptions 1–11 are made, price is not only equivalent to cost but is solely determined by cost. (2) Pragmatic, or empirical cost means the amount of one commodity given up when the other is produced. The cost of the extra half unit of Alpha is clearly the half unit of Beta which is no longer produced. In the same sense, the cost of the two units of Alpha that were formerly produced was the one unit of Beta which might have been obtained in place of the Alpha units. This cost is not determined by price, but is determined by the value of the productive service in the other use. Two Alpha units will cost one Beta unit, and the Alpha units will not be produced unless their price also is in a 2 to 1 ratio. Their cost in terms of Beta units determines their supply as well as the price at which their supply will be sold; while the cost of Beta units in terms of Alpha units determines the supply and price of Beta units. Under the restricted assumptions of this case, the cost principle may be stated thus:

> The number of units of any commodity B which exchange in the market for one unit of any other commodity A must be the number of units of B which are sacrificed in production in adding the last unit of A to the total produced.

In this simple case, the cost in terms of resources used is the same as cost measured in terms of alternative product sacrificed ("potentiality cost," as Whitaker names it), and both are the same as price.

It is to be observed that we have resolutely avoided any attempt to determine the meaning of cost in philosophical or "real" terms. Any such proposal is outside the sphere of the economist's rightful jurisdiction, and the many attempts at determining such costs are additional witness to the fact.[1] Money cost, on the other hand, explains very little; looked at from an individual point of view, what is cost to one man is income to another. Looked at from a social point of view, money-cost, being a mere summation of individual money-costs, means even less.

II

We can now relax some of our original assumptions, not in an attempt to reach complete reality, but so that we may get somewhat nearer that reality in our analysis.

Assumptions 1 and 4

Productive services are not homogeneous, but are infinitely varied and dissimilar; which is to say that they are specialized and not freely transferable without an effect on cost.[2]

We can study the effect of this heterogeneity on cost under two conditions. In the first place, limiting ourselves to a purely "pecuniary" apportionment of productive resources; in the second place, extending our analysis to include that definition of "pecuniary" which takes into consideration the large volume of production and consumption which is on a margin of indifference between market and non-market uses of productive power.[3]

(a) It is evident that specialized resources will not be equally suitable to all occupations. As demand for two commodities shifts, that for Alpha increasing at the expense of that for Beta, and as resources are transferred from the one use to the other in response to the price change which is the visible evidence of the change in demand, the resources are progressively less suitable to their new use (psychologically and physically).[4] Price of Alpha will rise with expanding demand for it (even when there is no lag in production changes) due to the fact that the particular resources taken into the expanding industry are successively less suitable to it than to the production of Beta. With each successive transfer, it is the final incremental, alternative-product cost of a unit of productive service ["resource"] that is equal to and determines the price of the incremental unit of product; but this cost does not solely determine total price of the total product, that being determined by supply and demand and their relative elasticities, in relation to specialized resources.

The surplus of marginal price ["cost"] per unit over total price ["over cost"] per unit is a payment made to specialized resources ["and the earlier units of unspecialized"]. In the occupation to which they are best adapted that payment is a "differential rent," which will increase as demand expands and price rises. The payment made to resources in the occupation to which they have been transferred, and are less adapted, is in the nature of an increasing alternative-product cost. With each successive shift, the alternative product given up becomes greater and more out of proportion to the amount of product gained in the other use. The price of the marginal unit in the expanding use is directly determined by this alternative-product sacrificed. With each shift there comes to be a greater proportion of more mobile resources in the expanding industry, and a greater proportion of less mobile in the declining industry. Parallel with this, the specific productivity of less mobile factors decreases, that of more mobile factors are price-determining, the greater the relative immobility of factors, the steeper will be the curve of increasing in price with expanding output.

Heterogeneity of resources also raises a problem of imputation or distribution. For it follows, not logically but nearly inevitably, that if resources are dissimilar, more than one kind will be used in the production of any given commodity. Varied productive resources will be combined in some proportion, presumably that best adapted to yielding maximum returns (In such a case, it is impossible to measure in physical units the contributing amounts of different resource services used in making any unit of output). With expansion of demand for one commodity and consequent expansion of output, the proportion in which heterogeneous resources are combined may change, this breaking down that combination which was most suitable, and setting up some new, less favorable proportion. That is to say, as well as ["besides"] the rising cost due to employment of resources in less suitable occupations, there is a rising cost due to their being combined in less suitable proportions. This is an <u>additional</u> [deleted] source of a permanent rise in price, of "specialization rent," of increasing marginal alternative-product cost, and an additional source of price-determination-costs. {"X" in the margin alongside last sentence.}

(b) Passing on now to that other category of limitations on the freedom of transfer arising out of heterogeneity of resources, we come to an analysis of the efforts on cost of the existence of non-pecuniary alternative uses of productive resources. This is the other plane of less restricted pecuniary alternative uses of productive resources. This is the other plane of less restricted pecuniary uses, more in accord with the realities of our economic organization. As has been said already, the very existence of a non-pecuniary alternative, even if it is on the margin of indifference with a pecuniary use, is a destruction of the original assumption of complete homogeneity ["necessarily?"]. But it is possible to <u>consider</u> change along this margin of indifference as constituting free and costless transferability. Having merely noted, then, that the existence of the problem is, in strictest logic, already limitation on free transferability, we pass on to consider the effects of such a margin of indifference on cost, and the factors acting to prevent the creation of such a margin.

One of the most important assumptions on whose validity the existence of a margin of indifference is postulated is that the functional relation between the quantity of productive capacity ["expended in pecuniary use"] and the remuneration to it does not change with any shift of a productive resource from one occupation to another.[5] If this functional relation or proportion does change with a transfer, if the productive services are on a margin of indifference between producing a given commodity with a given inseparable combination of pecuniary and non-pecuniary uses, and producing some other commodity where the inseparable combination of those uses is different, then the transfer is accompanied by costs that do not determine price. For cost is determined by price in so far as a transfer of productive capacity involves a change in non-pecuniary remuneration of the instrument transferred. Furthermore, all instruments whose use is divided between the pecuniary and the non-pecuniary uses but which are not on a margin of indifference between them, are "specialized" instruments earning rent. {"?" in margin alongside last sentence.}

Due primarily to inseparability of the resources involved, and a necessity of choice between large complexes – "bundles of advantages and disadvantages" – involving especially non-pecuniary elements, a transfer frequently does mean a giving up or receiving of other considerations than those directly expressible in money-price, and because of this there is limitation on the establishment of a permanent margin of indifference. Some element of non-pecuniary, "differential rent" must have entered price, and more than constant alternative-product is at stake in the transfer.

On a parallel to the increasing cost already mentioned as arising out of the less suitable combination of heterogeneous resources, after an expansion in the industry, we must point out that an additional increasing cost is occasioned when a shift in the proportions of those heterogeneous resources involves redistributing their productive capacity into new proportions between pecuniary and non-pecuniary uses.

In other words, this case b) is but a special case of what has been already stated under a): when the assumption of constant alternative-product cost is broken down, cost, to the same extent, is price-determined, not price-determining.[6]

Assumption 2

As we have already found it convenient to speak of production of more than two commodities, and by implication have found that that did not alter our principle. However, we must extend this assumption to state that we continue to assume that productive resources are wholly used up in the totality of their uses, otherwise we are embroiled in the great complex of unemployment of resources, i.e. the trade depression.

Assumption 3

We now drop this assumption, so that there are producers for each of the many commodities we are considering. We continue to assume that books are kept. However, certain difficulties arise when we relax this original assumption. Whereas, on the consumer's side, marginal utilities of equal price-units of any two or all commodities are equal to every consumer of both or all, on the production side no significant parallel to this is possible. For different productive activities are seldom directly compared as to subjective quality by the same individual.[7] A single productive resource is practically never in a position where its owner can establish a relative disutility of one employment as this employment is progressively combined with other employments. Such a comparison could only be drawn for owners who were actually on a margin of indifference between two occupations, but in that case there would again be free, costless transferability, and no problem for cost theory. Because this can scarcely ever happen, there is an almost inevitable limitation on free transferability, and increasing cost is usually the result of expansion of any one industry.[8] Likewise, there is on the side of production no parallel to the consumer's spending of a fixed total fund of purchasing power; for the producer's choice among different uses of resources involves a more or less different total income.

Assumption 5

While we can enlarge the concept of total demand so as to cover more than two commodities, without affecting the cost principle, it is necessary to retain the assumption that total demand for all commodities does not change, for that would be a change in the Historical sense. Likewise it will continue to be assumed that demand must clear the market, otherwise we are again involved in the trade depression. Demand for one commodity continues to be considered as the reciprocal of demand for all other commodities.

Assumption 6

Giving up this assumption requires that we make additional qualifications. Marginal utilities of equal price units of two or all commodities are equal to every consumer of both or all. But to the extent that all commodities are not consumed and compared by all consumers, strict logic required a qualification on the demand side of price. There is some modification of the importance of this qualification, in reality, as any individual consumer divides his expenditure among an indefinitely large number of products, and any small number of commodities are likely to be compared and their marginal utility price established by an indefinitely large number of consumers of both (Whitaker has pointed out the same of the producers side even more of a mystery).

Assumptions 7, 8, 9, 10 and 11

These are related, as they are the basic assumptions on which any attempt at a rational analysis of economics is postulated.[9]

NOTES

1. In an assumed case, somewhat analogous to the "Crusoe economy" we have just set up, A. C. Whitaker – having given an excellent summary of alternative-cost theory – drops it from mind altogether, to "take the springboard leap into darkness" and attempt a philosophical explanation of cost. However, he takes one step which it is impossible to grant him. He takes up the case of an assumed "Crusoe economy" (one man, who is at the same time producer and consumer, etc.), but soon takes a step where it is impossible to follow him, for there enters that peculiar phenomenon of classical economics; instead of assuming one, homogeneous supply of productive services, of any nature whatever, he assumes one "factor of production," labor; and instead of considering is as a purely economic resource, he treats it in a humanitarian way, cloaking it with feelings of pleasure and pain. Now the essential thing to assume, for the "Crusoe economy," is one single homogeneous productive resource, and one might, if he felt so disposed, name such a resource, labor. But that a resource should acquire, by this arbitrary appellation, any special attributes, is clearly illogical. This, however, is exactly what Whitaker does; and he proceeds in this realm of "real" cost to define that cost as disutility, or "costliness," and to state as the theory of value, that the disutility of the last unit of labor offered or exerted is equal to the utility produced by it.

We might grant him, for the sake of discussion, that he is trying to describe another kind of Crusoe economy, in which labor, human labor, was the only resource of production. But this raises two additional criticisms. First, even when the comparison of utilities and disutilities is made in the mind of the same person, while we can admit with only slight qualification that that person might be capable of comparison, even of some commensuration of different utilities, or of different disutilities, it is a much greater assumption to make the claim that

that individual has any basis on which to base comparisons between utilities and disutilities (Naturally, when this case is extended so as to cover two men, any comparison of utilities between them, is impossible; but any comparison of utilities in one group and disutilities on another is the sheerest sort of guesswork). Second, absolutely no extension of this sort of Crusoe economy is possible, so that to bring in any other productive resource spoils the whole theory (Whitaker tries to leave capital out of account, not even making any comments about the "pain-cost" of capital. "In this essay we will arbitrarily set aside the problem of abstinence cost") It is not long, however, before he is discussing capital again. If he has broken his own assumption without telling us, he suffers; on the other hand, if he means his theory of capital to have no relation to a pain cost of abstinence, then he is in the same predicament that he has so ably explained Ricardo to have been in: of the added difficulty for any such theory of the unequal proportions of capital and labor (In short, it seems that Whitaker's powers of analysis failed considerably, or else that he was inordinately rushed, during the last 30 pages of his book).

2. At the same time, heterogeneous services do not group themselves spontaneously into four classes – the "factors of production" of classical economics – nor can they be made to fit any such classification or any modification of it. A theoretical analysis arranges them, rather, along a ["or many? What variable trait?] continuous scale, a series of indistinguishable points. In actuality, the discontinuities of these points are, of course, more evident and the continuous scale more difficult to establish. But the principle that all productive services, instruments, resources are of the same nature, from an economic point of view, is absolute. Such a principle is not denied by the complexities of actual economic life, nor is it altered by our admittance [admission] that, in the light of legal and humanitarian causes, one subdivision of productive services into those that are, and those that are not, human is sometimes of advantage in economic analysis. In all that follows in this paper, it must be realized that this distinction, important as it is, is not being carried further – for want of time. Instead, productive instruments will be considered in their fundamental nature; only statements true of both classes ("capital" and "labor") will be made.

3. Any discussion of the purely "non-pecuniary" use of productive resources, never entering the market and never affecting any choices that do enter the market, is of course beyond the scope of the pragmatic economist. By purely "pecuniary" uses, we mean uses that may be bought and sold in the market without giving any consideration whatever to such subjective influences. Because even subjective influences do enter into market choices, affecting the "margin of indifference" between market and non-market uses, it is necessary to make this separation in our analysis between purely "pecuniary" uses, where there is no possible alternative, marginal, non-market use, and "pecuniary" uses, where there is such an alternative, marginal, non-market use.

It is important to notice that any such non-market alternative entering into pecuniary choices can arise only when we have previously relaxed the assumption of strictly homogeneous resources. Such elements, by their very presence, destroy the essence of homogeneity which would require that resources have no subjective attachments, or else that their subjective attachments were of the same degree with every resource.

To put this another way: giving up the homogeneity assumption creates two planes of specialization and restriction on free transferability. One, our definition of purely pecuniary heterogeneous uses; the other, our definition of pecuniary expanded so as to include marginal, non-market alternative considerations. This is the separation that will be followed under (a) and (b) below.

Non-pecuniary uses, especially when in connection with human instruments, are often referred to as "leisure" uses.

4. Some resources will be so evidently specialized to one occupation that they present no problem to the understanding of this point. But it must be realized than this is not the important thing, from an economic point of view. The first unit of a heterogeneous supply that is transferred on a rise in price, is, to the degree that it is less suitable to the new than to the old use, specialized to the old. This specialization is of the same nature, even though of infinitesimal degree, as that other more obvious kind.

5. The whole problem of the amount of productive capacity is raised at this point, and must be largely passed over. It will be enough to assume that the total amount of productive capacity in any resource is a fixed sum (although that total is in reality itself a function of remuneration, which is another factor breaking down the determination of price by cost) but that it is subject to allocation, and that one of the kinds of allocation is between pecuniary and non-pecuniary uses. If any resource earns less than the maximum pecuniary income potentially available among the most remunerative of a alternative employments, then the difference must be assumed to be made up by the production of non-pecuniary utilities.

6. It is a paradox that at times there seem to be decreasing costs ["not costs of equal values!"] in connection with such transfer of resources involving change in non-pecuniary remuneration. An example is that at times it is said that there are values as well as costs in transfer of resources: "habit-formation vs. boredom, the familiar vs. change, uncertainty and risk vs. the love of the gamble." However this paradox is solved, it would appear, by pointing out that to the extent that there are such improvements accompanying transfer, such resources were not used previously in their most productive use, violating our assumption (cf. Note 5).

7. Individual, i.e. any owner of "capital" or any self-owner of "labor," the distinction between the two is extremely important, but leads far afield into capital theory and wages theory, and can not be discussed here.

8. It is in connection with this point that we find one of the weakest points in Whitaker's theory. In discussing comparability among producers he says, "all I can be supposed really to know is that if I ere in occupation A, I would suffer more discomfort than if I were in occupation B. If I am a person of average (i.e. typical) constitution, I may infer legitimately that this is true also of average person." And, "We may conclude, then, that there is a perfectly legitimate sense in which we can compare the subjective costliness of commodities produced in society by entirely different groups of persons (pp. 181–182).

Admitting that he is talking in philosophical rather than empirical terms, this is still an extreme example of he way even the best of economics can talk foolishness about this tricky subject.

9. In assumption 9 we have abstracted from any possible time-lag. If we should give up this assumption, the effect on cost theory would be the same as giving up any other assumptions as to free and costless transferability. The elements of time and time-lag are as much disruptive of constant cost and as much causes of price-determined "rent" as any other form of specialization.

NOTES AND OTHER MATERIALS FROM FRANK H. KNIGHT'S COURSE, CURRENT TENDENCIES, ECONOMICS 303, UNIVERSITY OF CHICAGO, 1933–1934

Edited by Warren J. Samuels

Published below are the course reading list and student notes taken by F. Taylor Ostrander in Frank H. Knight's course, Current Tendencies, Economics 303, at the University of Chicago during Winter term of the 1933–1934 academic year. The reading list is surprisingly casual and uneven in detail among items. The notes are assumed, as usual in these volumes, to be a reasonably accurate summary account of what Knight said.

As such the notes provide important insights into the mind and thinking of a great economist; indeed, the notes are an unusually rich source of materials. Even when one senses limits, puts in perspective or criticizes what Knight is reported to have said, one is impressed with the topics he considered, the depth of his thinking, and the candor of his speech. To argue with Knight, even if only in one's own mind, is to learn a great deal. His insights are important because of their relations to some of the deepest concepts and issues in economics whether narrowly or broadly defined. Especially interesting to this editor is that Knight's course encompassed new publications up to the academic year in which the course was given (see Warren J. Samuels, Kirk Johnson, and Marianne Johnson, "What We

Documents from F. Taylor Ostrander
Research in the History of Economic Thought and Methodology, Volume 23-B, 87–140
Copyright © 2005 by Elsevier Ltd.
ISSN: 0743-4154/doi:10.1016/S0743-4154(05)23102-4

Learn from the Problem of Recent Economic Thought," in Samuels et al., *Essays on Fundamental Topics in the History of Economic Thought*, London: Routledge, forthcoming).

Knight (1885–1972) was nearly 50 when Ostrander was his student. Not only are the notes from Economics 303 exceptionally rich in materials for understanding Knight intellectually, they contribute to our understanding of him as an emotional person. He was darkly pessimistic. He was cynical as to the behavior of others and the consequences of that behavior. He repudiated many doctrines held by other economists while affirming other doctrines and several of his own. He knew too much too deeply to unqualifiedly follow any approach to any question; he loved economic theory but stressed its limits. He often adopted the role of devil's advocate, perhaps in part because he loved controversy but especially because he felt the limits of "the other side" needed emphasis. His criticisms of the ostensible ethics of capitalism were deeper and more powerful than those of many anti-capitalists. So far from assuming given individuals (as well as given tastes) Knight emphasized that the chief product of the economic system was not goods but man. The one thing Knight accepted, as a relatively absolute absolute, was freedom; his arguments for capitalism were so idiosyncratic as to be contrary to those of most other Chicagoans. Not only were many of his ideas idiosyncratic, he did not refrain from voicing them. He particularly opposed the basing of policy on the religious or moral teachings of any church. His candor was matched by his own opinionated views. As an economic theorist he could juxtapose pure theory and the search for the conditions of equilibrium to actual arrangements and the limits of equilibrium theory. He could defend pure theory against institutionalist critics and criticize institutionalist arguments, and vice versa. He affirmed the importance of institutions but not what institutionalists made of them. Neoclassicism, especially its emphasis on price as a function of demand and supply, was relativist, but he was as uncomfortable with relativism as he was with absolutism. Neoclassicism's emphasis on existential scarcity and the consequent ubiquitous necessity of choice was but one source of the tragic nature of the human predicament. Knight sugar-coated nothing.

The notes refer to Mortimer Adler and "Adlerism." Adler, a brilliant and prolific author across the spectrum of the learned disciplines, was an important but controversial figure at the University of Chicago. He was widest and best known as an advocate of "Great Books" curricula and multi-disciplinary programs and for his influence on the *Encyclopedia Britannica*, as well as for his arguments favoring the generation of social policy based on moral and, eventually, religious ideas. He and Knight were opponents on many issues, most notably the basing of policy on what Knight considered to be questionable, ambiguous, even meaningless, and wishful-thinking "principles" of morality. Ostrander recalls having attended the series of debates between Adler and Knight during his academic year at Chicago, 1933–1934. He remembers their arguments as sharp and relentless, and that Knight "would end each debate by saying something like 'You haven't convinced me; I disagree with all your argument[s], and the only way you can win is by hanging me,' or 'burn me at the stake' " (Ostrander to Samuels, February 3, 2003).

Ostrander's notes from Knight's lectures indicate that Knight made several key points in the first few minutes of the first class meeting – points that help clarify his version of neoclassical economics and relate it to other versions and to other schools of thought, however only implicitly. These key points include the following: (1) The purchase of consumer goods under a tension between conformity and distinction. Here we have tension, interpersonal utility functions, and other-directed learning of preferences. To this is added; (2) the emergent quality of preferences and ends; the individual literally does not, in Knight's view, know what he or she wants. Whereas "Classical economics assumes that the <u>end</u> (a utility function) is known to the individual," Knight believes that "there is, in reality, no <u>known</u> end-realizing action" and that "A point of the end of action always is to find out what is the end of action."

In a later discussion it is clear that the reach for Knight of these evolutionary facets of economic behavior, however, is limited:

– Technical economics gets along all right so long as we talk about wants and the technical machinery and means for satisfying them.
– But <u>economic</u> theory has to assume that wants are reflected in behavior – they want what they get.

Knight says we are interested in people's wants and, especially, their interests. But we are unable to get beyond their existence as physical agents and beings and their communicative methods, symbols and words. Moreover, "As soon as you begin to talk about any social phenomenon – social psychology, institutions, sociology – you begin to talk about <u>history</u>. You must ask how existing cultural phenomena 'got that way.' " Also, "no theory of consciousness has built any bridge from 'physical' world to the world of <u>real interests</u> (psychology). We can find physical measurements to any degree of accuracy – but there is no possibility of establishing <u>preciseness</u> for the social-mental world. . . . Even in the physical world there is no precision in the ultimate analysis."

There are several problems with this position. First, this leaves much of economic theory, especially that of consumption, as circular reasoning; as Knight himself puts it, "they want what they get." Second, it is not necessary to penetrate the minds of people to examine the social forces and processes at work. Status emulation and consumption patterns based thereon have analytically meaningful content. But, of course, Knight rejected not only going beyond utility theory but also behaviorism.

Third, this position has not prevented later Chicago economists from jumping from empirical, behavioral evidence to maximizing rationality. Fourth, it has not kept economists from taking positions erected on genetics and evolutionary psychology, so-called.

As for utility (returning to the discussion of early points); and (3) it is a purely conceptual category, notwithstanding its quantitative character: "Utility is motivation viewed as pure quantity – how far utility is pure quantity is another problem." The conjecture is offered that economists' comfort (training and experience) in working with utility as a purely conceptual quantitative category, prepared them for the then-emerging transformation of the neoclassical paradigm. The transformation was from Marshallian economics, a construct "of what we take to be 'real' objects, persons, institutions and events," to what Shackle called "the logical or mathematical construct or machine, a piece of pure reasoning, almost of 'pure mathematics', able to exist in its own right of internal coherence, as a system of mere relations amongst undefined thought-entities" (G. L. S. Shackle, *The Years of High Theory*. New York: Cambridge University Press, 1967, p. 294).

Knight seems, therefore, to have stood astride (at least) two paradigms of neoclassical economics. One is the world in which markets are a function of institutions; the other is the world of pure conceptual a-institutional abstract markets. One is relatively concrete; the other is abstract. Both in fact are a mixture of the concrete and the abstract, but one is much more in the direction of the concrete and the other is much more in the direction of abstraction. Much of his work was at the concrete end. But not all; his handling of uncertainty and of utility and other topics indicates the infusion of the abstract (see my introduction to Ostrander's notes from Knight's Economics 301, Economic Theory, in this volume). Knight was not of the new paradigm but he contributed to it and its research protocol, the quest for unique determinate optimal equilibrium solutions.

Other early key points, both distinctively Knightian, are the agency of the individual economic actor, in juxtaposition to pure behaviorism and historicism and to physical science; and a definition of "economy" that embraces the approaches of both Lionel Robbins and Paul Samuelson.

The most important early discussion, however, comes under Knight's heading of "Social Control." To appreciate his argument, one has to understand that Knight's social theory is developed within a tension between: (1) his knowledge that social control is both inevitable and necessary; and (2) his correlative desire for individual autonomy. One *could* add to that a hatred of social control, some of which *is* relevant. But what Knight dislikes is, first, selective elements of existing social control and, second, change of social control, e.g. change of the law by law, except for those changes of the law that remove the selective elements he dislikes; i.e. Knight is not opposed to all change of social control. In any event, the problem of social control is also for Knight (as it was for Pareto) the problems of social change and of the status of the status quo.

For clarity, one could postulate Knight's alter ego. Let S be the totality of existing social control; K that part Knight dislikes; R the part Knight likes but the alter ego dislikes; dK the change in social control Knight would support; and dR the change the alter ego would support. The differences between Knight and his alter ego are: (1) that which they like and would not change; and (2) that which they dislike and would change. Neither is totally for existing social control, neither is for changing everything. Neither can claim the anti-social control high ground. Nor can either claim the anti-legal change high ground – though given existing social control, change of law (or other rules) by law is the point at issue. Some writers, e.g. Bruno Leone, define coercion as legal change – not the law already in place. If the law in place is L1, the new body of law is L2, and the difference between L1 and L2 – legal change – is dL, it is impossible for me to see only dL as coercion (I surmise that no revolutionary wants to change everything – though it may appear that way to both some of them and some of their opponents; at least that is the record, as I read it, of historical revolutions. This has not prevented conservatives from talking about total as opposed to incremental or marginal changes).

(For exposition of several of the foregoing themes, see Warren J. Samuels, Review of Gordon Tullock, ed., *Explorations in the Theory of Anarchy*, in *Public Choice*, Vol. 16 (Fall 1973), pp. 94–97; "Anarchism and the Theory of Power," in Gordon Tullock, ed., *Further Explorations in the Theory of Anarchy*, Blacksburg: University Publications, 1974, pp. 33–57; and "The Concept of 'Coercion' in Economics," in Warren J. Samuels, Steven G. Medema and A. Allan Schmid, *The Economy as a Process of Valuation*, Lyme, NH: Edward Elgar, 1997, pp. 129–207).

Thus, in the notes, Knight is reported to have repudiated what he called the idea that is "the intellectual element in social enterprise," namely, the "Idea that society is my chariot for me to drive." This, he says, is "Analogous to [the] making of rules for a game – with [the] aim of a better game." He has two problems with this. It involves, first, "Making rules for yourself, not for others;" and second, "Aiming at a good game, but . . . aiming at winning the game." It is "childishness," he claims, of those "who want social control, but don't see the elementary fact that what they want is a society that would be their plaything; that with would work only if they controlled."

Knight goes one step further. He says that "aiming at winning the game . . . is not analogous to economic theory," that "The step from understanding of economic theory to social policy is a tremendous one," that people "Have enough to understand economic theory – and it is essential," and that "talking about social control on a basis of that understanding is overpowering" (Interestingly Knight seems to have a view, close to that of Vilfredo Pareto, of elites competing for control of the masses; here he says, "And the masses love to lie down before the Juggernaught car, if it's done with right technique").

This additional step raises further questions. Why is economic theory the test? Would economists, or economic theorists, then not become the rule-making authority for others in society? Would not this mean that economists – or some economists – would become the otherwise maligned chariot drivers? On what basis are they to have this exalted position? And which economic theory, and which economists, would control the rule making? (Knight himself would put it, which rules = whose rules). Further, as above, inasmuch as there must be rules, the question is not whether or not but which (= whose) rules. And since Knight interposes economic theory, hence economists, against changing the rules, why is his argument not self-referential, or self-reflexive (Knight, in a critique of Mortimer Adler, uses the phrase "talk turns back on itself")? Or is this but the guise in which one group seeks to become in fact but not in name the chariot drivers of society?

What about the objective of being a chariot driver? Is this anomalous or aberrational, as well as wrong? Knight seems to think so? For whatever it is worth, Adam Smith thought it distinctively human:

> The desire of being believed, the desire of persuading, of leading and directing other people, seems to be one of the strongest of all our natural desires. It is, perhaps, the instinct upon which is founded the faculty of speech, the characteristical faculty of human nature. No other animal possesses this faculty, and we cannot discover in any other animal any desire to lead and direct the judgment and conduct of its fellows (Adam Smith, *The Theory of Moral Sentiments*. New York: Oxford University Press, 1976, p. 336).

Smith is wrong about other animals; whether he is wrong, in part or otherwise, about human nature is another matter.

Knight is absolutely descriptively correct when he talks about "Aiming at a good game, but . . . aiming at winning the game." Legislation is not written and enacted neutrally, but by interested parties using law as a political means to their economic ends. Objections to legislation sought by others, as class warfare or as violations of non-interventionism or as putting the government on the backs of the citizenry, are not forthcoming from the same objectors when it is their legislation on the table. Such is the predicament – call it plutocracy – targeted by John Rawls's notion of a veil of ignorance; Rawls shared with Adam Smith a concern for the welfare of the least, or lowest, among mankind. Knight may have considered rule making with a view to winning as a violation of economic theory. But George Stigler, a student at Chicago at the same time as Ostrander, much later argued that people pursue their self-interests in politics no less than in economics, or in political no less than economic markets.

Some of the foregoing is echoed in and supported by the following. In an article in the November 27th, 2002 issue of *Business Week*, entitled "Biting the Invisible

Hand," Martin Fridson, the chief high-yield strategist at Merrill Lynch, is quoted for making a critical distinction apropos of the Enron scandal. His first point was that the Invisible Hand, in his view a metaphor for harnessing individual self-interest to serve the general well-being, is a powerful principle. His second point was that it is a "very convenient cover story for people who are actually trying to stack the deck in their favor" – for people who preach the virtues of competitive capitalism but practice the crony variety.

The same point is made by users of rent-seeking theory who invoke it to condemn all change of law – except those changes they believe needed to correct existing wrong law. Assuming the ubiquity of rent seeking (= aiming at <u>winning</u> the game), such does not render normatively repugnant *all* efforts to change the law; nor do the critics of rent seeking perceive it in their own agendas.

Here we have, rather, arguably empirical support for the general theory of business control of government, of ideology as a system of preconceptions, of capitalism as predatory behavior, and so on.

Predictive power is generally not very powerful in economics, but, absent a desire to predict precisely who will act in a predatory manner and precisely how they will do so, Knight's ideas, like Thorstein Veblen's theories, for example, predict the fact of these types of behavior very clearly. The likelihood of business-oriented because business-dominated government, for example, being complicit in arguably numerous ways is successful prediction.

If the problem of social control, therefore, is also the problem of social change and of the status of the status quo, the operative problem is, who decides? Knight understood that very well and lamented the fact that a group less than a unanimous whole made such decisions. But this is an impossible requirement, one seemingly calculated to retain the status quo and its *other* modes of change.

Knight's notorious, pessimistic low regard for the intellectual abilities and independence of most other people are evident in the following. In his discussion of those for whom society, in their view, is their chariot to drive, he is recorded as remarking, "And the masses love to lie down before the Juggernaught car, if it's done with right technique." In his wide-ranging discussion of behaviorism, he laments that the "Human race is inherently wonder-hungry, salvation-hungry." The human race, he says, "Wants a verbal formula to solve the world's problem" and that "Any good salesman can capitalize this market." It is an open question whether Knight knew (or remembered) Pareto's concept of *derivations* – beliefs that are used to manipulate others' psychic states and to mobilize political psychology – or Sigmund Freud's concept of belief in the omnipotence of words – that invocation of words can manipulate reality/nature/God (Pareto would say that at least two and possibly all three of these terms were derivations).

In a remarkable discussion of the alternative (now, opportunity) cost theory, Knight first says, "The trouble is that resources can be on an indifference margin at different levels of pecuniary occupation." He next adds that "What is involved is a lack of freedom of choice" and "If there were pure freedom, unequal non-pecuniary uses of resources would be impossible." The conclusion – which resembles but goes beyond the welfare economics of A. C. Pigou – is that "Cost theory holds only to the extent that all individuals are equal and the same." Knight concludes that the "Problem really leads to a restatement of economic theory." Numerous economists have held that economic equality in some form is necessary for the proper working of markets (an "ought" statement in "is" form) but for Knight to be recorded as saying this is striking.

Knight had the habit, or else had adopted the pedagogical device, of making strong, even extreme, *ex cathedra* declarative statements – especially about what constitutes economics. One example: "If you assume away division of labor, you have assumed away the economic system." Just as he narrows economics to neoclassical economics, he narrows the economic system to one driven by the division of labor. He offers no cognizance of either a different economics, however uninteresting or dangerous to him it might be, or of a different economy, no matter how much less productive of goods and narrow economic welfare it would be. Another example is in the immediately preceding line, in which he asserts the "impossibility of any individual working at more than one occupation." Still another is, "You can never state a problem in advance, the statement of it and the solution of it are a simultaneous process, not a time sequence." As is often the case, Knight's statement covers his point in question but, because of its blunt, even brusque, Delphic character, seems to extend to more blanket coverage.

These statements are in partial juxtaposition, if not in conflict, with Knight's general position, namely, the complexity and difficulty of working things out in an explorative and emergent manner. Still, on a variety of issues Knight had strongly felt positions.

Whatever else one might think of such strong or extreme statements, they undoubtedly contributed to the success he had in making converts to the Chicago "free market" belief system (Ostrander comments, on reading the foregoing: "I doubt he intended to convert them this way." My response: I wrote only "undoubtedly contributed," nothing about intent – though I do now note the Chicago "style" of taking strong or extreme positions).

Another such statement comes soon after the foregoing two: "One is not free to make marginal choices between the non-pecuniary aspects of two expenditures of resources – they are in different worlds originally, and, even if not, are seldom directly compared at one instant, and might as well be in two worlds." What, one wonders, would Knight make of Gary Becker's extension of marginalism into non-pecuniary areas, not merely non-pecuniary aspects of expenditures.

One does not know whether the following excerpt is an extreme statement or a statement seriously, if pessimistically, questioning the meaningfulness of the rationality assumption:

> – Money is the worst devil in economics – people never will understand it, or act sensibly about it – perhaps only solution is to get back to barter terms.

In these lectures, too, Knight does not refrain from using strong language: foolishness, lunatic point of view, criminal imbecility, and so on. He also makes somewhat strong, even outlandish, statements hoping to persuasively prove a point; for example, "can't draw a picture with yourself in the picture" in a discussion of different positions of the individual in the physical and social world, the latter being "a part of the world you study." Grand pronouncements are made, e.g. that terms from the physical sciences "can not be used in social sciences."

Knight's thoughts are rarely far away from two topics: economics as a science, and non-interventionist government policy. In a discussion entitled in the notes "How does economics become a science?," Knight combines the two. He recalls the period when "people gradually got the idea that competitive market operations could take care of the needs of life without any definite, positive <u>policy</u> by governments." The result was, "Policy becomes negative – lack of it was wanted." His next point is a typical Knightian exercise in candor: "Policy of lack of policy is a propaganda as much as the propaganda of positive policy." This is eminently correct. But also correct, and not brought up by Knight, is that positive policy is analytically equivalent to negative policy, given the change of interest protected. If government regulates Alpha in the interest of Beta, and if this becomes seen as positive policy (leading to the cry, get the government off of our backs, etc.), then doing away with the regulation – commonly called deregulation – means that Beta's interest is now in the same position as was Alpha's hitherto. In the case of mercantilism – the practices in question – government action that promoted the interests of some economic actors (businesses) in effect constrained the interests of other economic actors (businesses and consumers); eliminating mercantilism (something never fully done) promoted the latters' interests. Positive and negative are a function of point of view, of whose ox is being gored.

Knight carries his analysis the further step: "<u>When</u> policy becomes negative – you have a science."

Knight next returns to social control. His argument is, first, that "Social control – ought <u>not</u> to mean <u>somebody</u> controlling others," and second that "True social control" takes place when actors mutually control each other, "where the actor is acted upon to the extent, at least, that he acts." This latter is a system of free association, economic freedom, and limited government. Policy is negative, in his view, and is so because economics has supported freedom.

Turning to the question of inequality of income and wealth, Knight makes two unforgettable points. The first is the "know nothing" position:

–We find we know nothing about the question.
    –It's better not to know so much, than to know so much it isn't so – John Billings. [Pseudonym of Henry Wheeler Shaw, 1818–1885, American humorist and lecturer: "The trouble with people is not that they don't know but that they know so much that ain't so." *Josh Billings' Encyclopedia of Wit and Wisdom* (1874)].

The second is a position using Veblen's theory of pecuniary emulation, conspicuous consumption, and the pecuniary standard of living:

–Corporate heads are not paid for services, but so they can live the way it "befits" a corporate head to live – their salaries are window-dressing.

This position is not quite the same as, but likely includes, Martin Bronfenbrenner's (a Knight student) view that upper-level corporate executives are paid more than they are worth in order to serve as a stimulus for mid-level management to over-work in order to secure promotion. Altogether, the lectures seem to this writer to be less technical economics than what Knight himself likely would call a propaganda for existing inequalities.

Further as to inequality, we read that Knight thinks there is about as much sense in talking about equalizing money incomes as equalizing baseball scores.

Knight goes on to ask

> What is the diminishing utility of high scores? Is there any? A great deal depends on how the game is scored – as when handicaps are used so that there is a fight all the way.

Knight surely would appreciate the degree to which the rules of professional football (and baseball) are changed in order to stress offense over defense, producing more "big plays" and higher-scoring games, generating greater excitement for marketing purposes.

Apropos of rule making, Knight asserts that, "This is the sort of question society has to answer." One interesting point about that assertion is not the social construction of rules, which is not arguable, but the invocation of "society," presumably as more than or different from the mere sum of individuals comprising it. Another interesting point is his degradation of changing the rules – though he knows this has to be done.

–What effect does intelligence <u>have</u> on the social problem?
  –What is the social problem?
    –How is society to solve it by itself?
      –The players changing the rules.
    –Or are you and your gang going to change it.
      –Discussion of this sort of thing is absurd.
  –What is the intellectual basis of discussion.
  –We think intelligently about the external world, but talk nonsense about the social world.

Knight loves and again projects the idea of free exchange, concluding, in part, that

–In a pure exchange economy there would be no . . . distribution problem.
–We are not born into the world on contract terms.
  –Our "inalienable rights" put the individual in a handicapped position with reference to other owners of property.
–If you believe in progress of civilization – you must realize that the <u>private exchange economy</u> is the greatest invention working towards that end.
  –It provides social cooperation without any political guidance – in subconscious.

This presentation is like a model in which some of the important variables are omitted. Absent here are the legal determination of the conditions of access to, use of, and intergenerational transmission of interests protected as property, the putative need for changes therein, the manipulation of the law of property and other laws by the already powerful to further enhance their position, and "guidance" of economic performance by government in both the preceding and other ways. This view takes, selectively to be sure, as part of the natural order of things what is a function of past government action, and would render present government impotent – all largely under the fiction or illusion of a nonpolitical socioeconomic system. Unless one is playing the role of high priest, selectively seeking to reify the status quo, one cannot accept Knight's limited story, or model, even if one is relatively wealthy and agrees that, in principle, "the <u>private exchange economy</u> [perhaps in contrast to its capitalist version] is [one of] the greatest invention[s]" of mankind. One further interesting aspect of the problem is raised by the caveat about the capitalist version of the market economy. Even if we all agreed with Knight on the basic propositions that the market economy is vastly desirable and that its institutions/working rules should not be frequently changed, we would still disagree as to both the details of its institutionalization and of their change.

Knight perseveres, first affirming non-deliberative over deliberative decision making – " Any kind of self-conscious process disintegrates the whole thing – is anarchy" – and, second, tying democracy to laissez faire – " Knight thinks that the nearest approach to democracy has been when it was allied to a laissez-faire system of economics." The problem with the former is that it misrepresents so-called non-deliberative decision making, neglecting the deliberative critique of received arrangements that Menger said is the task of every generation. The problems of the latter are, first, that some consider democracy the prerequisite to a free market economy and others vice versa ("laissez faire" this writer considers a sentiment, not a meaningful term; see Warren J. Samuels, "Laissez-faire," *Encyclopedia of Economic Sociology*, Jens Beckert and Milan Zafirowski, eds., London: Routledge, forthcoming); and, second, that others consider both a function of political and other pluralism in society. The overriding problem is whether Knight is prepared to be self-referential in applying his "any kind of self-conscious process" dictum to himself; what is he doing if not engaging in conscious, deliberative discussion?

Knight is happy with neither "big government" nor deliberative legal change. He desires escape from the burden of deliberative choice. He supports the image of and movement toward his idealized economy because such movement is perceived as neither governmental nor change. One example of this concerns the money and banking system, specifically credit-creation – money creation – by private banks (the idea that commercial-bank credit creation is money creation was only slowly then being widely understood; Ostrander writes me, "Henry Simons took this from Knight and ran with it.") Knight was unlike those of similar disposition who favor "free banking," saying

–By prohibiting the strictly defined coinage of money by individuals.
–It has come to restrict note-issue, and even deposits.
–But it still leaves the creation of credit in private hands – and enterprise economy can not work on this basis.

Another remarkable discussion includes Knight's preference for minimal government: "People want to do what they want to do – this seems to be the basic assumption of what Knight wants government to be." Then Knight is recorded as posing a central, or "real," problem: "Does the sovereign make law, or law make the sovereign?" He then gets to his main point, how a consciousness of law as man-made *policy* is a truly "real" problem. For an advocate of laissez faire, Knight seems remarkably concerned with the effectiveness of law as social control:

–Problem of law – the respect for law as law <u>vs</u> the disrespect for law as law.
  –Taking away the divinely ordained quality of laws puts society up against a <u>real</u> problem.
  –When man realizes that laws are made by other men like himself – we have a <u>real</u> problem.
  –Most men think the majesty of law is something <u>other</u> men ought to follow.

–As with religion – most educated classes and religious leaders have always believed that religion was necessary for the mass of the people – the opiate – to preserve society.

–When a religion ceases to believe it is descended from god – other religions from the devil, then it is dead as a religion – especially when its advocates <u>argue</u> for it – and on the basis of its utility, rather than "burn the unbelievers."

This is exceedingly perceptive of Knight, a skeptic or cynic about both organized religion and the state. Presumably the modicum of law he recognizes as necessary must have the aura of the transcendent to give it absolutist legitimacy. As for religion, Knight felt it had a social control role to play but tended, like the state, to overdo it. His fuller, if still terse, position is rendered by the epigrammatic, "Religion is the opium of the masses and the sedative of the [upper] classes." One interesting aspect of Knight is that he was quite learned on that which he disliked; for all his skepticism or cynicism towards religion, for example, he was a deep-thinking sociologist, or student, of religion.

Knight was sensitive to the linguistic problems of economic discourse and in this was well ahead of others in his time. He brilliantly states that

–The terms we use in describing society have usually physical penumbra attached to them, and we haven't any other concepts – but are set, applied from physical to social.

He cites "an urge to reduce things to physical analogy," but emphasizes, "Analogy breaks down." The key point is that use of a borrowed term by analogy can lead to economic interpretation that mistakenly imports from the physical what does not apply to the social. A subordinate point (implicit in his space, time and mass example) is that a physical model may be only one way of modeling the physical world and economists must be concerned with which model is taken over by way of analogy – or by some other figure of speech, such as metaphor or simile.

Knight continues in a manner indicating his appreciation of two points, the possibility and definition of the domain receiving the analogy, and the role of human agency.

–In social phenomena – we think in physical terms.

–And don't usually recognize the real problem – can there be a concept of society? or social science?

–In physical level you have to abstract from the fact that you are part of the physical world – this doesn't cause much trouble.

–But on social level you <u>cannot</u> abstract from the fact that <u>you</u> are a <u>part</u> of the <u>social</u> world.

–But when you become a part of the world you study – there is an impasse.

And,

–Any statement you make that is about society is never true after it has been said – for society has been changed by its being said.

The last two points again suggest the question of whether Knight was self-reflexive, prepared to apply the implicit caveat to himself as well as to others. The same question applies to the later statement, "What happens is that everyone gets a vested interest in some one theory and discussion is then out of the question."

Neoclassical economics posits given preferences. This is a methodological assumption, necessary to generate unique determinate results. It is so salient an assumption, however, that many interpret it to constitute actual description. Knight, for one, knew better:

–When you shop – you don't know what you want, you shop to <u>learn</u> what you want.

This position brings Knight close to existentialism (although existentialist ideas had been articulated for some time, to my knowledge existentialism as a school of philosophy did not yet exist 1933–1934). He says,

–Motivation has a large provisional reality; but in the end motivation has no reality.
–It is the experimental drive towards finding out what to express and at the same time expressing it.

At any rate, when Knight is indicated as saying, "Another term of analogy is to seek an <u>idea</u> in history, to forget the human beings," he can be seen as straddling the conceptual reality/actual reality dichotomy.

Given Knight's more or less respectful hostility toward Thorstein Veblen and his ideas ("Only real institutionalist in U.S. was Veblen. . . . Knight can find no contribution."), it is striking to read him validate Veblen's portrayal of the mainstream conception of the economic actor. Says Knight,

–Ordinary economic theory treats of <u>behavior</u> as <u>motivated</u>, but of the <u>motivation</u> as <u>mechanical</u>.

Compare Veblen's famous language:

> The hedonistic conception of man is that of a lightning calculator of pleasures and pains who oscillates like a homogeneous globule of desire of happiness under the impulse of stimuli that shift him about the area, but leave him intact. He has neither antecedent nor consequent. He is an isolated definitive human datum, in stable equilibrium except for the buffets of the impinging forces that displace him in one direction or another. Self-imposed in elemental space, he spins

symmetrically about his own spiritual axis until the parallelogram of forces bears down upon him, whereupon he follows the line of the resultant. When the force of the impact is spent, he comes to rest, a self-contained globule of desire as before. Spiritually, the hedonistic man is not a prime mover. He is not the seat of a process of living, except in the sense that he is subject to a series of permutations enforced upon him by circumstances external and lien to him (Thorstein Veblen, "Why is Economics Not an Evolutionary Science?," originally published in the *Quarterly Journal of Economics*, vol. 12, July 1898, pp. 373–397; reprinted in Malcolm Rutherford and Warren J. Samuels, eds., *Classics in Institutional Economics*, Vol. I, *Thorstein Bunde Veblen*, London: Pickering & Chatto, 1997, pp. 19–20).

Knight – as Ostrander's collection of notes taken in four of Knight's courses (and various of Knight's own writings) abundantly show – knew better. Here he is speaking of "ordinary economic theory," not real people. Just like Veblen (Also see below).

Knight is recorded as saying,

–Self-legislation <u>or</u> social legislation.
–This latter is the only <u>meaning</u> of social control.

These statements typify Knight's aggressive affirmation of his point of view. He does not try to establish *why* this is the only meaning of social control. His analysis, affirming self-legislation over social legislation, makes no provision for that which requires collective action, such as determining and reforming rights.

Another statement in the same lecture, "Alteration of behavior by discussion means freedom on a social level," contemplates inter-individual discussion but not government as a mode of discussion (just as the idea of social self-regulation tends to exclude government, especially democratic government, as an institution of social self-regulation (following either Hobbes or Locke)). Knight new better; he was engaged, however, in working out what he called a propaganda for economic freedom (See below regarding democracy and deception).

Knight continues a tradition that commenced with John Stuart Mill, in saying, "Yet if we get away from the science of competition, we are outside the range of economic science – and talking ethics." A Knightian critic of Knight might say, "There is absolutely no reason for this to be the case; it is based on narrow, even foolish, notions of what economics is all about." First, it presumes that the object of analysis is unique determinate equilibrium optimal solutions, for which competition is widely believed to be necessary – which is not the only possible research protocol. Second, the meaning of "competition" is left open; without a definition, one does not know what to make of the proposition; and there are many definitions. Third, if by ethics one means the normative, a positive economics is possible under non-competitive conditions. Indeed, both Knight's theory of profits due to the successful bearing of uncertainty and Schumpeter's theory of monopoly profit are examples of such positive analysis; the ethical penumbra and implications of each is a different story. Thus, Knight says, in a discussion of imperfect competition theory, "Pure competition for the system is conceivable, hardly realizable."

As mentioned elsewhere, Knight anticipated the core argument of the characteristics theory of demand, that people demand not goods but certain of what they believe are the characteristics of goods. Knight:

–What do people do when they buy commodities? What do they buy?
–<u>Commodities are not human wants</u>.
  –A list of one and a list of the other would have no word on both lists.

And as to what drives the demand for those characteristics, according to Knight,

–We want mainly <u>distinction</u> and <u>conformity</u>.

Status emulation through conformity and distinction – distinctive conformity to what Veblen, in his *Theory of the Leisure Class*, called status and pecuniary emulation under the aegis of the canons of invidious comparison, pecuniary standards of taste, and conspicuous consumption and leisure. Veblen, again.

Included in the discussion of imperfect competition theory is the following:

–The art of living is the art of <u>fooling</u> everyone else, and getting away with it.
–Yet, one would assume that the deceives, once they knew this, would refuse to
 be fooled.
–In fact, Barnum was right, they love to be fooled.

One could object that this is what Pareto called rule by fraud, that it is cynical, that it ultimately derives from Knight's deep pessimism, and so on – and one would be right. Except for one thing, Knight is right. He is trumped only by Winston Churchill's admonition that, democracy is a poor type of government, but all other forms are worse ["Many forms of government have been tried, and will be tried in this world of sin and woe. No one pretends that democracy is perfect or all-wise. Indeed, it has been said that democracy is the worst form of Government except all those other forms that have been tried from time to time" Winston Churchill, speech to House of Commons, November 11, 1947].Returning to the problem of unique determinate solutions, we find the notes saying, "Mathematical economists try to set up equations to show that the science can be made determinate – the exact reason why Knight lost faith in it." Knight did in fact straddle both deterministic and non-deterministic research protocols. While he preferred and largely practiced the latter, Marshallian one, he did, as we have seen, partake of the former, albeit with obvious misgivings.

A particularly interesting irony is found in Knight's point that the word "'political' was added to the word economics, in order to turn its attention away from state policy, to welfare." That is not strictly accurate. Some used "political economy" to emphasize the importance of government willy nilly; some, to call attention to their policy program; and some, despite the fact of their non-interventionist sentiments, used the phrase as a term of art.

Knight is reported to have said critically of Othmar Spann, that he believed "The group is prior to the individual, the individual finds his expression only in the group. [The group] is more real than the individual." This priority was anathema to Knight. But analytically, and without quibbling over the use and precise meaning of terms, does not Knight's statement, "We want mainly distinction and conformity," amount to the same thing? It is the group, in my view, that has ranks of distinction and patterns of conformity, maintained by individuals' learned behavior, through which individuality and its expressions are formed.

One possible exercise to perform on these and other notes from Knight's courses (and his writings) is to identify how much, and what, comes for him under the heading, "Anything dangerous cannot possibly be true." Several examples have already been mentioned in the present notes, e.g. that labor and capital, each within quotation marks in the notes, are no more ethically than theoretically different. A few lines later, Knight takes up "class."

A minor but not uninteresting point: Knight uses "classical" to refer to the mainstream of economic thought and not solely to the school of David Ricardo and his followers. John Maynard Keynes did likewise, and was seen by some as wrong in doing so. The usage has died out but may have been more or less common in the 1930s.

Knight is recorded as saying,

–As long as people think of life in terms of achievement, of efficiency, the type of theorizing of the classical school seems to be borne out by the behavior of people.
–This is not to say that they are right.

Here are two interesting problems. First, is Knight saying the theory is confirmed by behavior, or that the theory is confirmed only so long as people are induced to behave that way, or, apropos of "seems," the confirmation is problematic? Second, apropos of "not to say that they are right," to whom does "they" refer, the people or the classical theorists?

Shortly thereafter, Knight says,

–Rationality – the world is tired of this – wants "action," wants to think with its blood, wants experimentation, not thinking.

Knight's rhetoric, not surprisingly, is situational. When Knight wants to combat deliberative decision making he invokes the superiority of non-deliberative decision making. When he wants to combat social experimentation (especially legal experimentation) he invokes deliberative decision making, i.e. "thinking."

Knight's view of the "Idea of individual liberty" was that it was the "Perfect formula for every individual manipulating every one else" (a view close to that of Pareto) but also that "The more intelligent the individuals become the more manipulative becomes social action, until it breaks down." His view of the liberal economy resembles a benevolent plutocracy, in that "Individual liberty is a personal property competition under liberalism – i.e. the respect for property comes first – then comes 'individual liberty' – then principle of mutual consent in exchange." Thus, "In certain lines an individual taste and choice is possible," but "In social lines, especially political, it is not possible." In the political domain, "Life [is] according to rule." There, reverting to the topic at the end of the second preceding paragraph, Knight's intellectual armor now includes discussion plus, when it comes to voting, unanimity – which, like his student, Buchanan, he finds "essential." What is odd, also likely an incongruity or inconsistency, is Knight's correlative condemnation of what he calls the "Adlerist" and "Thomist idea that the first person to get the rostrum shall define the terms and lay down the rules." The principle of unanimity privileges those who first established the rules." All this is introduced by a pronouncement recorded by Ostrander: "Knight doesn't believe in freedom of thought." This, especially when one considers the next line, "The society that tolerates Communism and Jesuitism is crazy," suggests that Knight was endeavoring to get his students to think. As for Knight, while he hated those systems that offered only absolutes, he suffered great unease as a result of the conflicts touched on in class and in the notes. He, too, sought absolutist legitimacy of his system, but for him these absolutes were, as he put it, only relatively absolute absolutes.

As for the assumption of consumer rationality, Knight acknowledges the influence of advertising ("Did not take into consideration the power of propaganda") but his more important point was, again, "the fact that people live, not to fulfill purposes, but to make life interesting." Unclear is whether Knight had something deeper in mind in his next recorded sentence: "Even economics is written about not for the consumer, but by economists for their own interest."

When Knight is recorded as saying, "Competitive system takes wants as given," it is the economists' model and not the actual economy that does so. It is in the latter that "Business uses as much money to create wants as to supply goods." For the economist, given wants is due to either belief or methodological limiting assumption. For business, existing wants are a double opportunity or target, one to supply and the other to alter.

It is fascinating to see Knight, both in the notes just commented upon and throughout the entire set, weaves together the wants and behavior of people in markets and in government, respectively. Some of the problem with government is due to how government is organized and controlled. But part of it is due to the fact that zero-sum problems and their attendant conflicts require a venue in which to be addressed, and in the modern world government is the principal such venue. Markets, on the other hand, are the domain of positive-sum behavior and issues. One link, of course, is that market agents seek to redraw the rules in their favor – which is one source of zero-sum problems in government. It is in government wherein Knight locates the conflict between the principle of unanimity and the prior-appropriation principle with which in practice reformers have to deal. The irony here is that Knight for all his snickering about and hostility to reform, has his own agenda of reforms.

Which brings up once again the problem of how self-referential or self-reflexive was Knight. That Knight argues,

–Intellectual life is itself competitive.
  –Mainly for personal recognition.

is no surprise. It was Adam Smith's deepest principles of human nature that people sought social recognition and moral approval. The question is whether Knight saw it in himself – and, then, if he did, how did he differentiate himself from others. For Knight went further, leading Ostrander to write,

–How separate the promotion of truth from the promotion of the thinker of it.

and

–Promotion of social reform can't be separated from promotion of social reformers.

and again

–Propaganda is primarily aimed at getting the propagandist into a position of power
  – Thomism or Communism.
  –Virtue, social reform, etc. are <u>sales</u> talk.

"Propaganda" did not then have its current pejorative meaning. It was a synonym for "education." Knight used it to describe the objective of his friend Lionel Robbins's book on classical economic policy – a propaganda for economic freedom. Surely Knight would have concurred that his lectures constituted propaganda = education, too. No, the issue is not, "I educate, he propagandizes." The issues are self-promotion and self-reflexivity (Ostrander is displeased with and questions the use of "propaganda," saying "He [Knight] hated Nazi propaganda." But Knight, writing as a sociologist of language, does use the term).

And one other issue: Power, through control of government. Here the contenders are politicians, social reformers, businessmen, religious leaders. "Which type of dictator would be best?," Knight asks – world conquerors or moral fanatics. Frank William Taussig had written (in his *Inventors and Money Makers*, New Brunswick,

NJ, 1989, p. 129) the choice was between the "blood-guilty Napoleons of history" or the Napoleons of industry; by 1933, a new type was on the scene, Napoleon as Hitler or Stalin. Now the alternatives are plutocrat and religious terrorist.

In a discussion of cost theory and resource allocation, Knight once again borrows from Veblen, saying,

> one knows that people value things according to their price.
> —And sellers put up the price accordingly

In a discussion of social well-being, Knight makes the following point, without affirming it as a matter of practice:

—Law breaking is an essential <u>value</u> of human existence.

As to why people want freedom, Knight finds incorrect the answer, "in order to get maximum economy." His answer is that "One wants to control his own resources, and be free of dictation from others (also to <u>dictate</u> to others – the <u>immorality</u> of it)." This brings Knight to one of the great questions with which he wrestled throughout his intellectual life, the relation of freedom to power. Here he is recorded as saying,

> —Is it possible for the individual to be both free and without power? No.

Another of his great questions follows from his perception that individuals want both to be free from dictation by others and to be free to dictate to others. This is a conflict between freedom from and freedom to, one rephrased as Alpha's freedom from being correlative to denying Beta's freedom to, and vice versa. Somewhere in society, in some institutions, the ongoing resolutions of such conflicts take place. Whatever else those institutions may be called, therein is the state, government. Knight brilliantly and admirably identified, from his own perspective, much of what is involved in such great questions. But he was never able to do the impossible, namely, to provide an unequivocal formula for their solution. As attractive as his sentiments and his applications of them may be, his social philosophy was ultimately inconclusive; it could not unequivocally solve the issues to which it was addressed. He understood this, I think, and this understanding either helped cause his pessimism or reinforced a pessimism whose ultimate origins resided elsewhere.

On efficiency, Knight is recorded as saying, "Economic action is maximizing behavior, efficient, so far as it is thought about." Several points: (1) Knight seems here to intend to describe the world as it is, to define reality, not to state a limiting methodological assumption; (2) The identification of behavior as maximizing behavior requires that the observer know the end and the relation of means to the end, something whose possibility Knight in other contexts (see *supra*) denies; and (3) Knight here maintains that maximizing behavior is deliberative behavior.

Given what Knight has to say about deliberative in relation to non-deliberative behavior (again, *supra*), this does not leave much room for maximization.

Efficiency is not only a matter of the relation of input to output; it is also a matter of whose interests count in the choice of output as end and of input as means. Knight identifies the basis of equilibrium in input and output markets and then raises a major problem:

–Based on property and exchange – impossible to define those accurately.

"Property" is a primitive term. Distinguishing between the institution of private property and the rights that constitute property in practice, there are various possible forms of private property and various possible bundles of rights. Statements about property, especially those about government and property, tend to beg not only the questions of institutional form and bundles of rights, but also the fundamental role of government in adopting and reforming property in both regards. Especially the changes: they cannot be defined once and for all time.

The question of property also arises in the penultimate topic recorded in the notes, in connection with the then-recently published, now classic, study, *The Modern Corporation and Private Property* (1932), by A. A. Berle, Jr., and Gardiner C. Means. Part of their argument is the separation of ownership and control, therefore the changed nature of property in the form of corporate equity securities. Ownership no longer meant control for most stockholders; practically, ownership meant only a claim on dividends – and only if and to the extent that such were declared by management. The later Chicago School, especially George J. Stigler, found that argument to be wrong and dangerous (it opened the door to, indeed invited, government control) and tried to finesse it, in part, with the idea of a market for corporate control. The idea did not go too far, in the light of insulated upper-level management and their self-interested decision making, e.g. the Enron case. Knight, too, settled upon the separation of ownership and control. His view was more moderate albeit in the same direction. He acknowledged, contrary to legal theory, that "The only element of stockholder control is when he buys the stock." As for the man who buys the stock, "In every case he delegates authority and trusts to luck – law can do something, not much."

The final page of notes commences with a reference to Veblen by Stigler, then another student in the class. It is difficult to distinguish between what Stigler and Knight say, though the dividing line is probably where the notes read "Knight can't find out" what Veblen means by Darwinian or Darwinism. After a beginning by Stigler on Veblen's dichotomy of making money versus making goods – vendability versus serviceability, selling versus making, speculation versus production – the discussion turns to Veblen's use of Darwin's analysis. Of particular interest is

Knight's seemingly implicit acceptance of a conflict between institutions that work to inhibit change and those that promote change. The notes read,

–Pre-industrial society has a definite aversion to <u>change</u>.
–Western, post-Renaissance history has changed to a belief in the improvement of life through change.

The notes conclude with two points of perhaps wider interest. One has to do with the dialectic:

–Dialectic – theory of the way people change their minds.
   –Does not <u>seem</u> to be at random.
     –For social historians <u>seem</u> to be able to bring order in to the pattern of social behavior.
   –But what <u>is</u> an intelligible theory of economic change?

The final point is spectacular but, alas, undeveloped:

–The whole <u>content</u> of economics is of institutional origin.

The implications are that institutions matter and that economics – "the whole content of economics" – is derived from, an emanation of, the institutions of the market economy. Economics is, indeed, like markets, a function of the institutions that form and operate through it.

READING LIST

READING LIST ECONOMICS 303 CURRENT TENDENCIES
General Surveys
Boucke, O. F., Development of Economics
Boucke, O. F., Critique of Economics (especially 1st and last chaps).
Hayes, E. C., (Ed.) Recent Developments in the Social Sciences (essays by J. M. Clark on Economics and Ellwood on Sociology)
Homan, Contemporary Economic Thought
Ogburn & Goldenweiser, The Social Sciences in their Interrelations
Tugwell, R. T., (Ed.) The Trend in Economics (esp. essays by Mitchell and A. B. Wolfe). Also review by A. A. Young, Q. J. E. Feb. 1925

<u>Method</u>
Bagehot, "English Political Economy," in Economic Studies
Cairnes, Character and Logical Method of Political Economy
Dickinson, Z. C., Am. Ec. Rev. Supp. March, 1924
Keynes, J. N., Scope and Method of Political Economy (esp. Chaps. VI–X)

Mill, J. S., Essays on Unsettled Questions, Essay V
Mill, J. S., System of Logic, Book VI
Mitchell, W. C., "Quantitative Analysis in Economic Theory," Am. Ec. Rev. Supp., March, 1925

Psychology, etc.
Boucke, Critique of Economics, Chap. I
Dewey, Human Nature and Conduct
McDougall, Wm., Paper on Soc. Psych. & Instincts, Am. J. Sociol., Vol. 29 (1923–24), pp. 657ff.
Cooley, Human Nature and the Social Order; Social Organization; Social Process (See references to Chap. I of B. M. Anderson's Value of Money)
Frank, L. K., Am. Ec. Rev., March, 1924
Kantor, Jacob, Am. J. of Socio., 29, 674ff. Ibid., 27, 611ff.
Snow, A. J., in J. P. E., Aug., 1924
Dickinson, Z. C., Economic Motives
Dickinson, Z. C., Article, Q. J. E., May, 1919
Parker, Carleton, A. E. R., March, 1918, Motives in Economic Life
Clark, J. M., Economics and Mod. Psychol., J. P. E., Jan. & Feb., 1918
Knight, F. H., Q. J. E., Aug. 1924 and A. E. R., June, 1925
Tugwell, R. G., J. P. E., June, 1922
Viner, J., Marginal Utility and its Critics, J. P. E., Aug–Sept., 1925

Institutionalism & Historical Point of View
Homan, P. T., Book above under Surveys, and Articles, Q. J. E., May, 1928, J. P. E., Dec. 1927 [added in longhand: "and Encyc. of Soc. Sc. article"]
Leslie, T. E. C., "The Philosophical Method in Political Economy," in Essays in Political and Moral Philosophy
Anderson, B. M., See above under Psychology
Hamilton, Walton, J. P. E., Jan., Mar., Apr., 1918; A. E. R., 1919 (and note following "Discussion")
Commons, J. R., Legal Foundations of Capitalism. Article in Yale Law Journ.
Ogburn, W. F., "Social Change," Art. In A. E. R. Supp., 1919
Veblen, The Instinct of Workmanship; Theory of the Leisure Class; Place of Science in Civilization, esp. essays on Marginal Utility, Clark's Economics, and Schmoller's Economics.

Mathematical Econ. & Prob. of Max. Welfare
Cassel, Theory of Social Economy, Chap. IV [added in longhand: "minimum requirement"]
Pigou, Economics of Welfare, 1st 200 pages

Pigou, Economics of Welfare, 2nd ed., App. III, Secs. A–D
Young, A. A., Rev. of 1st Ed. of Wealth and Welfare, in Q. J. E., Aug., 1912
Robertson, D. H., Ec. Journal, March, 1924
Sraffa, Ec. Journal, December, 1926
Pigou, Ec. Journal, June, 1927, and June, 1928
Shove, Ec. Journal, June, 1928
Schultz, Henry, Statistical Laws of Supply and Demand, Treatment of Demand and General Theory portion on Supply
Bowley, Mathematical Groundwork of Econ. [added in longhand]

[Added in longhand on back of Reading list:]

Robinson, Joan	Econ. of Imperfect Competition
Chamberlain, Ed.	Theory of Monopolistic Competition
Robbins, L.	Nature and Significance of Econ. Science
Fraser, L. M.	
Sauter, R.	[sic: Ralph William Souter] *Prolegomena to Relativity Economics* 1933] ["answers" and arrow to Robbins, *supra*]
Memorial to Wieser	[*Gesamtbild der Forschung in den einzelnen Landen*, Wien: Julius Springer, 1927; for other volumes, see this annual, Archival Supplement 2, 1991, p. 8, n1].
Viner	Cost Theory – Soc. Sci. Encyclopedia
	Comparative Cost
Knight	J. P. E., 1927 – Simplification of Cost Theory
	Risk, Uncertainty and Profit
Wicksteed	
Jaffé, W.	"Les theories economiques et social de Thorstein Veblen" 1924 [sic]
Knight	Q. J. E., "The Ethics of Competition," 1923
Harris, A.	Dialectical Institutionalism

F. TAYLOR OSTRANDER'S NOTES ON FRANK H. KNIGHT'S COURSE, CURRENT TENDENCIES, ECONOMICS 303, WINTER 1934

Conformity }–to what other people have.
Distinction }–from what other people have.
  –What is wanted when commodities are bought.

Classical economics assumes that the <u>end</u> (a utility function) is known to the individual.

–Edgeworth pointed out that we need only to measure <u>relative</u> quantity, not absolute quantity.

  –But there is, in reality, no <u>known</u> end-realizing action.

    –A part of the end of action always is to find out what is the end of action.

    [Double vertical lines alongside in margin]

<u>Utility is motivation viewed as pure quantity</u> – how far utility <u>is</u> pure quantity is another problem.

–Mechanical causality – behavior is <u>caused</u> by the situation in which the organism finds itself.

  –The other, contrary view, is that behavior is motivated, with the end in view.

  –Or, that behavior is motivated but the end is also to be sought for <u>in</u> the behavior.

  –Probably a mixture of all three.

–"Historical" point of view.

  –Forget the human being.

  –There is a recognizable <u>order</u> in the data, or progression of events.

–Economy means the accomplishment of <u>given</u> ends with the most efficient use of the means to hand.

–Knowledge of the physical world is (superficially) very different from knowledge of society.

  –Being a <u>member</u> of society, we have to deal with it in a different way than the physical world.

  –Even when interest in another person is completely hard-boiled, one's relation to him is sentimentalized to a larger extent than anything in physical world.

  –Fellowship relations enter in.

    –But also enter into relationships to physical world.

  –Knowledge of human beings; knowledge of society – What is it?

    –cf. Cooley – Human Nature in Society.

  –How far do we get in trying to <u>understand</u> each other?

    –cf. Laredo Taft's statue [grave of Henry Adams's wife in Rock Creek Park] – two men fastened back to back. Human understanding – a vague feeling.

Social Control

  –cf. Tugwell, worst and thus best example.

    –Idea that society is my chariot for me to drive.

> –And the masses love to lie down before the Juggernaught car, if it's done with right technique.
> –What is the intellectual element in social enterprise.
>> –Analogous to making of rules for a game – with aim of a <u>better</u> game.
>> –Making rules for yourself, not for others.
>> –Aiming at a good game, but if aiming at <u>winning</u> the game then it is not analogous to economic theory.

The step from understanding of economic theory to social policy is a tremendous one.

> –Have enough to understand economic theory – and it <u>is</u> essential.
> –But talking about social control on a basis of that understanding is overpowering.
> –The childishness of Tugwell and Slichter, who want social control, but don't see the elementary fact that what they want is a society that would be <u>their</u> plaything; that with would work only if <u>they</u> controlled.

I – Human Behavior
(1) Descriptive – statistical approach (actual behavior).
(2) Mechanics of nature – classical approach (want satisfaction) (economic behavior)
(3) Value Realizing – explorative-experimenting.
> –#2 abstracts from error – conceives of unknown ideal.

II – "Culture" – "Historical-Sociological approach"
> –Like grammar – no concept of individual (word) behavior.

[Referring to above:]

1. Behaviorism allows no mistakes, cause and effect fixed.
2. Error of not achieving end.
3. Error in end set up.

Talcott Parsons – The Sociology of Marshall – Q. J. E.
Halévy – History of English Utilitarians
Stephens – The English Utilitarians
Bagehot – Economic Studies (Postulates)
Murdoch – Psychologies of 1925

Pearl's a priori mathematics and aesthetically appealing curves.
Human race is inherently wonder-hungry, salvation-hungry.

–Wants a verbal formula to solve the world's problem.

–Any good salesman can capitalize this market.

Dictator: (a) Absolute proprietor of society, treating it like a pigeon farm.

(b) Wants to give man his due, benevolent.

Democracy: (a) Values realized in the process of realizing, life is to have as good a time as possible.

(b) Values realized in carrying out a destiny; life is to fulfill this.

Problem of <u>Agents</u>

–How to pick one – a doctor can not be intelligently chosen unless the chooser knows all medicine and more than the doctor, and must know how much each doctor knows.

–How to trust one, when he is picked.

–Why <u>should</u> the agent carry out his master's interests rather than his own?

–He does usually do it, but why?

–Relation between morals and legal sanctions.

–We have no laws against incest and cannibalism – and none of either.

–We have laws against stealing, can't enforce them.

<u>Adlerism</u> – embezzlement of words – criminal lunacy.

–Words get meaning attached to them and cease to be descriptive. (<u>Honorific</u> – Veblen) – when words have acquired color they can not be used in precise definition.

–It is technically impossible to talk about rationality.

–And actually impossible anyway, because such talk turns back on itself – it is always possible to be rationally irrational (Most of Adler is irrationally rational! F.T.O.).

–The foolishness of trying to lay down a formula for the future.

–Conceptions are conceptions and thus by nature are not adaptable to precision of definition.

Social control is a problem of political philosophy.

Intellectualism is itself a religion – and has been as uncritically accepted as any other religion.

Natural sciences get along without much problem of epistemology.

–All social phenomena are mental, all mental phenomena are social.

–"If it wasn't for differences of opinion, we couldn't have any horse races" – Puddinhead Wilson.

– Technical economics gets along all right so long as we talk about wants and the technical machinery and means for satisfying them.

–But <u>economic</u> theory has to assume that wants are reflected in behavior – they want what they get.

–But when we ask "what do they want," going behind the economic theoretical abstraction – we are in hot water.

–Do we know anything about human beings? – aside from the communicative methods and symbols?

–Do we know anything about them aside from their existence as physical agents and beings?

–But these things do not interest us. What we are really interested in is their <u>interests</u>, especially as they pertain to social living.

–We have got to know what is <u>in</u> other people's minds before we know <u>what</u> is <u>in</u> our own – this brings us squarely against the problem of communication, epistemology, symbols, words.

–We must <u>separate</u> our interest in physical things from our interest in human beings.

–And our two <u>methods</u> of dealing with those two subjects – which methods are not and, can not be, similar – i.e. "Social <u>science</u>" is God.

–How are we to check or verify our knowledge of <u>communication</u>?

–We have no technique of enforcing any single verification, any objective standard, about human thought.

–Nor any technique of attaining such an objective standard.

–Yet we can not get along without <u>some attempt</u> to deal with interests, intents, etc. from an impartial point of view.

–As with law and law suits.

–So with economics.

–As soon as you begin to talk about any social phenomenon – social psychology, institutions, sociology – you begin to talk about <u>history</u>. [Triple vertical lines alongside in margin]

–You must ask how existing cultural phenomena "got that way."

–No philosophic thinker has <u>ever</u> shed any light on the mind-body problem – philosophers are still in the trees – no theory of consciousness has built any bridge from "physical" world to the world of <u>real interests</u> (psychology).

–We can find physical measurements to any degree of accuracy – but there is no possibility of establishing <u>preciseness</u> for the social-mental world. [Single double line in margin alongside "establishing" on]

–What is the relation of <u>truth</u> and education?

–Even in the physical world there is no precision in the ultimate analysis.

Cost Theory
 –Position held in 1927 paper [cf. Knight, "A Suggestion for Simplifying the Statement of the General Theory of Price," *Journal of Political Economy*, vol. 36 (June 1928), pp. 353–370] was essentially Wicksteed's.
 –The <u>essential</u> element of the classical pain-cost theory was an <u>alternative-product</u> ["cost" written above "product"] theory.
  –Their philosophy of pain-cost doesn't make any difference to <u>this</u> matter.
 –Effective cost (price determining) is the alternative product sacrificed.
  –Resources not distributed between competing uses don't <u>count</u> (though they follow the essential lines of classical rent theory).
 –The trouble is that resources can be on an indifference margin at <u>different</u> levels of pecuniary occupation.
  –Alternative-cost theory breaks down unless resources are freely transferred at the <u>same</u> level of remuneration
  –Problem really leads to a restatement of economic theory.
  –What is involved is a lack of freedom of choice.
   –If there <u>were</u> pure freedom, <u>unequal</u> non-pecuniary uses of resources would be impossible [Single vertical line, alongside these three lines, in margin].
  –Cost theory is subject to almost infinite reservation – because of the inevitable monopoly of the individual.
  –Cost theory holds only to the extent that all individuals are equal and the same.
  –Concept of productive capacity.
   –What can we consider as an invariant?
   –We <u>have</u> to assume that an individual equalizes the return to an invariable productive capacity.
  –Assumption that one producer owns all the resources (<u>my</u> assumption) – gets rid of <u>competitive</u> assumption [Double vertical lines alongside in margin].
  –Problem is less for consumption side than for production side.
   –Due to fact of division of labor – impossibility of <u>any</u> individual working at more than <u>one</u> occupation.
   –If you assume away division of labor, you have assumed away the economic system [Single vertical line alongside this and preceding line, in margin].
  –But the same difference [in space above: "problem"] does hold on utility side – lack of freedom to combine <u>all</u> alternatives at will in any proportions at a <u>simultaneous</u> time.
   –Problem of time is the devil all the way through [Single vertical line alongside in margin].

–You must either go into the utilities (subj[ective]) on both sides – or stop
with pecuniary measure on <u>both</u> sides.
  –As Pigou does not do, in going to utility on work side but stopping with
  money on wages side – "equal wages for equal work."
    –"Equal wages" – to take in utility – would have to be equal <u>real</u> wages,
    or equal (real expenditures and resulting) real income resulting from
    wage.
–One is not free to make marginal choices between the <u>non-pecuniary</u> aspects
of two expenditures of resources – they are in different worlds originally,
and, even if not, are seldom directly compared at <u>one</u> instant, and might as
well be in two worlds.

Production	Money	Consumption

Money-making
Division and coordination of
productive resour<u>ces</u>
  a. between modes of
    money-making
  b. between money-making
    and other uses
  c. between different
    owners.
		Money-spending
		Divison and coordination of <u>uses</u> of
		money in procuring utility for each
		individual.

–Greatest weakness in economics is to assume that people usually <u>want</u> goods and
services.
–Fundamental difference between the two sides is, that on consumption side only
an individual, who varies one resource among different uses, is concerned. While
on production side, <u>many</u> individuals are concerned, usually each owning one
resource, and varying its use between different uses. ["–a plurality of resources"
written below the line under "each owning one resource"]
  –i.e. every individual is carrying on a Caruso economy, at the same time that he
  is entering into a social economy (in his activities as consumer and producer).
  [Double vertical lines in margin alongside this point]
    –This also brings up the problem of use of resources which do not go through
    the market.
      –As personal services.
      –What of book assets (in rural life).

–Do not make <u>any</u> call for money.

–In a perfectly organized economy – every book credit would carry interest – on both sides.

–Money is the worst devil in economics – people never will understand it, or act sensibly about it – perhaps only solution is to get back to barter terms.

–Short run – resources are concrete things.

Long run – resources are a capital flow.

–<u>Basic economic realities are intensities</u>, are flows <u>without rate</u>, only intensity.

–Not like water – if it stops flowing, <u>it</u> still exists.

–But like electricity, if it stops flowing – there is nothing there.

–It can only be measured by taking <u>intensity</u> of flow, over a period.

–Satisfaction is like that – it is <u>something happening</u>, with an intensity over a time.

–When it is not happening, nothing is there.

–There is <u>no</u> magnitude to remain static between events.

–Distinction between <u>labor</u> and all other productive resources in society.

–One does not often keep a capital account for himself (to some extent, one does keep an account with himself – to that extent, he is capital).

–One can not mortgage himself – i.e. he is not capital – to the extent that one <u>does</u>, [e.g.] wage contract, (baseball player), one is a piece of capital).

–Any thing one can not increase or decrease is not capital.

–The owner gets one return, the instrument gets another.

–Owner gets rent – instrument gets a wage or an interest.

–Can one replace, or does one merely maintain, the individual.

–A piece of capital is thing with which one (can)(does) keep an account.

–Businesses vary greatly, in practice on this point.

–Increase of capital <u>must</u> be considered – from an accounting sense, as income; but it can not be consumed as such without eating capital.

Concepts having to do with <u>capital</u>, in an advanced society.

1–Genesis: <u>Some</u> productive instruments

a–Have been produced

b–Must be maintained

c–Can be, and are, increased; alternatively to consumption goods

d–Can be, and are, exchanged; especially for consumption goods

2–<u>Where</u> some capital goods can be and are produced additionally, and can be and are exchanged

a – Then all productive instruments that are exchangeable are "capital" "value"

3–The amount of capital in any instrument is (a) always capitalized value of a yield at the market rate or (b) is cost of construction cumulated at same interest rate
–Two ways to capitalize an instrument of limited life.
    1–Capitalize it for its limited period (although there will always be something to sell for salvage, or a carcass expensive to be disposed of. Capital never behaves like the One Horse Shay).
    2–Or capitalize it for perpetuum – the realistic way.
–But whether you consider capital to be perpetual or limited – is entirely a technical matter – any piece of capital may be considered in either way.

4–Time
    a–Period of construction is one dimension in cost – no information re amount of capital
    b–Service life from >0 to infinity – no information re amount of capital
    –Impossible to say that production of capital is production, and that production of income from that capital is also production – yet this is what accountants must do.
        –Get lunatic point of view, says modern psychiatry, and everything else he says and does will be perfectly logical and clear.
            –Apply this to Mill – "demand for goods is not a demand for labor" – criminal imbecility.
            –But if you get his point of view – it follows easily.
            –If we are no longer interested in anything but goods – we will not have a demand for capital producing labor. (?)

How does economics become a science?
    –Between Renaissance and end of 18th century – people gradually got the idea that competitive market operations could take care of the needs of life without any definite, positive policy by governments.
        –Policy becomes negative – lack of it was wanted.
            –Policy of lack of policy is a propaganda as much as the propaganda of positive policy.
        –When policy becomes negative – you have a science.

Talking about intelligence and social management is bound to be pessimistic.
    –Most "revolutionary" leaders don't do anything to society – their actions allow them to preempt a great acreage out of future historiography – that is all.
    –The person who sets out to control society sets himself, by definition, outside it – which is impossible – it is anyway an infinite conceit.
    –Social control – ought not to mean somebody controlling others.

—True <u>social control</u> is a problem of free association – we know ourselves-in-society.

—What do we really mean by <u>intelligent association</u>.

    —What society has taught us to conceive as intelligent action – is somebody manipulating some<u>thing</u>.

    —What can <u>we</u> accept as intelligent <u>action-in-society</u> – where the actor is acted upon to the extent, at least, that he acts.

What <u>is</u> the relation of economics to social control? cf. Knight (Ethics of Competition), Laidler (Socialism in Thought and Action).

—What is a definition of poverty?

—What is inequality of distribution of wealth? – economic power? – If so – the definition of economic power is tremendously more unequal in its distribution in Russia today than in U.S.

—Mob follows leader as long as he leads where they want to go – this is pol[itical] life and leadership.

    —The thing that has <u>meaning</u> for inequality is <u>income</u>, not wealth.

        —But that does not measure much; <u>consumed income</u>? If so, we are getting away from the excess of inequality – no <u>rich</u> man consumes 1% of his income.

        —We find we know nothing about the question.

            —It's better <u>not</u> to know so much, than to know so much it isn't so – Josh Billings [pseudonym of Henry Wheeler Shaw, 1818–1885, American humorist and lecturer: "The trouble with people is not that they don't know but that they know so much that ain't so." *Josh Billings' Encyclopedia of Wit and Wisdom* (1874)].

            —Corporate heads are not paid for <u>services</u>, but so they can <u>live</u> the way it "befits" a corporate head to live – their salaries are window-dressing.

        —What would be the inequality of income if all <u>money</u> incomes were equal? [Single vertical line alongside in margin].

            —Most of the things that account for the inequalities of <u>expenditure</u>, as big estates, big cars, servants, would <u>not</u> be <u>on</u> the <u>market</u> and many things would be on the market at <u>very different</u> prices than now.

            —All <u>preferential goods</u> which would still exist and <u>could</u> not <u>go round</u> – <u>who</u> will get them, and <u>how</u> – i.e. most high values are really scarcity values.

            —What of champagne land – would it raise as much wheat as Minnesota land – no, this would be sheer loss.

            —What of skilled craftsmanship, would it be set to dishwashing? – this also sheer loss, other better dishwashers.

   –If these things didn't happen, we still have the fact of sure inequalities
     of real income – and problem of how and why.
   –One solution is change a society so that preferential goods could be
     Communistically or cooperatively enjoyed – big estates made clubs,
     etc.
–Knight thinks there is about as much sense in talking about equalizing
   money incomes as equalizing baseball scores. What is the diminishing
   utility of high scores? Is there any? A great deal depends on how the
   game is scored – as when handicaps are used so that there is a fight all
   the way.
   –This is the sort of question society has to answer [Double vertical lines
     alongside in the margin].
–Diminishing utility has absolutely no meaning in the question of
   distribution of money income.
   –D[iminishing] U[tility] holds only for the budgeting by one person of
     his income.
     –Jevons saw this.
     –Austrians never have.
   –We can not extend this argument to a diminishing utility of total income
     for two persons in their total budgeting [Single metrical line alongside
     in margin].
   –We don't know anything about total utility anyway
     –Bosquet [Bernard Bosanquet, 1848–1923?] – says we have to
       consider total income as always the same, whatever the shifts in
       money income (and expenditure).
     cf. Slichter – last chapter.
       Homan – Economic Stabilization in an Unbalanced World –
       last chapter.
   –Is graduated income tax really dependent on principle of D[iminishing]
     U[tility]? – Not at all, theoretically.
     –Or on McCloud's [likely Macleod] principle of the fiscal system –
       "to get as many feathers from the goose as possible with as little
       squawking."
     –Or on human aesthetic interest in a nice-looking formula.
   –Question of whether we can have a historical economics that is any
     different from economic history – leads into same morass of philosophy.
–What effect does intelligence have on the social problem?
   –What is the social problem?
     –How is society to solve it by itself?

–The players changing the rules.

–Or are you and your gang going to change it.

    –<u>Discussion</u> of this sort of thing is absurd.

–What is the intellectual basis of discussion[?]

    –We think intelligently about the external world, but talk nonsense about the social world.

–Economics discusses the organization [of] a two-stage market – owners of productive capacity sell them to intermediaries (business units or enterprise) and buy productive services (goods and services) from the intermediaries.

    –Problem of relation to this of political policy.

    –In a pure exchange economy there would be no pecuniary cost of production nor any distribution problem.

    –We are not born into the world on contract terms.

        –Our "inalienable rights" put the individual in a handicapped position with reference to other owners of property.

    –If you believe in progress of civilization – you must realize that the <u>private exchange economy</u> is the greatest invention working towards that end.

        –It provides social cooperation without <u>any</u> political guidance – in subconscious.

–Any kind of self-conscious process disintegrates the whole thing – is anarchy.

–Knight sure that society will never reorganize itself – especially in the direction of anarchy.

    –Even if a group steps in to reeducate society in this direction – what of the problem of its turning over control, after reeducating society for disorganization.

    –Knight thinks that the nearest approach to democracy has been when it was allied to a laissez-faire system of economics.

        –You can imagine an exchange economy without money.

        –You <u>can't</u> imagine an enterprise economy without money.

            –No enterprise economy has allowed this framework of money-circulation to be in private banks.

                –By prohibiting the strictly defined coinage of money by individuals.

                –It has come to restrict note-issue, and even deposits.

                –But it still leaves the creation of credit in private hands – and enterprise economy can not work on this basis.

–The intellectual problems involved in <u>cheating in games</u> games are of great import to democratic society.

–There is no <u>theoretical reason</u> why any detail of our civilization <u>should</u> change if we replace free enterprise by socialism. – Some changes would probably occur – what?

–People lay down rules for themselves – act accordingly – "self-legislation individuals."

–Would a socialist economy break down because of irresponsibility of government servants, without usual incentives, or because of the <u>over</u>-responsibility of its servants? – Probably the latter – as in Russia.

Rational management <u>depends on</u> maximizing total return to factors – <u>whatever</u> is done with the actual product.

–This can not be forgotten (re interest) even in Russia.

Mathematical Economics

–Walras – the system on which most subsequent analysis is based.

–<u>Productive efficient</u> [sic] = input of a given factor per unit of output. [Alongside in margin: "<u>Better concept</u>"]

–<u>Distributes</u> the product and allocates the resource.

–Partial derivative of the product with respect to the factor.

–i.e. small change in product resulting from small change in factor – other sides remaining the same.

–The rate at which the effect varies when the cause is varied at a given rate.

–Mathematics is a way of describing what goes on in the world – though never <u>taught</u>. [Alongside in margin: "<u>Real income</u>"]

cf. Fisher – Measurement of Utility, in J. B. Clark memorial

Frisch

Schultz J. P. E.

Bowley

The whole organization of society bristles with problems that cannot be solved by free competition.

–The <u>real problems</u> of economics, <u>vs</u> the things economists talk about.

–cf. Herbert Spencer – dignity of man; limitations on government, etc., one of most <u>ardent advocates</u> of laissez-faire.

–People want to do what they want to do – this seems to be the basic assumption of what Knight wants government to be.

–Problem of democracy – will people vote what they want, or differently.

–Does the sovereign make law, or law make the sovereign.

–Problem of law – the respect for law as law <u>vs</u> the disrespect for law as law.
   –Taking away the divinely ordained quality of laws puts society up against a <u>real</u> problem.
   –When man realizes that laws are made by <u>other</u> men like himself – we have a real problem.
   –Most men think the majesty of law is something <u>other</u> men ought to follow.
   –As with religion – most educated classes and religious leaders have always believed that religion was <u>necessary</u> for the mass of the people – the opiate – to preserve society.
      –When a religion ceases to believe it is descended from god – other religions from the devil, then it is dead as a religion – especially when its advocates <u>argue</u> for it – and on the basis of its utility, rather than "burn the unbeliever."
   –People are willing to accept a lower return on the ownership of <u>land</u> than in other forms of ownership, of securities, etc.
      –Social prestige and land in England.

–There is <u>no</u> principle on the production side to balance the principle of diminishing utility on consumption side.
   –Utility acquires a <u>social objectivity</u> – even when every man does not consume <u>every</u> product – but every man consumes a <u>wide</u> range of products.
–In order to say that one dollar's worth of two things has the <u>same cost</u> – what do we assume?:
   –That sacrifices are the same for two individuals.
   –That all productive capacity is expended <u>within</u> the pecuniary system.
–The problem – what do we mean by comparing a quantity of pecuniary and some additional non-pecuniary dimension <u>with</u> a <u>different</u> quantity of pecuniary plus some additional non-pecuniary division?
–Problem of time division in comparison –?
   –Same hours per same year.
–In old formulation – we had to <u>assume</u> that when two people work for the same wage in same occupation – that then their sacrifices were equal – but this is too much to assume.
–Problem of total productive capacity?
–How estimate <u>availability-utility</u> as separate from use-utility?
–These must be answered to be able to say that we <u>equalize</u> productive capacity.

Institutional or Historical Approach
   –Motive of intellectual pursuit – what craving drives us to think about the world in a scientific way.

–We have to act in our environment – we <u>have</u> to live in an environment.

–Are led to think about that environment to try to <u>know</u> it, with aim of interfering in it.

–The world would be knowable – if there were no time-change (natural, physico-chemical, biological, social).

    –The terms we use in describing society have usually <u>physical</u> penumbra attached to them, and we haven't any other words – showing that we haven't any other concepts – but are set, applied from physical to social.

–In physics we separate out space and time as the dimensions of change – mass as a constant (?).

    –Then we explain physical things in terms of motion of mass in time and space.

        –Herbert Spencer's <u>The</u> Utilitarian – utterly naive, but completely important for understanding that philosophy. Jevons and Menger made utilitarianism more analytical.

        –<u>Mach</u> – <u>The Conservation of Energy</u> – very imp[ortant].

    –Explain change by getting rid of it.

–There is an urge to reduce things to physical analogy. Conservation of momentum is far more important than conservation of energy. But these principles break down forcing us to go to higher, <u>less empirical</u>, less intellectually satisfying principles.

    –Energy breaks down – potential energy, radiation loss.

–In social phenomena – we think in physical terms.

    –And don't usually recognize the real problem – can there be a concept of society? or social science?

–In physical level you have to abstract from the fact that <u>you</u> are a part of the physical world – this doesn't cause much trouble. [In margin, with arrow pointing to this and the next point: "can't draw a picture with yourself in the picture."]

–But on social level you <u>cannot</u> abstract from the fact that <u>you</u> are a part of the <u>social</u> world.

    –But when you become a part of the world you study – there is an impasse.

    –In social science you can not talk <u>about</u> phenomena without talking <u>to</u> them, and with the aim and end of <u>changing</u> them.

–Social science is itself a social phenomena [sic].

–Any statement you make that is about society is never true after it has been said – for society has been changed by its being said.

–Possibility of a science?

Historical Economics – as separate from economic history
  –Must have some <u>attempt</u> to explain history in intelligible terms, i.e. a philosophy or system of history [Double vertical lines in margin alongside this point].
  –This brings up the problem of <u>what terms of analogy</u>:
    a–<u>Physical</u> terms – can not be used in social science – no possibility of measurement where different individuals are concerned. We assume that human action does not violate physical laws. But there is violation of them in life.
    b–<u>Biological</u> terms – Veblen thought he was a Darwinian – impossible to see how – no connection between the two.
      –Adaptation in the social sciences – we don't know – there may be the opposite!
      –Either kind of explanation of society is intellectually satisfying – but utterly false.
        –Veblen doesn't discuss either <u>variation</u> or (selection) adaptation in his attempted biological explanations – What is his criterion of variation? his principle of selection.
        –Henri Sée – Theory of Economic Determinism.
        –Seligman – The Economic Interpretation of History.
    c–You do not have to assume intelligence in historical economic behavior – as long as individual experiments at random and a random selection of satisfying results is sought.
  –What happens is that everyone gets a vested interest in some one theory and discussion is then out of the question. Why do people want to hold theories?
    –All knowledge is rooted in intelligence of <u>social</u> life. Even behaviorism has a metaphysics.
      –Human beings do not <u>want</u> the complete answer to all their wants – they want relative goods – relative to their environment.
      –When you shop – you don't know what you want, you shop to <u>learn</u> what you want.
      –The nature of motivation is that there is an end, the consciousness of some end.
        –The problem is to get some content into it, getting more and more precise in putting in that content – but still the <u>end</u> always <u>remains</u> to some extent[:] find out <u>what</u> is the end?
          –Light, sand, matter, muscle-skeleton – these are our only external relations with the external world.
          –But we have to be unconscious to complex function of use of those – or we can't use them.

–Motivation has a large provisional reality; but in the end motivation has no reality.

 –It is the experimental drive towards finding out what to express and at the same time expressing it.

 –Living is an art – science is an aid to art, but never takes the place of art.

 –You can never state a problem in advance, the statement of it and the solution of it are a simultaneous process, not a time sequence.

c– Another term of analogy is to seek an <u>idea</u> in history, to forget the human beings.

 –(German), culture-entity.

 –Language is the primary tool of civilized life – and it is used unconsciously – except in <u>professional</u> life. The only effect of civilization on language is to simplify it.

 –<u>Conscious motivation</u> doesn't apply to language.

 –To study language historically you study it as an entity – not as a mode of human behavior – does one think of grammar as behavior? What <u>does</u> one think of when one thinks of grammar?

 –What imagery does one use in thinking of it?

 –<u>How explain linguistic sequence</u>? After you've explained this – you have something to work on in history.

 –<u>Language is the perfect example of Institutionalism</u>.

[In margin at top of page: "Babbitt – Rousseau and Romanticism" and "'intellectual stupidity is a worse crime than immorality' K[night]"]

–The product of an unmotivated environment, in which the aspect of human behavior is subordinated to some <u>historical order</u> in phenomena.

–Economics <u>can</u> be explained in terms of such culture-entities.

–"Language as a historical product."

–Rice Case Book – Method in Social Science

–Marshall – "Natura non facit saltum" – Nature does not proceed by jumps.

 –Not so – modern physics stresses discontinuities.

 –But in <u>history</u> – there is continuity, it can be plotted in curves (usually by mechanical analogy).

Approaches to our study:

 a–<u>Factual</u> – historical – statistical.

 b–Motivation

 –Mechanical

 –"Control" – political

c–Historical

  –In economics we plot the quantities having to do with production and consumption – but make an arbitrary definition of those.

  –Problem of simplicity – a subjective fact – on the basis of it, you predict confidently.

  –Make a "simplicity" out of chaos by plotting.

  –Theory of errors in observations.

  –Proposition that a simple curve is more probable than a complex one – base on faith.

  –Empiricism vs rationalism

  –You can argue on the basis of a simple pattern – but you either assume there are no other facts, or that they are unimportant.

   –Commonsense doesn't question simplicity.

  –The two things are not so different, run into each other [Double vertical lines alongside in margin].

  –In relativity theory, there is no absolute measurement – the answer depends on where you measure from – what is a straight line in relativity theory?

  –Empirical principles come to be held or felt, as a priori. And the a priori principles have a solid empirical basis.

  –The two come out at nearly the same point.

 –There is an absolute contradiction between the proposition to treat a subject scientifically and the one to change a thing – unless you are yourself outside the subject – not true in social science. [In left margin: "cf. Bergson."]

  –Our observation is a process of communication.

  –The social science is of its own subject matter.

 –Ordinary economic theory treats of behavior as motivated, but of the motivation as mechanical.

 –Alteration of behavior by discussion means freedom on a social level.

 –Control implies individual freedom and some social freedom for the social mind.

  –Self-legislation or social legislation.

  –This latter is the only meaning of social control

Calculus – what to do about velocity when its rate is not uniform? How treat it?

 –This is the function of the calculus.

 –By studying the velocity at a point! – via the concept of the limit.

  –Simple velocity is a first degree, linear function.

  –i.e. Newtonian – constant force is one that will create a uniform velocity.

[Diagram with unlabelled axes, with ascending curve starting at origin, labeled "Kt^2 total velocity (increasing)," and straight line from origin intercepting rising first curve from below, labeled "2kt increase of velocity (uniform)"]

–A mass, in its natural state, has the natural tendency to keep moving at the same velocity. The Greeks thought the natural principle of mass was <u>rest</u>.
  –Unless a contrary or additional force affected it – as friction,
    –Velocity produces heat.
–Marginal utility is on this principle.
–Mathematics as a substitution for the inaccuracies of verbal words.

Imperfect Competition – whether the seller has any control over price, or not, is the test – for competition, and the individual seller.

[Diagram, with quantity saleable on vertical axis and price on horizontal axis. One vertical line from and perpendicular to horizontal axis at point X. To right of it is negatively inclined curve at second point X, not crossing perpendicular line]

–At any price above the market price, indefinite quantities could be sold.
–The problem is to define competition, it is realized that it will be a limited case.
–Transportation and selling cost are the main causes of imperfect competition.
–Even making a better product – spending more money on it and selling more – is only the same as advertising more, or being further away from the market.
–Yet if we get away from the science of competition, we are outside the range of economic science – and talking ethics.
–We need to make a dimensional list of meanings.
–What do people do when they buy commodities? <u>What</u> do they buy?
  –<u>Commodities are not human wants</u>.
    –A list of one and a list of the other would have no word on both lists.
    –We want mainly <u>distinction</u> and <u>conformity</u> [Single vertical line in margin alongside].
      –We buy discomforts as well as comforts, we buy to shorten life as well as to lengthen life.
      –"Men travel far to seek disquietude."
–Pure competition for the system is conceivable, hardly realizable.

[In margin on top of page: "cf. Q. J. E. May 1918 (effort to increase monopoly) S. Bell"]

–Pure monopoly for the system is not conceivable – no man ever has had or could have that power.

–Yet partial monopoly is always with us.

–Mrs. [Joan] Robinson assumes a perfect fluidity with no historical conditioning; thinks she can rearrange the whole thing, reshuffle the cards, at her own will.

  –You can not confuse the size of the plant and the size of the firm.

    –The first is possible to be talked about – by technical considerations.

    –The second we don't know anything about – it gets into the human problem, about which we can't generalize – there is something mystical abut the whole thing.

      –cf. the problem of Generals and War.

        –It is an organizational activity of the same kind.

        –In which the mystical element stands out largely.

      –Perhaps this is the sort of thing a real social science would talk about – if there is anything to be said on the subject.

–Problem of strategy and deception.

  –The art of living is the art of fooling everyone else, and getting away with it.

  –Yet, one would assume that the deceivees, once they knew this, would refuse to be fooled.

    –In fact, Barnum was right, they love to be fooled.

  –Mathematical economists try to set up equations to show that the science can be made determinate – the exact reason why Knight lost faith in it.

[In top margin: "Antonelli (Mathematical Economics)]

–Cassel's formula – doesn't take in the fact that demand is a function of income.

–This determinateness doesn't prove anything anyway.

–Even three linear equations don't have to intersect, at a point.

–Unless you get to counting equations and counting functions – the whole thing is meaningless.

–Try to write equations which picture consumers and producers choices.

–Domestication and expansion of the sphere of economics.

–Economics – has always dealt with state policy.

–Great change of attitude – from concept that trade was robbery, cheating, to concept that it produced mutual advantages.

–Division of labor – of a very different sort than Plato's.

–Freedom of exchange, of markets in which sellers and buyers compete.

–Classical system – one of mechanism of exchange (economics – in Greek – (household), but a better translation is <u>estate, plantation management</u> – a closed unit.

   –"Political" was added to the word economics, in order to turn its attention away from state policy, to welfare.

The idea of an <u>institution</u> is very important.

   –American institutionalists don't know anything about the term.

  –Particularly the <u>individual as an institution</u> – bishops, priests.

  –And the <u>state as an individual</u>.

   –<u>Spann</u> – romantic fascist.

    –The group is prior to the individual, the individual finds his expression only <u>in</u> the group – is more <u>real</u> than the individual.

    –<u>Bentham</u> – as crazy as Spann, but on the other side.

    –Traditional Anglo-Saxon view – that the state is a mutual aid society of individuals.

   –The group must speak through an individual.

[Notes on a] <u>Lecture</u>

<u>Modern Non–classical Economics</u>

  A. Socialists

   –Marx got surplus value from William Thompson – acknowledged it in 1st edition only.

    –Thompson had been anticipated by the "Ricardian Socialists."

     –There is no more ethical than theoretical difference between "labor" and "capital" income – all education is capital.

    –Sismondi, Saint Simon, Fourier.

    Proudhon

   –<u>Class</u> has no meaning unless there is a <u>fairly</u> limited number, to which entrance is extremely limited, and in which membership is fairly homogeneous.

   –In U.S. – more classes than individuals (each one belonging to several classes).

   –Marx uses fatalism for exhortation.

   –Every religion has a theory of history and a cosmology – as selling talk – having <u>nothing</u> to do with the religion's scheme of living.

  B. Nationalists

   –Spann (Vienna), builds on A. Müller – is one of the intellectual godfathers of Hitlerism.

–U.S. has a paradox of external laissez-faire, and of great nationalism, especially tariff.

–List – fairly moderate protectionist

–Early and late Marxism.

    –Marx 1848 – a supreme democrat.

    –Late Marx – justifies Lenin's interpretation.

        –After communist [Paris uprising?] of 1871, and example of universal suffrage in U.S. – since 1840.

    –Dictatorship <u>by</u> the proletariat.

        –Means dictatorship <u>over</u>, and <u>above</u> the proletariat.

C. Historical School

–Older[:] Roscher, Hildebrand, Knies.

    –Connected with development of <u>culture</u>-history.

    –Idealism from Kant to Hegel.

    –German Romanticism – of the people, <u>folk</u> lore.

    –Out of which came <u>Historical Jurisprudence</u>.

    –Produced plausible propaganda for the study of economic history.

–Late school – <u>Schmoller</u>.

    –Contemporary of Menger – who brought back the theoretical, marginal utility approach.

    –Schmoller refused any theoretical approach, even in economic history – the time too young for generalization.

    –Ingram (disciple of Comte), Cliffe Leslie – English historical school.

        –Against a science of economic theory.

        –William Ashley.

D. Mathematical or Pure Theory

–Completely cuts out all matters of policy.

E. Statistical School – Mitchell

–His methodological discussion not important.

–But his subject-matter of attention, is the big hole in classical approach – equilibrium does break down.

–The business cycle – what reaction on <u>general</u> classical theory does his conclusions have? Any?

–The actual shape of demand and supply scales.

–Mills, F. C. – important work on crises – what comes out of it is a mystery.

–Sargent Florence.

F. American Institutionalism – means anyone who studies non-classical economics.

   –Only real institutionalist in U.S. was Veblen.
      –Did he mean any of it – did he think he was a satirist (Wesley Mitchell) or did he think he was a social scientist.
      –Knight can find no contribution.
   –As long as people think of life in terms of achievement, of efficiency, the type of theorizing of the classical school <u>seems</u> to be borne out by the behavior of people.
      –This is not to say that they are <u>right</u>.

   [In margin at top of page: "Substitute competitive politics for competitive business."]

   –It is a study of <u>economic behavior</u>.
      –It does not destroy its force to say that actual behavior is not wholly economic.
      –It went out of fashion, broke down, because of its policy implications.
         –Laissez faire – an anachronism now[,] untrue then.
         –Rationality – the world is tired of this – wants "<u>action</u>," wants to think with its blood, wants experimentation, not thinking.
         –It assumed a rationality of the consumers' choice.
            –Did not take into consideration the power of propaganda.
            –And the fact that people live, not to fulfill purposes, but to <u>make life interesting</u>.
            –Even economics is written about[,] not for the consumer, but by economists for their own interest.

[End of Lecture]

–What the group superiority means <u>is</u> essentially, l'etat cèst moi.
–Christianity was in origin an escape from the most thoroughly regimented world ever organized – with slave intellectualism.
–French Enlightenment (doctrinaire), English Utilitarianism (Bentham was a freak)
   –Idea of individual liberty.
      –Perfect formula for every individual manipulating every one else.
      –The more intelligent the individuals become the more manipulative becomes social action, until it breaks down.
      –Pretty much what has happened.
   –Knight doesn't believe in freedom of thought.

–The society that tolerates Communism and Jesuitism is crazy.

–Liberal society has not educated people to meet the problems liberal society must face in the near future.

–In certain lines an individual taste and choice is possible. In social lines, especially political, it is not possible.

  –Life according to rule.

–Individual liberty is a personal property concept under liberalism – i.e. the respect for property comes first – then comes "individual liberty" – then principle of mutual consent in exchange.

–Majority rule is a way of finding out the group's mind.

  –No sane society tries to impose anything on a considerable minority.

  –The principle of unanimity is essential.

  –Without it society goes to pieces.

–Liberalism had a naive faith that people would agree.

–Problem is a matter of discussion.

  –Approach to discussion is vital.

    –With thought of <u>agreements</u>.

    –Instead of <u>winning for one's own</u> belief – survival of the fittest.

      –Instead of Thomist idea that the first person to get the rostrum shall define the terms and lay down the rules. –Adlerism.

Competitive system takes wants as given.

–Business uses as much money to create wants as to supply goods.

–Persuasion is as much or more, the issue of politics.

–Intellectual life is itself competitive.

–Mainly for personal recognition.

–How separate the promotion of truth from the promotion of the thinker of it.

–You can't tolerate people who deny tolerance – Communists and Jesuits.

–Promotion of social reform can't be separated from promotion of social reformers.

–Propaganda is primarily aimed at getting the propagandist into a position of power – Thomism or Communism.

  –Virtue, social reform, etc. are <u>sales</u> talk.

–People will <u>not get what they want</u>.

  –From exploiters, <u>or</u> from benevolent social reformers.

  –Which type of dictator would be best?

    –Ruthless world conqueror, or moral fanatic.

–Older order was based on <u>personal</u> contact – with religious, customary behavior tradition.

–But with new means of communication of printing press, and radio, so that man can: (1) compete with other men for the law of larger numbers; and (2) he can extend his voice and word to the ends of the earth faster than the sun.
  –Gives a possibility, with backing of military power, of a despotism never before known.

<u>Cost</u> – Alternative cost – in <u>classical</u> theory it was in terms of pain – equal cost for equal pain.
  –In <u>Austrian</u> theory it is in terms of (agency) (labor) time.
  –What if labor does not receive the same reward in equal occupation at equilibrium?
  –<u>If</u> labor has had a free choice of pecuniary and non-pecuniary alternatives, then the above may hold true and still bring equilibrium.
    a. Money income.
    b. Alternatives to work – "leisure," "not working" – will be mixed with work in varying degrees.
    c. Work
  –Money income is different in different occupations.
  <u>Work</u> is different in different occupations.
  "Not working" is different in different occupations.
  –We don't know what money income is.
  –<u>If</u> we could assume: (1) that net money income is the same in two occupations; (2) that is <u>means</u> the same; and (3) that not-working is the same, then the difference between occupations, for this same individual, is dependent on the working.
  –<u>If</u> we know what money income <u>is</u> in two occupations, if it has the same utility curve in both, and if the individual can vary the non-working proportion and the working proportion,
    –Then is it correct to say that at the same money income the two occupations will be on a margin of indifference.

The principle of economizing: the maximizing of return from limited resources.
  –Human being – an electric battery – a current goes out and comes in.
    –The capacity of the cell is apportioned among alternative (ways of expenditure) channels, with a view to maximizing return.
      –Electric current is in inverse proportion to the resistance.
        –Every channel open will get a current inversely proportional to its resistance.
      –The economic cell is like this in the expenditure of a given money income.
        –Apportionment of resources to maximize utility.

–Buying the product of greatest utility down to the point at which the last increment becomes equal to the lost increment of utility of the next greatest utility.

–Battery takes all the paths open, conditioned by a resistance law.

–Economic cell takes paths down to the point at which the utility of the last increment of one path, taken, becomes equal to the last increment of another path, and all other paths.

–Problem of the division of labor and the use of productive capacity.

  –No distribution among different occupations.

    –The division off labor prevents this.

    –Technical impossibility of getting away from one line and closer to another line.

  –In a Crusoe economy, even with division of labor, he would apportion productive resources on the basis of equal sacrifice – sacrifice of alternative products.

–In a complex economy, all that is out.

  –Equalization of utility only within each individual's expenditure.

  –Within every enterprise the allocation is similar to individual consumption.

  –There are other considerations – other than maximum pecuniary return.

    –On property side.

    –Even more on labor side.

    –An alternative between work and doing what you do when you don't work – as long as life is chosen.

    –If one works, there is a different margin of alternatives than if one does not work.

    –Also money income and non-work uses off time are complementary goods.

    –The individual is not on a margin as between two pecuniary incomes.

      –Although in lower ranks of labor, this is not so true – work for what they can get – money is all they can get.

    –But, to have alternative cost theory – you assume that all laborers are the same in all occupations, and within each occupation.

      –If utility concept enters – (personality) into the conception of a resource – allowing psychological differences, then any cost concept is impossible.

      –Division of labor and allocation of resources – [are] not done in accordance with "equal disutility."

–There's no comparison of different alternatives.

–Only pecuniary cost considered: at the margin cost is alternative product.

–Psychological, <u>individual</u> cost theory.

    –Labor produces everything, more labor exerted with higher price.

      –Utility and disutility equal, for the individual.

      –But <u>no</u> connection with the prices of commodities for two products.

    –Smith and Ricardo – "pain cost" is alternative cost theory – allocation of resources with a view to pecuniary maximization.

    –How much is left of cost theory?

      –Resources apportioned have different owners.

      –Money income changes with alternative shifts in apportionment of resources.

      –Due to the <u>mores</u> of a <u>competitive industrial</u> society – its folkways.

        –Money income is in the great majority of cases the <u>sole</u> end of action.

          –Salaried intellectual work (law professors, other professors) is the main exception.

            –Production is mainly an elaboration of a few main materials.

              –In the stages of production – little concern for final product.

            –Time dimension – if a person is free to work as little or much as possible, why should he not offset varying irksomeness of work, by varying the intensity of work?

        –Yet it seems as if the essential thing is to work the same amount, regardless of irksomeness – if non-work is given up, the result is work, of about the same intensity per hour per one, as per another, wage.

          –Only money income per day will be increased by working longer (cf. Sir J. Stamp – essay on Stimulus).

      –What is a day's work? – it seems to be largely a matter of what people think it <u>ought</u> to be, which varies.

        –Husking corn, a day's work? The concept increased by about three times in a generation.

      –Opposing proverbs: mobility of labor – hardest species of baggage to move = man? as easiest? – both.

      –Idea of natural gait.

–Dividing the time stream laterally or vertically.
–Mental effort is attention (cf. Ladd – Attention).

–How much will a man work?
a = non-work utility
b = utility of income from working
C = capacity (total) of <u>work</u> <u>and</u> leisure
r = amount of leisure given up.

$U=a^i,b^j$ $\qquad\qquad i+j=1$ (*i* and *j* are fractions)
$\qquad\qquad\qquad\qquad\qquad$ C is divided between *a* and *b*.

–Economists have said, generally, that one would work harder with increase in wages.
–Knight argues that one would work less with inc[rease].

$U = (C-r)^i \times (pr)^j =$

–*p* disappears with differentiation!
–There is <u>some</u> complementary relation between leisure and work.

$U= a^i + b^j$

–Where the relation is additive, it is harder to work out.
–*p* does not drop out in differentiating this – one knows that people value things according to their price.
–And sellers put up the price accordingly.
–Buyers can't be said to miss getting their money's worth.
L. Dennett – what America needs is a large class of artistic loafers, expensive loafers.

Ultimate economic question: What can the "science" of competitive economies tell us about social well-being? [Double vertical lines alongside in margin]
–cf. Pigou – magnitude, distribution, regularity.
–Is welfare increased with an increased income magnitude?
–Absolute equality of distribution of income is an enormity – would certainly not bring greatest total income (Democratic people tax themselves to have something to look up to).
–Regularity of income, especially regularity of rational marginal utility alternative choices – is a perfect formula for boredom.
–Life can't be put in a straight jacket.
–We live by putting up rules, then breaking them.
–Law breaking is an essential <u>value</u> of human existence.
–If we make a formula for making and breaking laws – we must break that.
–Economics <u>begins</u> with notion of economy (efficiency).

–Form is fairly well defined, content exceedingly thin.

–In electricity – law of conservation of energy means that efficiency of a light bulb is always 100%.

–In order to get any other concept of efficiency – some value judgment must be made.

–Once it is made, the proportion of usefulness to waste is a precise mathematical function – in form; in fact it is as precise as the possibility of measuring usefulness!

–Is economic efficiency desirable? Why?

  –One does economize effort – even in attaining inconsequential ends.

    –Is there not a supreme urge to economy? Physics – law of least action.

    –Why do people want freedom?

    –In the interest of economy? – probably not.

    –It is not correct to say that one wants freedom in order to get maximum economy.

    –One wants to control his own resources, and be free of dictation from others (also to dictate to others – the immorality of it).

      –Is it possible for the individual to be both free and without power? No.

–Economic action is maximizing behavior, efficient, so far as it is thought about.

  –Concept of efficiency – was not thought of in Greek economy, or in Middle Ages – was an offshoot off the invention of the steam engine.

    –Which was known before Watt – he made an improvement in efficiency – began speculating about it – Cournot formulated it in scientific terms.

  –What has been the psychological effect of the introduction of mathematics into common education – in 16th century[?]

    –Capitalism and accounting would be unthinkable without Arabic notation (cf. Lipson, Robertson).

    –Its effect on rational thinking and notion of economizing?

      –Arabic system came from India, where it had been brought from Persia.

    –Which is cause and which is effect – in this introduction of mathematics and accounting?

      –Arabs never put it to practical use.

      –As Chinese never put the invention of gunpowder to use except as firecrackers.

      –The need for accounting was existent.

      –Adam Smith had first concept of a demand curve.

–Accounting came in <u>after</u> the development of a competitive market in which goods and services acquired price tags.

–Picture of economic society is of an aggregate of Crusoes.
  –A summation of individual equilibria, into a total social equilibrium.
  –Based on property and exchange – impossible to define *those* accurately.
  –Yet Crusoe must be born and brought up in society.

Re <u>Berle and Means</u> (cf. Carver, New Industrial Revolution [sic; Walter Meakin, *The New Industrial Revolution*, New York: Brentano, 1928?; William Jett Lauck, *The New Industrial Revolution and Wages*, New York: Funk & Wagnalls, 1929)] cf. E. L. Haney (an intellectual crook), Corporations [Lewis H. Haney, *Business Organization and Combination*, New York: Macmillan, 1914, 1920].
  –The important point is why do people invest in stock? What do they expect to get out of it? Most discussion on such matters is mere "wagging of ears."
  –The only element of stockholder control is when he buys the stock.
  –The whole thing goes back to a change of attitude toward the whole question during the war.
  –The problem of delegated authority and guaranty of original rights.
    –Is a vital one for society – pervades it.
      –Man in barber [shop] chair, man who hires a lawyer or doctor, man who buys stock.
        –In every case he delegates authority and trusts to luck – law can do something, not much.
    –How get effective industrial units of any size without getting delegated authority and compromised ownership?
      –Only alternative to those is to go back to small, local, community concerns.
      –But size works as a form of protection which [one] does not get in case of small local unit.

<u>Veblen</u> by Stigler
  –Vendability and serviceability (business <u>vs</u> industry)
    –Two terms are based on an ethical approach and rely on an appeal from their ethical meaning for their acceptance by others.
  –Selling and making.
  –Speculation and production.
  –What does Veblen mean by "Darwinian or Darwinism"[?]
    –Knight can't find out.

–"Survival of the <u>fittest</u>" found in Spencer with an elaborate develop[ment] of idea of <u>ameliorative</u> progress.
  –If Darwinism means anything it <u>must</u> mean amelioration – it is the fittest who survive.
  –When carried over into social science:
    –The only thing we <u>know</u> survives is improved technical efficiency.
    –Though this is distinctly modern and Western.
    –Pre-industrial society has a definite aversion to <u>change</u>.
    –Western, post-Renaissance history has changed to a <u>belief</u> in the improvement of life through change.
–Dialectic – theory of the way people change their minds.
  –Does not <u>seem</u> to be at random.
    –For social historians <u>seem</u> to be able to bring order in to the pattern of social behavior.
  –But what <u>is</u> an intelligible theory of economic change?
  –The whole <u>content</u> of economics <u>is</u> of institutional origin.

NOTES AND OTHER MATERIALS FROM FRANK H. KNIGHT'S COURSE, ECONOMICS FROM INSTITUTIONAL STANDPOINT, ECONOMICS 305, UNIVERSITY OF CHICAGO, 1933–1934

Edited by Warren J. Samuels

In his *Intelligence and Democratic Action* (Cambridge, MA: Harvard University Press, 1960) Frank H. Knight, in a brief discussion of capital and its yield in the context of limitations on knowledge, concluded that "[t]his...makes the whole rationality assumption rather fantastic if taken literally" (pp. 81–82). This view of a principal assumption of mainstream neoclassical economics by a leading neoclassical economic theorist might seem odd – except that Knight follows this statement with the following admission:

> ...when I am talking with an orthodox economist who expounds all these economic principles as gospel, I am a rip-roaring institutionalist, and when I am talking to an institutionalist who claims the principles don't make any sense at all, I defend the system, the "orthodoxy" that is treated with so much contempt by followers of Veblen and others who wear the institutionalist label (p. 82).

It would appear odd, that is, except that Knight agreed that the issues, or the topics, raised by the institutionalists were exceedingly important, even if he disagreed mightily with the positions on those issues or topics taken by the typical institutionalist.

Documents from F. Taylor Ostrander
Research in the History of Economic Thought and Methodology, Volume 23-B, 141–192
Copyright © 2005 by Elsevier Ltd.
All rights of reproduction in any form reserved
ISSN: 0743-4154/doi:10.1016/S0743-4154(05)23103-6

As to the issues themselves, Knight was critical of both the institutionalists' substantive analyses (or their absence) and those who misunderstood the limits of his beloved price theory (of whatever school). In notes distributed to students, Knight wrote:

> Note unfortunate use of "institutionalism" to refer to everything except price mechanics. Also that controversy over method is chiefly a phenomenon of "attack" on latter due to inability to see or refusal to consider its problems. Conversely, most "classical" economists have done "institutional work in all senses. But exposition of price-mechanics without constant emphasis on relativity to other considerations does tend to give wrong impression (On author also?).

One of the issues in disagreement involved the inevitability of change, the determination of what proposed changes would be adopted, and the resolution of the conflict between continuity and change. Knight distinguished between deliberative and non-deliberative modes of change and believed that the latter predominated. Thus, as a member of a panel on institutional economics, Knight argued that

> Human society must always be largely of the original institutional character; custom and habit must rule most of what people feel, think, and do. Institutions . . . are more or less explained historically rather than scientifically and are little subject to control (Frank H. Knight, "Institutional Economics-Discussion," *American Economic Review, Papers and Proceedings*, vol. 47 (May 1957), p. 20).

That session was part of a program organized by the institutionalist, Edwin E. Witte, as President-Elect of the American Economic Association. Three years earlier, Witte had written the following:

> Institutions cannot be taken for granted, as they are man-made and changeable. Changes in the working rules are possible and occur frequently, although normally only slowly. Institutional economists . . . do not rule out the possibility of changes in institutions (Edwin E. Witte, "Institutional Economics as Seen by an Institutional Economist," *Southern Economic Journal*, vol. 21 (1954), pp. 131–140, reprinted in Malcolm Rutherford and Warren J. Samuels, eds., *Classics in Institutional Economics* II, London: Pickering & Chatto, 1998, vol. III, pp. 288, 289. Witte acknowledged that when Knight organized the program, seven years earlier, "he gave more opportunity . . . to institutional economists to present their views than had been accorded them for many years" (p. 287n.4). Witte obviously reciprocated).

The conflict between deliberative and non-deliberative decision making as the basis of socio-economic change is not the only issue, but it is obviously an important one. Even within institutionalism, Thorstein Veblen is often seen as emphasizing non-deliberative decision making and John R. Commons, whose disciple Witte was, emphasizing deliberative decision making. The truth is both Veblen and Commons and both Knight and Witte had room, in their complete system, for both modes of decision making (The same issue affects the interpretation of Friedrich von Hayek's ideas; see Warren J. Samuels, "Hayek from the Perspective of an Institutionalist

Historian of Economic Thought: An Interpretive Essay," *Journal des Economistes et Des Etudes Humaines*, vol. IX, Juin–Septembre 1999, pp. 279–290).

Thus the notes have Knight saying, "Any <u>fixed</u> economic institution is a part of legal background." The problem, as Knight well knew, is that the tension between continuity and change applies to law itself; no economic institution is literally fixed (one has only to examine the statutes enacted by legislatures and the decisions reached by courts to see this as an empirical matter). Moreover, when Knight (like Hayek later on) carries the argument further, saying, "Historical jurisprudence looks at law as an institution growing by itself, aside from human efforts," he neglects both the deliberative and incremental nature of legal change, however much legal change is characterized by the principle of unintended and unforeseen consequences – which are largely due to human action. One can always emphasize, apropos of historicism, either the fact of change or fact of the gradualness of change; but without human action there is nothing. Knight, in my view, wrongly identifies the difference between historical and analytical jurisprudence as made versus grown (= found). The difference is that historical jurisprudence (Legal Realism in its U.S. form) emphasizes the exercise of choice in the making or changing of law, whereas analytical jurisprudence posits law as a deductive system proceeding from given first principles or legal norms. Knight does get it right when he notices the rationalizing element. He also gets it right when he says that economic "institutions do get embodied in law – almost a branch of law."

Be all that as it may (and I will return to several themes below), Knight appreciated the issues raised by the institutionalists because he, like them, knew that the problem of organization and control was not only important but paramount. One manifestation of this appreciation was the course offered by Knight at the University of Chicago, Economics 305, Economics from Institutionalist Standpoint.

Taylor Ostrander had this course as one of three that he took from Knight during the summer term of 1934, during the year that he attended the University of Chicago (see the biography of Ostrander in volume 22B of this annual). Ostrander's materials from the course are of three kinds: the Syllabus, nine pages; Ostrander's notes from Knight's lectures, 31 pages; and Ostrander's notes from his readings, 15 pages.

The notes are edited in the same manner as the other Ostrander materials. In the Syllabus, items within square brackets are materials added by Ostrander when Knight went over it and orally added new items. Minor corrections have been made. The idiosyncrasies of the Syllabus largely remain as in the original.

The syllabus – designated "Main Topics and Notes on Literature" – reveal that Knight had a deep understanding of the several dimensions of institutionalism. His knowledge of the literature encompassed both U.S. and European sources. For example, he appreciated the role of belief system, inclusive of the dominant socio-economic philosophy of life provided, in part, by religion and studied by Werner Sombart, Max Weber and Richard Tawney. In this regard, Knight followed in the footsteps of Adam Smith as a sociologist of religion – but in a period and place that permitted both deep study and open critique of the institution of religion. Inter alia, religion helped provide people with the fundamental postulates forming their definition of reality and legitimizing the structure of power. For many believers, such was to be neither questioned nor secularly studied; not so for Knight in either regard (An opponent of religion, Knight nonetheless could readily cite or quote scripture – based on his upbringing).

The foregoing notwithstanding, the notes reveal that the lectures dealt much more with Knight's view of the world, especially the philosophy of economics, than with the literature and ideas of institutional economics, the nominal subject of the course. Presumably the readings provided that material (specifically on institutionalism) for the student. Yet the philosophy of economics may well be what the students needed to hear from Knight. A partly alternative view is that Knight devoted most of his class time (but not all; he does connect to institutionalist ideas) to the large issues *qua* issues that impinge on institutionalism, or on which institutionalism impinges or takes a position. Be all that as it may, the course was pure Frank Knight; it was a course on Knight – and his view of the issues.

Much of what Knight lectured about comes under the heading of social control, or deliberative collective decision making. But Knight did not welcome the exercise of deliberative organized social control. Linguistic escape was provided by using "control." The problem of organization and control, while recognized as paramount, was also largely anathema, for the same reason. But institutions, he understood, were important – even their change, indeed especially their change, since that was the agenda of current politics. Whereas institutionalists stressed holism and evolution of institutions, in the mimeographed document, "The Development of Economic Institutions and Ideas" he used in the course, Knight used a different construction: "Economic study of the patterning of conditions and the changes in those patterns, is Institutional economics." The conditions are the conditions of individual choice, the conditions under which individual choice is undertaken. "[T]he most important condition is the whole situation." The conditions and their patterning, he says, "is an institutional problem." "It" is all there, the formulation is different; hence "it" is not all there. The "it" is the institutionalist formulation which stressed institutionalism as: (1) a critique of the form that the market economy has taken (the hegemony of capitalists) and of market economics; (2) an approach to problem solving; and (3) a body of knowledge. The astute reader will see Knight's critique of all three of these meanings of institutionalism.

The crux of his critique is on deliberative social control. Like Friedrich von Hayek, therefore, Knight needed a category of the non-deliberative. Thus, like Hayek, Knight identifies language as a "primary social institution." He says that there is "No individual choice in the matter" and that "Such an institution determines the behavior of humans, and is practically an ultimate cause." Off stage, however, are the deliberative elements in the formation and change of language.

The foregoing is given effect when, in his mimeographed handout, he is recorded as saying, "Freedom of choice and Institutionalism are antithetical ideas, but freedom of choice is itself an institutional fact." On the one hand, there is individual choice necessarily taking place within institutions, and itself an institution ("an institutional fact"); on the other, there is an approach that would deliberatively, if selectively, change institutions – that is what is antithetical, even if inevitable. Thus, "Civilized" men deliberate [and] choose much more than traditionalized primitives, who had narrow possibilities." Primitive men have a narrow range of choice, modern man has a much wider range. That is the source of the problem that disturbed Knight the most: the difficulties and often adverse consequences of social choice. He was conscious of the ubiquity of social choice and was chagrined about it. That the wider range of social choice more or less parallels political democracy and social pluralism, means that skepticism about the former translates into skepticism about the latter, i.e. the anti-democratic element of modern conservatism.

Knight proceeds to query, "Is it possible deliberately to change institutions? – by breaking or defying them?" and points out that "Legislation changing them is itself an institutional product." The second point is correct but insufficient to prove a negative answer to the question of possibility. Changing the law of torts, the tax law, the law of property, etc., will be done through legislation and legislation will be a function of institutions (= power structure) and thus an "institutional product." The deliberative element is ineluctable. Knight is recorded as saying, "Means and ends and efficiency are characteristic of most human activities; it is not the main characteristic of most activities (artistic creation, etc.) but it is the main characteristic of economic activity." Given that politics (legislation, litigation, lobbying) is an economic alternative, the argument against, or limiting, political action falls – except as an argument per se.

The materials are loaded with points of interest to both historians of economic thought and institutional economists. Two are in the very first section of the Syllabus. First, Knight says that Max Handman is the only "true example" other than Thorstein Veblen of American institutional economics. If Knight is correct in this judgment, he is alone; Handman is not so viewed by institutionalits.

Max Sylvius Handman (1885–1939) was born in Romania. He received his bachelor's degree from the University of Oregon in 1907 and his Ph.D. from the

University of Chicago in 1917. Handman taught, as economist and sociologist, at the Universities of Chicago, Missouri, Texas-Austin, and Michigan. He was a student and colleague of Veblen's at Missouri, a close member of Veblen's group (see Elizabeth Watkins Jorgensen and Henry Irvin Jorgensen, *Thorstein Veblen: Victorian Firebrand*, Armonk, NY: M. E. Sharpe, 1999, pp. 147, 154); according to Joseph Dorfman, he supplied Veblen with expensive cigarettes (Joseph Dorfman, *Thorstein Veblen and His America*, New York: Viking Press, 1934, pp. 305–306). Clarence E. Ayres eventually replaced Handman at Texas.

A search on JSTOR indicates that Handman published one review article, five other journal articles, and five reviews; that he participated in four American Economic Association panel discussions and had eleven cites in journal articles and reviews of others' books. His own publications appeared thrice in the *Journal of Political Economy*, twice in the *American Economic Review* plus four in the *Supplement* thereto, five in the *American Sociological Review*, and one in *Political Science Quarterly* (total of 15). He also published pieces on European and American Indian cultures in contact (in Edward B. Reuter, ed., *Race and Culture Contacts*, New York: McGraw-Hill, 1934) and on the pillage economy (*Louisiana Historical Quarterly*, vol. 14, no. 1, January 1931).

Handman seems to have been a colorful, even Bohemian, character. His work is occasionally cited in works on the history of sociology (e.g. Nicholas S. Timasheff, *Sociological Theory: Its Nature and Growth*, New York: Random House, revised edition, 1966, p. 318); especially his "The Sociological Method of Vilfredo Pareto," in S. A. Rice, *Methods in Social Science*, Chicago, IL: University of Chicago Press, 1931, pp. 139–153; see Harry Elmer Barnes, ed., *An Introduction to the History of Sociology*, Chicago, IL: University of Chicago Press, 1948, p. 567; Harry Elmer Barnes and Howard Becker, Social Thought from Lore to Science, Boston, MA: D.C. Heath, 1938, vol. II, 1015, 1023; and Don Martindale, *The Nature and Types of Sociological Theory*, Boston, MA: Houghton Mifflin, 1981, p. 101n2.

Handman was a conspicuous and moderately productive scholar but, while well known, not a major figure (e.g. Robert C. Bannister, *Sociology and Scientism: The American Quest for Objectivity, 1880–1940*, Chapel Hill, NC: University of North Carolina Press, 1987, pp. 57, 58, 116–117, 120, 126). It is not clear to me why Knight ranked Handman alongside Veblen.

Second, Knight seems to parallel the often-criticized John Maynard Keynes in his use of "classical" to refer to the mainstream of economic orthodoxy rather than only to David Ricardo and his group in the early 19th century.

Shortly thereafter, Knight says in the Syllabus that as a "quantitative or statistical economics," Wesley Mitchell is "properly at [the] opposite pole from institutionalism, but usually included in the movement." This is not correct. Mitchell followed Veblen in trying to study actual, not pure conceptual economies, and, in part, to do so through the study of price formation and behavior; in Mitchell's case, the study of business cycles in a pecuniary culture. Veblen's own quantitative work was later included in a collection of important contributions to the *Journal of Political Economy*; see Thorstein B. Veblen, "The Price of Wheat Since 1867," *Journal of Political Economy*, vol. 1 (1892), pp. 68–103, 156–161, reprinted in Earl J. Hamilton, Albert Rees, and Harry G. Johnson, eds., *Landmarks in Political Economy*, Chicago, IL: University of Chicago Press, 1962, vol. 1, pp.1–44).

When Knight next says that Mitchell "[h]as, like most economists, written some things of a really institutionalist character," he seems to support the findings of Malcolm Rutherford that many economists of the interwar period were cosmopolitan and did not distinguish between parts of their work as belonging to two different schools; they did economics, period.

The class notes commence with Knight's remark that Werner Sombart "knows nothing about economic theory, [and] is proud of it." One tends initially to think that this is the now legendary but wrongly conceived canard that institutionalists have no theory as do the neoclassicists. But this implication is immediately negated by Knight's recognition of a non-neoclassical type of theory worthy of the name, when he is recorded as saying, "The problem of Institutional Economics is that of developing a <u>historical economics</u>, as distinct from <u>economic history</u>, i.e. a <u>theory of history</u>."

Knight queries, with obvious criticism intended, "Have either Marx, Sombart, or Weber produced a theory that <u>explains</u> economic change? – so that we can predict future change?" Here he makes or assumes too much of the importance of prediction. As such disparate economists as Friedrich von Hayek and Charles Wilber have insisted, prediction is impossible in such matters, in which we are dealing with structural or process models, or pattern models, or arrays, not unique specific outcomes.

Ostrander heard and recorded epigrammatic statements by Knight in this course as well as in his history-of-economic-thought course. Shortly after his remark about Sombart, Knight is recorded as saying, "Any great <u>literary</u> acclaim or position usually depends on a conceited interest in one's own verbosity."

As part of his discussion of what today would be called either rationality or motivation, Knight says, "<u>Problem-solving</u> – the essence of the motivation is that the end is not known in advance – the nature of the activity is the defining of the end." Several aspects of this statement are noteworthy. First, one defining term of institutionalism is precisely "problem solving." Second, instead of assuming given wants, Knight argues that "the end is not known in advance – the nature of the activity is the defining of the end" (Such is one problem to be solved – though not once for all time). In both respects, Knight is more akin to the institutionalists than the neoclassicists, though it is only accurate to say that neoclassicism has a spectrum of views on the second matter. Knight's profound view was that life is explorative and emergent.

Thus the notes have him arguing, "this kind of purposive activity is universal. There is no purposive activity which does not redefine the end in the process of achieving it" and that he "is opposed to pragmatism... for it assumes the end is known, and Knight sees most behavior as explorative." By "this kind of purposive activity" Knight is referring to activities in which the "ends are not given; the desire for an end is the end" and nature of the activity, problem solving is that "the end is not known in advance" and "the nature of the activity is the defining of the end." Knight may well be mischaracterizing pragmatism, but his argument that life is explorative and emergent is apposite.

Both institutionalists and their critics have decried the enormity of materials pertinent to a meaningful institutional economics. It matters not a whit, therefore, whether it is Knight or Ostrander who is responsible for the statement "(We need five years for this course)" found in the notes. My guess is that it is due to Knight and he appreciated, not criticized, its importance – as, for example, when he says, according to the notes, "Any kind of complete explanation, or the mere concept that one is possible, is antithetical to the concept of a rational, choosing, planning sort of life." One wonders what Knight would have made of the rational expectations hypothesis in light of his claim that "Any statement about Social Science which is true at the time it is made is rendered immediately false the moment it is made." More immediately, one wonders if he was prepared to be self-referential about it.

Many of the foregoing issues are echoed in subsequent passages.

Another issue that arises is that of the use, and definition, of "society" as distinct from the "individual." Quite apart from the problem of the socialized or institutional individual, the issue raised later by followers of Hayek is that society is not an independent entity, only the sum of individuals. But Knight himself uses the juxtaposition, when the notes have him saying (paraphrased), "Individual changes himself by thinking about himself. Society changes itself by thinking about itself. This is an antithesis to the whole setup of natural science, with observer and observed. Instead society talks to itself about itself with the intuition of changing itself." What does all this mean, or is it only metaphor?

Knight's pessimism is mentioned in my introduction to the notes from his history-of-economic-thought course. It arises here in numerous ways, for example, in his discussion of motivation, when the notes report him saying, "We can not even define what we mean by a problem, nor by a solution – we have to take them for granted..." A non-pessimist who emphasized the explorative and emergent nature of choice and behavior would not put the point so darkly (especially when the quoted phrase is read in context). But that is Frank Knight, whose pessimism was not absorbed (at least publicly) by later members of the Chicago School – who did apply it, of course, only to government action.

One of Knight's brilliant aperçus is reported thus: "Most economists treat uncertainty as a cost; Knight is sure it is the primary reason for interest in life." Another: "Poetry is the deepest social science." Still another: "Most people are poseurs."

Knight was skilled in asking pointed questions. In a discussion of history, he queries, "What is the explanation of anything?" and "Do we prefer a mechanical explanation?" In a discussion of evolution theory, he asks, "Is it progress or retrogression to get more wants than can be satisfied?"

Knight's thinking was often very deep. This is illustrated by his statement, "Social problem of the ethics of educating people."

Knight was nothing if not candid – a candor mixed with satire if not worse. Several examples stand out. After stating that Herbert Spencer's system was "The political rationalization of utilitarianism," he stated that Spencer's work was "written at just the time that Jevons, Walras, Menger were making an economic rationalization of utilitarianism." After identifying a number of social domains as games (ecclesiastics, military, commerce), he says, "Neither preaching nor teaching can be sold under an honest label – people won't buy either on the basis of a commodity," that "Education can be put over on a slightly more honest level," and that he "doesn't know what truth is but is sure he knows what lying is." In a discussion of the nature of social science, he remarks, "Difference between figurative and literal language – only one of degree."

Knight is especially candid with regard to the status of the domain of exchange and the price system. These, he is reported to have said, have their "roots in a theory of social policy arising from philosophy of individualism," to which apply no less than four forms of rationalization. Most of the content of such an economic system, in contrast with the abstract form deployed in economic theory, is "only relevant for social problem in terms of individualistic philosophy as far as this goes."

One of the Chicago School's principal tenets is already stipulated by Knight: "The element of monopoly does not keep the system from being competitive in principle." The market system, that is, is necessarily competitive. There is no question that members of the school believe this; but it is also part of what Knight once called a propaganda for economic freedom, i.e. the formation of a rationalizing ideology. This is evident, for example, in Knight's discussion summarized in Ostrander's notes on "The Development of Economic Institutions and Ideas," particularly the discussion of "Capitalism."

I conclude, somewhat, where I began. From Knight's, and Hayek's, and many others' point of view, a, if not the, basic social problem is how to bring knowledge to bear on change, or on decision making in general. Knight published *Intelligence and Democratic Action* in 1960. Already in these notes from 1933 we find Knight impressing upon his student that *Intelligence and the Social Process* is the main problem for Knight.

One problem with the Knight-Hayek position is its conventional application largely if not only to politics and not to the other principal sphere of action, business. Another problem is the difficulty of distinguishing the position as one of substance rather than a piece of conservative argument. Still another is the unequal application of the argument as between politicians they like and those they do not like, and between biases they share and those they do not share. At any rate, questioning the role of intelligence was in the air when Knight gave these lectures (but echoing a long conservative tradition insisting on the distinction between things seen and unseen). In 1933, the same year that Ostrander took Knight's course, Henry Louis Mencken wrote, in an article on the death of Calvin Coolidge, "We suffer most when the White House bursts with ideas" (quoted in Joan Acocella, "On the Contrary," *The New Yorker*, December 9, 2002, p. 133).

Knight's position on institutional economics is ironic. The Chicago School opposed much institutionalist theory and policy, defending and rendering the market system absolute whereas the institutionalists were critical of the capitalist-dominated form of the market system and demystifying it as well, the institutionalists saw regulation as a mode of changing property rights whereas the Chicago School saw it as redistributive and inefficient, and so on. The Chicago Economics Department during the pre-World War One and interwar periods had a number of institutionalists, including Veblen. Knight and other Chicagoans paid attention in their own way to institutionalist topics (See Malcolm Rutherford, "Chicago Economics and Institutionalism" (2002)).

That Knight knew institutionalism and why it bothered him, is indicated by what I have called the three elements of the historic meaning of institutionalism. First, it is a movement protesting both the capitalist-dominated market economy and the apologetic economics of that economy. Second, it is an activist approach to solving problems. And third, it has claims to constitute a body of knowledge. None of these sat well with Knight. His arguments about the limits of intelligence, he thought, applied particularly to the institutionalists. Apropos of the quotations from *Intelligence and Democratic Action* with which I opened this Introduction, Knight was more or less an equal opportunity critic. Interestingly, Knight seems to have said very little in praise of Veblen; much is said in criticism of what Veblen said and did not consider. Why did Knight expend so much time in reading institutionalist works and teaching this course? My guess is composed of three answers: First, he was bothered by institutionalism; he thought the institutionalists were not only wrong in what they said and tried to do, they were dangerous to society and to economics. Second, he appreciated that unlike the professional work of most neoclassicists, the work of the institutionalists dealt with important questions. Third, he appreciated that many things taken as given, or unexamined, by neoclassical economics were actually in a process of evolution – one of institutionalism's important claims.

One of those claims concerns the distribution of power in socioeconomic life, i.e. the problem of structure. If the principal problem is change of law through law, the principal subject (and object) of change is the distribution of decision making, i.e. of power. The theory of laissez faire and much economic theory focus on the ostensible automaticity of the market economy-price system. Thus the mimeographed document summarized in Ostrander's reading notes says, "Organization of modern <u>economic</u> life: an <u>automatic</u> system, production and distribution work through the medium of <u>prices</u>, which result from individuals' activities." Emphasis on automaticity functions to obfuscate, give effect to, and reinforce the existing structure of economic power. Institutionalists argue that it is not the market that automatically allocates resources; it is the institutions that form and operate through the market. Allocation being a function of structure, and there being more than one possible structure, there is more than one possible efficient allocation wrought "automatically." Automaticity is second to structure. Knight knew this very well. For one thing, it was stated forthrightly and emphatically by his long-time close friend, Clarence Ayres at the A. E. A. December 1956 session noted at the beginning of this Introduction.

ACKNOWLEDGMENT

I am indebted to Anthony Waterman for identifying the largely illegible phrase *cuius regio, eius religio*, found near the end of Ostrander's notes. Waterman writes, in explanation, apropos of Martin Luther:

> Lit. 'whatever of the king, so of the religion': it means that L. thought (being the Erastian he was), that the religion of a country should be that of its sovereign prince. Note: (a), the assumption, almost universal at that time, that there can be only ONE church in any Christian nation; and (b) the assumption, standard until the Scottish Enlightenment I should think (though people like Locke begin to chip away at it) that – as Louis XIV put it with admirable economy, 'l'etat c'est moi' (Waterman to Samuels, December 12, 2002).

SYLLABUS MATERIALS

Economics 305, Economics from Institutional Standpoint

Main Topics and Notes on Literature
(To be used with general, alphabetical bibliography)

I. <u>American Institutional Economics</u>
   1. <u>Veblen</u>, Th. (Perhaps the one true example, except Handman, who has written little.
      a. The Place of Science in Civilization. (1919) Collected Essays. "Why is Economics not an Evolutionary Science," 3rd paper, contains most of Veblen's position. For his criticism of classical economics, especially "Professor Clark's Economics" and "Limitations of Marginal Utility"; also three papers on "Presuppositions of Economic Science." For V[eblen]'s positive contribution, the title essay and second, on "Evolution of the Scientific Point of View" most important, to be followed with "Industrial and Pecuniary Employments," "Gustav's Schmoller's Economics" and papers on Capital, Marx, and Socialism.
      b. "Economics in the Visible Future," A. E. R., 1925 (<u>cf</u>. Discussion of J. M. Clark). Other works: Instinct of Workmanship, Theory of the Leisure Class, and Imperial Germany and the Industrial Revolution most important. Theory of Business Enterprise social-critical, on line of Industrial and Pecuniary Employments. Later books (Nature of Peace, Higher Culture in America [sic], Vested Interests, Engineers and Price System, Absentee Ownership, etc.) [m]ore satirical, and literary or controversial in appeal. [cf. Teggart, on Veblen]

   2. Handman, M. S.

   3. Commons, J. R. Legal Economist (Laws are not institutional in origin, but become institutions if long kept in force).
      a. Legal Foundations of Capitalism (cf. Reviews, Mitchell, A. E. R., June 1924, and Sharfman, Q. J. E., 1924–1925).
      b. "Institutional Economics," A. E. R., Dec. 1931 (Corres. Regarding same, ibid, June 1932) [Added in ink: "Copeland, Burns"].

   4. Mitchell, W. C. (Quantitative or statistical economist, properly at opposite pole from institutionalism, but usually included in the movement. Has, like most economists, written some things of a really institutionalist character.
      a. "Quantitative Method in Economics" (Presidential Address), A. E. R., 1925 (His main position; not institutionalistic).
      b "Prospects of Economics." (Leading Essay) In <u>Tugwell</u>, The Trend of Economics (Institutional only in sense of being more or less critical of the older classical economists).
      c. "The Role of Money in Economic Theory" (Institutional), A. E. R., 1916 Sup.; "The Backward Art of Spending Money," A. E. R., 1912; "Human Behavior in Economics," . . . Rev. of Sombart, Q. J. E., 1928–9; "Bentham's Felicific Calculus," P. S. Q., June, 1918.

    d. On Mitchell's main work on Business Cycles, see review by J. M. Clark, in Rice's Case-Book, with Mitchell's comment.

5. Copeland, Clark, Hale, Mills, Tugwell, Wolfe, etc., see Tugwell (Editor), The Trend of Economics. Sometimes treated as an institutionalist manifesto, but with several "black sheep." cf. Review of the volume by A. A. Young, Q. J. E. . . .

6. Other authors more or less sympathetic with the "movement," see Boucke, Clark, Edie (uses the word for all recent economics he approves of), Hamilton. [Added in ink with line to Edie: "Some Positive Contributions of the Institutional Concept," Q. J. E., vol. 41)" and, with line to Hamilton: "The Place of Value Theory in Economics," J. P. E., vol. 26" and "The Institutional Approach to Economic Theory," A. E. R., vol. 9, Supple."]

[Added in ink on back of first page of Syllabus:

Mitchell, W. C.	"Commons on the Legal Foundations of Capitalism" – A. E. R., 1924
	"Sombart's Hochkapitalismus," Q. J. E., vol. 43
	"The Rationality of Economic Activity," J. P. E., vol. 18
	"Quantitative Analysis in Economic Theory," A. E. R., vol. 15 (Am. Econ. Rev. Suppl. 1922 March)
Clark, J. M.	"Economic Theory in an Era of Readjustment," A. E. R., vol. 9, supplement
	"Economics and Modern Psychology," J. P. E. vol. 26
Ayres, C. E.	"The Function and Problems of Economic Theory," J. P. E., vol. 26
Copeland, M. A.	"Economic Theory and the Natural Science Point of View," A. E. R., vol. 21
Burns, E. M.	"Does Institutionalism Compete or Complement with Orthodox Economics?," A. E. R., vol. 26
Boucke, O. F.	(Chapter on Utility)]

Economics 305. Topics and Readings

[No Section I. is included]

II. Criticism of Institutional Economics

1. Eva Flügge, in Jarhb. F. Nationalökon. U. Statisik, LXII, 1927. Important; on relations to German Historical School Position.
2. Homan, P. T. Essays on Veblen and Mitchell in Contemporary Economic Thought. Also Paper, A. E. R., Sup., Mar., 1932, and Discussion following, by various members. cf. J. P. E., 1927 (Impasse, etc.), Q. J. E., 1928 (Issues, etc.).

3. Morgenstern, Schumpeter, Suranyi-Unger.

III. Earlier Historical Economics

1. Leslie, T. E. C. "The Philosophical Method in Political Economy" and "History of German Political Economy," in Essays in Moral and Political Philosophy.

2. Schmoller, The Mercantile System. (Example of and argument for the method. cf. Veblen's essay on Schmoller, under Veblen, above.)

3. Ashley, W. J. Trans. of Roscher Program; also "The Study of Economic History" and "The Study of Economic History after Seven Years," first two in Q. J. E., all in Surveys Historical and Economic.

4. Cohn, G., A. A. A., 1894 and Ec. Jour., 1905; Dunbar, Q. J. E., vol. I (and in vol. Econ. Essays); Keynes, J. N., in Scope and Method of Pol. Econ.; Ingram, in History of Pol. Econ.; Nasse, Q. J. E., 1886; Rae, in Contemporary Socialism, pp. 193–221; Seager, J. P. E., 1892; Wagner, Q. J. E., 1886.

IV. The Neo-Historical School in Germany, and Related Work

1. Parsons, T., "Capitalism in Recent German Literature: Sombart and M. Weber"; [J. P. E., vol. 36–7] best thing in English (For orientation see also Parsons, "Economics and Sociology" in Q. J. E., February, 1932).

2. Sombart, W., "Economic History and Economic Theory," Ec. Hist. Rev.; Nationalökonomie u. Soziologie, Kieler Vorträge; also in G. D. S., Vol. III.

3. Diehl, Carl, Life and Work of Max Weber, Q. J. E., Vol. 33.

4. Abel, Th., Chap. on Max Weber in vol., Systematic Sociology in Germany.

5. Weber, M., Protestant Ethic; and General Economic History.

6. Sombart, W., Die drei Nationalökonomien, Der moderne Kapitalismus.

7. Weber, M., Essays in Ges. Aufsätze zur Wissenschaftslehre, esp. on Roscher und Knies, and Objektivatät; finally, Wirtschaft und Gesellschaft (2 vols., in Grundriss d. Sozialökonomie).

8. Brinkmann, C., in Überbau etc., Schmoller's Jahrb., 1930; von Schelting, Zum Streit um die Wiessensoziologie, in Archiv. F. Sozialwis. U. Sozialpol., v. 62, 1930. And references in both.

9. Related work in other countries. Tawney, Religion and the Rise of Capitalism, and other work; Simiand, La methode positive daans l'economie politique (and French Neo-Positivism generally).

10. Another German movement closely related to neo-historism is the Universalistic economics of Spann. See in English his History of Economics. Also, C. Schmitt, Politische Romantik.

11. On the problem of Objectivity (Wertfreiheit), an essential issue throughout this movement, but especially under the influence of communism and fascism, see E. Spranger, Der Sinn d. Voraussetzungslosigkeit d. Wissenschaft. (1930) and references.

[In ink on back of page:

Weber, M., Protestant Ethic – trans. T. Parsons
           Spirit of Mod. Cap. – transl. F. H. Knight
Sée, H., Development of Modern Capitalism – trans.
Breasted and Robinson, Survey of World Hist. – 2 vols.
Wells, H. H., History
Thorndike (Hist.)
Cunningham, Western Civilization from Economic Standpoint
Marx, K., Das Kapital
Harris, Abram, J. P. E., Veblen and Marx, 1932, 1934
Knight, F. H., Development of economic institutions and ideas
Nussbaum, [*Econ. Institutions of Modern Europe*]]

V. ISSUES INVOLVED IN INSTITUTIONALISM

1. General Problems of Behavior (above bio-mechanics and chemistry and histology). Surveys, chiefly on level of physiology and animal behavior in Parmelee, Problem of Human Behavior; Allport, Social Psychology. cf. Metchnikoff, Nature of Man; Wheeler, Ants; Emerson, Termites. Psychology Symposia, Clark University, Psychologies of 1925, also 1930; also, The Unconscious, sponsored Mrs. E. Dummer. cf. Cooley, Dewey, Ellwood, McDougall, Sumner, Wallas. Survey of General Sociology, Park & Sturgess, Introduction. Sociology from standpoint of society as a unit, Spann, Gesellschaftslehre; from that of personalities in relation, Hornell, Hart.

2. History and Economic History. Müller-Lyer; Hobhouse, also Hobhouse & Wheeler; Gras; E. Gross. On Economic Interpretation of History: Communist Manifesto: Engels; Labriola; See; Seligman. (Hanson; Knight; Matthews) History and Historical Method: Adams, G. B.; Adams, Brooks; Barth; Bernheim; Cheyney; Flint; Fusster; Teggart; Rickert; Windelband. For Rickert-Windelband view of history, Chap. I of Park and Burgess, Sociology, with Bibliography. cf. Small, Origins of Sociology.

3. Institutions. Besides Sociology, see Anthropology, works of (esp.) Lowie, Goldenweiser; also Boas, Kroeber, Wallis, Wissler, etc.

4. Particular Institutions (all more or less economic in basis and function). Language: Sapir; The Family: Westermarck, Calhoun; Law: Commons, Pound, Jenks, Holdsworth, Maine, Maitland, Vinogradoff. Religion: Barton, Carpenter, Carus, Cumont, Harnak, Simkhovitch, Sohm, Laagarde, Walker.

5. Economic Institutions, Specifically. Bibliographies in Sombart, Der moderne Kapitalismus; use table of contents and index. Surveys of Economic History; Knight, Barnes and Fluegel, Economic History of Europe; H. See, Modern Capitalism (both with chapter bibliographies).

6. Methodology. See M. R. Cohen, "Social Science and Natural Science," in Ogburn & Goldenweiser (Ed.). The Social Sciences in their Interrelations; also most of the 33 papers in the volume, all with bibliographies. Rice, S. A. (Ed.), Methods in Social Science, a Case-Book; 52 papers, mostly analyses of particular works or groups of works from methodological standpoint. Keynes, J. N., Scope and Method of Political Economy.

7. Idea of Style and Culture-Pattern. Compare Wöfflin, Kunstgeschictliche Grundbegriffe; Sapir in Ogburn and Goldenweiser.

[In ink on back of page:

Knight, F. H. A. E. R. Facts and Metaphysics

Lord Acton]

Institutional Economics

1. Definition of Economics (By progressive exclusion).
   a. "Economy," use of given means to realize given ends, vs. other aspects of life-problems.
   b. Social Economy, the problem of <u>organization</u>. Excludes (1) all concrete physical aspects of the economic problem and (2) all private individual aspects; hence all concrete problems of both "production" and "consumption," also private personal relations (Private and technical nearly or quite coincident spheres vs. social).
   c. Science of social organization, excludes effort directly to affect the organization, i.e. exhortation or propaganda. Difficulty of being "objective" in the social sphere: (1) objective, and (2) subjective difficulties. A society discussing itself, "science" must have a very different meaning from "natural" science; cf. Law; criticism, pragmatism, historism; sham.
(Note: long-run vs. short-run organization problems, means being used to produce other means; "stationary" vs. "progressive" economy)

2. Basis of Types of Econoomic Science in types of social relations involved in social-economic organization. Main types:

    a. Price mechanics – "exchange" relations. Note roots in a theory of social policy arising from philosophy of individualism; three aspects of individualism – wants, resources, procedure; four rationalizations of individualism – "absolute," hedonistic, moral, negative (impotence of state).

    b. Institutional – "social habit" element.

      (1) In enterprise or price-mechanics system itself – most of its "content" in contrast with abstract form; latter only relevant for social problem in terms of individualistic philosophy as far as this goes.

      (2) The Family. A nearly pure "institution" and the real unit in economic relations, in large part from the stationary, and much more so from the long-run point of view.

      (3) The State. Various forms, "free" vs. "dictatorial." Relation between "politics" and "economics."

      (4) Church and other forms of association, insofar as habitual in character and economic in function.

    c. Power relationships, including monopoly.

    d. Negotiated relationships (bilateral monopoly).

Note high degree of overlapping in all these, a matter of elements more than of separate types. Power relationship is opposed to other three as cooperative, both really elements in every situation including exchange. Relationships originating in other ways tend to become institutions by habituation. cf. especially the state as to mixture of elements. Price-mechanical organization presupposes a "content" of other character, especially institutional.

Note unfortunate use of "institutionalism" to refer to everything except price mechanics. Also that controversy over method is chiefly a phenomenon of "attack" on latter due to inability to see or refusal to consider its problems. Conversely, most "classical" economists have done "institutional work in all senses. But exposition of price-mechanics without constant emphasis on relativity to other considerations does tend to give wrong impression. (On author also?)

3. Institutional Economics Defined: The study of socio-economic organization in its social-habit aspect or of the social habit element in it; two elements, the habitual and the social-economic.

    a. As regards the habitual character, the basis of habitualization of behavior is presumably the same as in any other field, a problem for psychology, physiology and non-social sciences of behavior.

    b. To speak of "habitual, economic" behavior involves thee concept of unconscious teleology, somewhat repugnant to many minds, since strictly habitual behavior is unconscious and economic behavior is purposive. But the concept of economy is used in Botany.

c. Passing over the psychological problem of habit formation, the field of discussion is that of the actual origin and growth of particular social-habits or institutions, those belonging in the field recognized as economic (first defined above). Both origin and change in habits are phenomena of non-habitual behavior. Problem historical.

4. Content of institutionalism. (A) Institutions in their economic and non-economic aspects.
   a. Language. Purest institution. Analogy of plant. In what sense an economic institution.
   b. Law. Institutional (historical) theory vs. others, including all possible theories of historical change. In a sense legal and economic institutions nearly synonymous.
   c. Religion. Individual vs. institutional views. How far economic.
   d. World-view, thought, taste, etc.
   e. Economic in narrow sense, largely legal. Wants, procedures.

5. (B) General Theory of Behavior (recapitulated to escape confusion). Necessity of a "hierarchy" of problems and sciences, à la Comte, cf. DuBois-Raymond, Giddings.
Physics-chemistry; biology (adaptation, survival); sociology (institutions in strict sense, cf. below); taste, sentiment, or interest, including comparison, i.e. economy; ethics; religion. Tendency to "reduce" higher to lower; limit in Lucretian world-view; how far physical and how far possible (energy, consciousness, social forms) why desirable (one form of wish-thinking); fact and meaning, uniformity, freedom, discussion; "realism" an epithet.

6. (C) The Problem of History: simpler views.
   a. Meaning or idea of history. The unique (Rickert), irreversible or cumulative change (Veblen): in either case non-repetitive. Also the meaningful?
   b. Historical explanation in simplest sense; mere tracing of continuity. (Economic History vs. historical or institutional economics)
   c. Theory of History in simplest form, an empirical formula. Role of cycles – a spiral? History, "progress," not necessarily improvement. Spencer's formula, and Bergson's criticism. Cheyney, law in History. Stage theories. Increasing self-consciousness? freedom, co-operation?

7. Explanatory (?) theories of history.
   A. Physical theories, so-called. Not "physical"; no conservation principle, mass, momentum, energy. "Cumulative" change. Not "dynamics" either, nor statements with parameters functions of time. Theories quasi-biological,

geography, race, etc. Consciousness ignored or treated as "epiphenomenal." Problem of analysis of cause where experiment impossible.

Darwinism; Lamarckism; Mutationism; Directed Evolution; all load the essential problem into unexplained changes, "metaphysical." Drift of controversy on Weissmanism and directed evolution.

How far theories applicable to institutions as such; nature of the group biologically, its struggle and survival.

B. Non-Physical theories, "Dialectical," "Spiritual," etc.
(1) Individualisstic. Power or "leadership"; (Carlyle)
(2) Social consciousness; Discussion.

NOTES: Theories giving consciousness a real role are not reducible to biological terms. Note that biological theories really tell nothing of <u>what</u> survives, to say nothing of why.

The "Economic" Theory (Materialistic Conception) either reduces to biological advantage, hence a biological theory, or else assumed economic interests distinct from biological, wherefore it becomes a metaphysical theory. If interests are taken as "data" as in the classical economic theory, they are either metaphysical dei ex machinae or built into the organism, which makes it all mechanical. Question as to sense in which a "spiritual" explanation explains.

On Economic Control. The terminology confuses two things; the word properly refers to individual domination, but what is meant (under democratic auspices) is rather a mutual-social process. Note relation of control problem to that of science vs. propaganda.

One fairly clear observation about history is the progress toward a more self-conscious, i.e. less "institutionalized" character, involving more individualism, more conscious co-operation, but also more conscious conflict.

On Statistical Economics. A method of study, to be used as a matter of course, where it is applicable, which is not generally in t he field of history. Prerequisite when objective is control.

PERIODS
For the Study of the History of Economic Institutions

1. Prehistory and Primitive Society.
   Economic and Institutional origins.
2. The Ancient East.
   Egypt, Babylonia, Aegeans and Phoenicians.

3. Greeks.

4. Hellenism.

5. Rome, Republic and Early Empire. (To end of 2d century)

6. Rome, the decline. Invasions. Teutonic states.
   The Church. Roman Law. (To Justinian & Gregory I)

7. Dark Ages (to 11 Century) The Franks, Holy Roman Empire.
   Norman Conquest. Establishment of Feudalism.

8. Byzantium and Saracens.

9. Twelfth and Thirteenth Centuries. The Medieval System (Feudalism and thee
   Manor, Church and Empire).

10. Fourteenth and Fifteenth Centuries. Transition.
    Trade and Towns. Law Merchant. Rise of National States and decline of
    Church. Commutation. Professional armies. Renaissance in Itaaly. Science,
    invention, exploration.

11. Sixteenth Century and Seventeenth to Gustavus Adolphus, and Richelieu.
    Renaissance and Reformation. Age of the Fugger. Religious (?) wars. The
    Commercial Revolution. Colonization. Early Mercantilism.

12. Later Seventeenth and Eighteenth Century.
    Dynastic and Trade Wars. "Liberal" mercantilism.
    The Enlightenment and English Empirical and Individualist Movement.
    Natural Rights and International Law. The Industrial Revolution. The French
    Revolution and European Wars.

TOPICS for each Period

1. General and political survey.

2. Character of economic life.

3. Economic Institutions (see outline).

ECONOMIC INSTITUTIONS. Outline

I. "SPIRIT"

Want-satisfaction vs. money-making)
Traditionalism vs. rationalism.) Sombart
Solidarity vs. individualism.
Beauty, creativeness, sport, power, etc.

II. TECHNOLOGY

Empirical vs. scientific (vs. traditional-instinctive).
Utilization of men (slaves), animals, plants, minerals.
Especial significance of transport.
Economic production and war.

Relation to "pure" science, hence magic, religion, etc.

III. ORGANIZATION AND CONTROL
(Sombart: Restriction vs. freedom; 2. Private ent. vs. pub. Ownership; 3. Democ. vs. Aristoc; 4. Compact vs. loose; 5. production for use vs. market; 6. individual concerns vs. social)

1. Politics and Economics
   Law and Order and the State. (Form of order, "constitution.")
   Maine's formula vs. Pound's three views of law and five stages.
   Tribalism (kinship) vs. territorial sovereignty.
   The City.
   German vs. English view of state in economic life.
   Military and economic organization. (Spencer)
   Feudalism, autocracy, democracy.
   Religion and the social constitution.
   Legal procedure.
   Economic and political units.

2. Division of labor (non-commercial)
   Men's and women's work.
   Medicine-man, priest, ruler. (Judge, law-giver)
   Large household (oikos); slavery.

3. Property (Freedom, "Competition")
   A. In consumers' goods, land, "capital: (cf. commerce), tools, money slaves, one's own labor, ideas, good-will, processes, "rights." Property and sovereignty (cf. #1).
   B. Property as usufruct, control, lease, sale, gift, bequest.

4. Commerce ("foreign," cf. #6 Local Trade)
   Theories of origin, primitive forms, "silent," gifts, etc.
   Exchange of surpluses vs. division of labor between groups.
   Princely & seigniorial trade.
   Trading class.
   Gild merchant.
   Fair, exchange.

5. Money
   Commodity, symbolic, prestige
   Coin
   Draft
   Bank note and cheque (Central banking)
   Govt. paper money

6. Trade (local)
   Market
   Competition (& monopoly)
   Advertising
   Speculation

7. Contract (Incomplete exchange)

8. Forms of Association
   A. Of Persons
   Gild
   Partnership
   Family concern
   Regulated Company
   Joint-stock Company
   Contract vs. Corporation
   Labor Union and Professional Association
   Employers' Association
   Cooperative Association Forms
   Friendly Society
   Church

   B. Of Factors of Production
   Hiring (a contract?)
   Lease
   Partnership (commenda), company etc. (See "A")
   Loan of "money" (capital)

9. Insurance

F. TAYLOR OSTRANDER'S NOTES FROM FRANK H. KNIGHT'S COURSE, ECONOMICS FROM INSTITUTIONAL STANDPOINT, ECONOMICS 305, UNIVERSITY OF CHICAGO, SUMMER 1934

<u>Knight</u>

Sombart – stimulating; knows nothing about economic theory, is proud of it. His economic theory is childish. cf. Die drei Nationalökonomien [1930] short, easy German – methodological.

The problem of Institutional Economics is that of developing a historical economics, as distinct from economic history, i.e. a theory of history. Has Marx got a theory of history?

Have either Marx, Sombart, or Weber produced a theory that explains economic change? – so that we can predict future change?

Marx takes economic change for granted, explains other things in terms of it. – what philosophy of economic change is this?

(Any great literary acclaim or position usually depends on a conceited interest in one's own verbosity)

Weber's Religionsociologie, his great work, is the antithesis of Marx. Took religion for granted, explained the rest of culture in terms of it. – how did religion get that way?

Other interpretations of history.

Marx – economic
Weber – religious
Veblen – biological ("Darwinian")
– intellectual
– aesthetic
– kinship
– literary
– romantic.
–All are right, no one makes the others wrong, they are complementary.

Relation between Institutional and Classical economics.
I. Human Behavior
1. –Behavioristic, "mechanical," physical
–Statistical economics (studies goods and services but leaves the human being out – there is no place left for the notion of economy – which is a value concept (efficiency-economy).
–Statistics is all right, if used critically, e.g. there is wheat and wheat – i.e. the mere definition of a commodity involves a value-judgment for economic wheat is not physical wheat.
–Principle of economy is necessary for valuation, choice of alternative means, and for classification.
2. –Categories of price theory economic; Economistic.
–Man motivated by wants (pragmatic) uses means to ends – efficiency.
–There is no analytical economics without a concept of the economicman.
– Actual behavior is not verbally [?] economic, but what of it. The same is true in mechanics, the frictionless machine is a necessary concept,

and is not invalidated by the fact that none exists. No amount of statistics of actual machines helps out the engineer – who is interested in the ideal.
–Man is rational in two senses. Our assumption is of errorlessness.
  a. Knows the ends.
  b. Knows the means.
  –In order to talk about this we have to assume that utility of end-achievement is a function.
  –There is a difference between economy and waste.
  [In margin at top of page: "an irrational passion for impassioned rationality." J. M. Clark]
  –Civilization depends on our knowledge of that difference.
  –In order to hold this we must assume that the end is reasonable, or is same function of something that is reasonable.
  3. –Other wants and motivations – i.e. ends are not given; the desire for an end is the end. Normative, human.
    a.–Social
    –Games
    –Symbols (we want goods and services as a symbol of something else, not as ends in themselves)
    b.–Problem-solving – the essence of the motivation is that the end is not known in advance – the nature of the activity is the defining of the end.
    c.–Creative self-expression – motive of inventor, artist, businessman, orator, etc.
  –Knight's point is that this kind of purposive activity is universal. There is no purposive activity which does not redefine the end in the process of achieving it [Double vertical lines alongside]
  –Knight is opposed to pragmatism, which is economic action (2) for it assumes the end is known, and Knight sees most behavior as explorative.
    –Says Dewey sees the problem – but is on the fence and usually falls to wrong side.

II. "Culture" – gives a historical or institutional economics.
  1. Language – is grammar human behavior?
    –Knight thinks of grammar as completely divorced from personality – i.e. study of history of language is made without reference to humans.
    [Alongside, in margin: "(Conscious effort has little effect on language.")]
    –Language authorities agree that language changes come about regardless of persons speaking it and that no one language is better than any other.
    –Like botany studying plants, taking the soil for granted – the soil is not absolutely irrelevant – neither is the human being absolutely irrelevant to language.

2. Law
   (We need 5 years for this course)
   –Any fixed economic institution is a part of legal background (Pound, R[oscoe])
   a/–Historical jurisprudence looks at law as an institution growing by itself, aside from human efforts.
   –Grows in response to essentially sub-conscious social forces.
   b/–Analytical jurisprudence looks at law as the creation of the sovereign.
   –This was the view held in U.S. law schools in 19th century.
   –Paradox – In France law is made – they believe in historical jurisprudence.
   – In U.S. law *grows*, the lawyer or judge "finds the law" – rationalizes by means of his search for precedents.
   –English common law can not be traced before Edward III – i.e. does not grow out of the innate wisdom of the German forest!
   –This problem goes back to the 13th century:–Does God will the law – and make it good – or does he will the law because it is good.
   –Insofar as one can talk about economic history in culture terms, you eliminate any possibility of discussion or action re controlling economic history.
   –Any kind of complete explanation, or the mere concept that one is possible, is antithetical to the concept of a rational, choosing, planning sort of life.
   –A challenge to the human mind:
   –The contrast between the intellectual craving for complete understanding – and the fact that one seems to choose, to plan, to alter.
   –Thus the cultural approach gets back to (I.1) the behavioristic, statistical approach – which likewise cuts out choice, planning, rational determination of action.
   –Yet the position of the behaviorist gives a clue to an escape:
   –The behaviorist is in a dilemma: talking machines vs. god-man.
   –It is the old dilemma of the "Map of England – or the box within the box picture.
   –It is impossible to find complete intellectual explanation of "ourselves in the world."
   –The position of the Social Sciences is akin to that of the person drawing a picture of himself drawing-a-picture.
   –The persons talking about Social Science is within what he is talking about, and changes the realm as soon as he talks.
   –Any statement about Social Science which is true at the time it is made is rendered immediately false the moment it is made. [Alongside in

the margin: "(Individual changes himself by thinking about himself).
(Society changes itself by thinking about itself)]
   –This is an antithesis to the whole setup of natural science, with
   observer and observed.
   –Instead society talks to itself about itself with the intuition of
   changing itself.
   –In a world which was completely explicable in intellectual terms
   there would benointellect – it would be lifting oneself by one's own
   bootstraps.
      –This has been the dilemma of modern philosophy.
         –Hegel thought he escaped it – but to change from realism to
         idealism does not change the problem, nor does it solve it.
      –Intelligent behavior becomes mechanistic behavior as soon as it
      becomes completely intelligible.
   –As soon as all the error is gone, all the motivation is gone.
      –Motivation is meaningless unless it is subject to error – or else
      it is no longer motivation, but just behavioristic action.
      –We can not even define what we mean by a problem, nor by
      a solution – we have to take them for granted, just as any
      conmmunication of experience is dependent on both persons
      having known the experience.

Relation of historical approach to classical economics – is really the relation of
content to form. Classical economics has no content, has only form.
   –Wants, technology, resources are given.
   –It does not discuss how much of these exist, only that they are present.
   –It makes formal propositions – such as $2 + 2 = 4$ – what? We don't find out
   by mathematics.
      –It is only in abstraction from content that $2 + 2$ are four, for content is never
      identical.
   –Economics makes formal propositions; if it seeks to make known the content,
   it must (a) look and see (economic history) [Commons said that this was his
   method] or (b) get into historical explanation.
   –Thus it is only stupidity to find a war between these two points of view – they
   are only complementary, two ways of looking at things.
      –Why, because economic phenomena have a history, must they be
      approachable only in terms of historical interpretation?
         –Why, because there is a body of formal propositions, should the historical
         approach be omitted?

–(a) To describe, is <u>economic history</u>.

–(b) To explain, to make the sequence of events fit into an explanatory scheme is <u>historical economics</u>.

   –<u>What</u> makes the sequence intelligible.

–There must be <u>something</u> in all the literature of historical interpretation.

Utility – only a tiny percentage of social action has a direct economic motive – what is the utility of a necktie.

   –Take a small income, $100 per month, make up a budget or expenditure – what proportion of that expenditure is for physical comfort and health?

      –Only an infinitesimal percentage (the amount necessary to keep a Soo [Sioux?] Indian in health).

      –Society itself imposes a certain minimum of expenditure.

J. R. Commons – is an institutionalist, in a different sense from Germans.

   –W. C. Mitchell is called, and calls himself, an institutionalist – but is not by standard of Germans, Veblen, or Commons.

Theoretical economics –

   –Income – a flow (electric).

   –<u>Utility</u> is the principle of alternative flow – the means by which income is allotted to the competing channels in expenditure.

      –Law of equal marginal utility is different from law of electric flow – which is that current takes all the paths open in inverse proportion to resistance.

      –Law of utility allocation – income takes one channel to a point where its "return" is equal to the "return" of <u>another</u> channel of <u>different resistance</u>.

      –Theoretical economics is the <u>mechanics</u> of the price system.

<u>Force</u> Motive Wants: a. current = consumption

          b. progress = saving

<u>Resistance</u> a. Resources (scarce)

          b. Technology

   –Putting in the figures to such formulae is a problem for <u>statistical</u> economics.

   –Finding how the formulae <u>got that</u> way is a matter of <u>historical</u> economics.

   –Deriving the formula is a matter for <u>theoretical</u>economics.

      [Preceding three lines accompanied for emphasis by triple vertical lines in near margin]

(cf. Deuteronomy and Leviticus – oldest and most authentic tribal records we have)

   –Difference between writing the history of economic action = economic history and explaining that history = historical economics.

   –History <u>is</u> somewhat intelligible.

–There is some kind of intelligibility in historical sequence.

–Yet we contrast this with the shallowness of all explanations we have.

–Getting ahead in Middle Ages was only possible in politics, war, or church.

  –No one in the community would have tried to "get ahead."

  –Outsiders, merchants, broke this down (perhaps they, being weak, tried to get ahead in the only way they could, by building on economic-legal institutions – the instruments of the weak [Alongside: "F.T.O."] [In margin: "i.e. inevitable connection of Christianity and capitalism."]

–Renaissance – national state, politics supreme.

–Marxism ties class struggle to modern society, which has less of it than any other society.

  –Really the "analysis" of class struggle is a preachment that one may exist; it is the fact that we haven't one, rather than that we have, that bothers the Marxist.

  –There are no classes in the U.S.

    –How define a "class?"

    –Agricultural reformers do the same thing – preach that there is a farmer class – in order to try to create one.

    –There are 50 times as many classes as there are people – at least for each individual.

  –The time when class struggle had any meaning – was a time when there was no economics, and could not have been an economic interpretation.

  –Tripartite division of productive factors was 100 years too old when Adam Smith took it up.

Science and History

  {–Only one History.

  {–Science deals with propositions independent of space and time – will be true everywhere and anytime.

  {–History is relative to individual uniqueness.

  {–Science is relative to universal propositions.

Kinds of Historical Theory

  –Any science can be made the basis of a history (mathematics, biology, economics).

  –There is a theory of history.

  –Is history an explanation or a question?

    –It seems as if people get satisfaction out of history as such, out of the mere sequence of events – but such does not explain.

      –Assumption that a thing is a cause if it precedes in a sequence.

–Empirical approach.

   [In margin: "Nietzsche – theory of history – theory of infinite"]

   –Positivist – Comte.

   –Statistical – (extrapolation).

   –No equation, passing through any number of points, tells anything about where the next point will be.

   –An equation $a + bx + cx^2 + dx^3 \dots$ can be made to fit any set of points a, b, c, d, but <u>any</u> next point e, can be added and the curve made to go through it.

–Yet nature tends to simplicity.

   –Simplest algebraic force is that of constant velocity.

   –Formula for falling bodies is the next <u>simplest</u> possible <u>algebraic</u> law.

   –Galileo started with the assumption of simplicity and made incredibly crude experiments.

   –Amazing thing is that the Greeks had no notion of <u>mass</u>

   – not even Archimedes.

   –Achilles and turtle – this paradox divides the space by one-half of its former distance, each time, and divides time by 1, i.e. time remains always the same for each division of space

     [In margin: "Argument could as easily be used of two runners approaching each other – if the space is divided by half, the time unit remaining always the same!"]

   –Why? Divide both by the same fraction – then Achilles easily catches the tortoise.

     [In margin: "Then you have merely a velocity of space per time."]

   –Is Newton's first law of motion a <u>rational</u> or an empirical truth?

   –Is it "<u>necessary</u>" or just a <u>fact</u>? – empiricists would say that a <u>necessary</u> truth <u>is</u> only a fact.

   –We do not doubt the axioms of geometry.

     –Our natural good sense prohibits us from doing so.

     –By the support that <u>our minds can not realize the contrary</u>.

   –Is there any truth in history than our natural good sense accepts, refuses to have violated?

   –What <u>is</u> a theory?

     –Is that the way things are?

       –Yes, if empiricism is true.

   –Mechanical explanations final – if you had formula for all things in the world, you would be able to fit all things into it.

   –Mechanics itself is a theory of history.

–Yet any other theory must violate it.

–Historical forces <u>do</u> act as though there were something like inertia and friction in them.

Theory of Uncertainty and Risk Taking is inseparable from theory of Savings and Investment.

–Most economists treat uncertainty as a <u>cost</u>; Knight is sure it is the primary reason for interest in life.

<u>The economic motive</u>

–Maximizing return

– [Blank]

The <u>problem-solving</u> motive

–The (whole) game is to find the end.

–Most economic behavior has the two motives mixed.

–What of <u>play</u>? Skill and luck mixed.

   –Game of all skill is not as interesting as game of all luck.

   (There is a law on the statute books somewhere in U.S. that if two trains meet at a station neither can proceed until the other has passed!)

–What has intelligence to do with social control.

   –The idea of <u>scientific</u> control has no meaning except in a dictatorship.

      –Society is not going to control itself as an outsider would control it.

      –Classical economics assumed people wanted liberty because they would be better off under it.

      –In games, intelligence for changing rules, intelligence for winning the game's end, intelligence of the poker kind.

–We do do certain things <u>more</u> or less intelligently – as setting up labor exchanges – but we cannot define what we mean by intelligence in this case.

   –Nor do we have any scientific test for measuring that better- or worse-ness.

   [Preceding three lines accompanied for emphasis by double vertical lines in near margin]

      –The test is a harder thing than the action.

         –Logic comes afterward; it involves reasoning of an advanced kind. Is not a first framework.

         –There is no such thing as thinking about thinking <u>in vacuo.</u>

         –All truth is a matter of <u>degree</u> of unanimity.

   –Hierarchy of sciences (Comte)

      –We answer questions in human fields <u>instages.</u>

         (3 <u>Methods in Social Science</u>, 1st paper on Comte)

      –Laws of mathematics.

      –Laws of <u>physics and chemistry</u> define a certain boundary to human action.

–Laws of <u>biology</u> – behaving is <u>adaptive</u>.
  –There are <u>limits</u> to how <u>un</u>adaptive an organism can be without
  destroying itself.
–Laws of psychology – not Comte's.
–Laws of sociology.
–Laws of history.
–Comte showed that there are different <u>stages of intelligibility</u> at the deepest
levels.

Knight – 3 lectures.
  –Since Middle Ages – have been created.
    1. Mathematics.
    2. Science.
    3. Freedom – of thought, of political action, to raise standard of living.
    4. Spread of culture.
    cf. H. O. Taylor – pp. 76–77.
Knight agrees with Max Weber in giving religious causes priority over
economics in the struggle for liberty – but <u>Commercialism</u> has been main force
making for tolerance.
  Liberalism has encountered one important counter-evolution – <u>Medievalism</u>.
  –But the greater danger is from <u>Nationalism</u> – to which form medieval method
  leads.
  –Nationality in pol[itics] and econ[omics] is under fire.
    –Political life is becoming more and more romantic, un-intellectual.
    (What of economic life?)
    –Rise of new type of leadership – unintelligible.
      –"Action for sake of action."
        –"Think with your blood." Hitler.
        –"Pull together." Roosevelt.
          –Experimentation – means follow the leader. [F.T.O. inserts (?)]

Democratic government only possible where no big problems presented
themselves – as on <u>frontier</u>.
–Room, exploiting of nature rather than usual exploiting of other men.
Religion – always the basis of <u>formal social order</u>.
    –Under liberalism – true, an unconscious sort of religion.
    –Negative – keep religion out of ordinary life.
    –Formal – archaic forms.
    –Positive – regard for human life and personality.
      –Property, i.e. liberty – or, i.e. power to do what one wants. – immunity
      from interference. Market-competition.

–Conflict gives way to mutual advantage.

–State as a main end, gives way to minimum interference.

  –Democracy.

    –Especially in U.S. – origin, frontier, reduction.

    –Connected with economic and religious laissez-faire.

    –Then a <u>new</u> "liberalism" of control, bureaucracy.

I <u>deal</u> state

  –A group of social scientists, no <u>one</u> of whom had a personal reputation.

  –Who express a <u>consensus</u> of opinion which is made available to the public.

  –<u>Darwinism</u> of the intellectual profession.

    –You write for publication, and a sort of <u>natural selection</u>, based on the public approval, determines the survival (since 1916 – when Knight began teaching – every economics textbook has been worse – a downward trend of 45°).

    –Academic life and classroom technique based on Supply and Demand – the consumer finally chooses what will be taught.

  –Problem of agency: –and leadership – leaders will either impose themselves by brute forces or will be chosen by those who are to be led.

–Then comes the problem of agency.

  –Most important book for social sciences is Tolstoi, <u>War and Peace</u>.

  –Knight thinks leadership has a good deal of force – but not of intellectual basis.

  –<u>Intelligence and the Social Process</u>

    –The main problem before Knight. cf. Wallas, <u>The Great Society</u> [1923]. cf. Cooley, <u>Human Nature and the Social Process</u>.

    –Education has backed "<u>independent thinking</u>'' but nine times out of ten that is wrong thinking.

      –The thing we need to push is <u>correct</u> thinking – above all else.

  –Marxism thinks people's ideas based on material bases! Absurd.

    –There is a personal vested interest in one's own ideas – wherever they come from – and God knows where that is.

  –Competitive campaigning <u>vs</u>. impartial thought.

A constant tendency in history towards technical improvement.

  –Not constant in rate.

  –Not always does technical increase result in <u>improved</u> product.

    –Seldom does it result in a changed product of the <u>same</u> product, but usually a change in the product produced.

  –You can explain as much history as you want in terms of <u>technical</u> improvement – but Knight doesn't see much in it, for we don't know what <u>technical efficiency</u> is, and <u>that</u> concept is essential to a historical explanation in technical terms.

–cf. Bagehot: <u>Politics and Physics</u>. Levy-Bruhl (of Durkheim school)
   –Truth is <u>entirely</u> a historical concept.
   –Knight doesn't go so far in this direction, but admits a lot.

The substance of history is <u>people's beliefs and values</u> (conscious or unconscious).
   –Efficiency = quality of result obtainable with given means.
   –Efficiency seems to mean <u>increased quantity of life</u> – survival value.
     –How define that? Is this <u>an</u> explanation, or does it call for explanation?
       –150 pounds of one biological species (mosquitoes) replaces 150 pounds of another species (humans).
         –Is the <u>quantity of life</u> increased or decreased? – Does life progress or regress?
         –What standards do we have for defining or measuring such a thing?
–Is there a <u>descriptive law</u> of history?
   –So that a "stranger" could always tell forwards or backwards in the time-stream?, could always place two events as being one or the other <u>first</u>.
–If we found such a descriptive law, would it <u>explain</u> history?
   [Double parallel vertical lines in near margin alongside preceding four lines]
   –<u>What is the explanation of anything?</u> – that quinine cures malaria?
   –Some answers <u>seem</u> to have meaning.
     –As mechanical actions.
     –But <u>organic</u> actions, and <u>social</u> actions are harder to explain.
     –Do we <u>prefer</u> a mechanical explanation?
   –This is a <u>real</u> intellectual problem.
   –Do we <u>impute</u> will into inorganic forms? Inevitably.
   –It does seem impossible to recognize any abstraction except in terms of "sense imaging."
     –There is no <u>absolute</u> reality which we can conceive as such.
   –Knight thinks we <u>strive</u> to get an explanation in <u>physical</u> terms, as a final step.
     –But 1934 physics is further from common man's vision than primitive spooks and demons.
–Is this itself a historical category (? F.T.O.) –Levy-Bruhl would say yes.
   –Greeks had no idea of <u>mass</u>, –without this a mechanistic explanation is impossible.

Theories of Evolution – Hegel, Spencer, Bergson
   –<u>Spencer</u> – First Principles, Inductions and Sociology, The Data of Ethics.
     –The political rationalization of utilitarianism.
   –Was written at <u>just</u> the time that Jevons, Walras, Menger were making an economic rationalization of utilitarianism.

–An empirical theory of evolution.

–From simplicity to complexity, from homogeneity to heterogeneity – but acquiring structure.

–Success in the human struggle seems to lead to biological extinction, not as Spencer thought, to survival.

–Darwin – progress in the animal world.

  –Excess reproduction.

  –Random variation.

  –Natural selection.

–What does Veblen believe Darwinism to be?

  –Biological survival, or social selection.

  –Do those ideas and social habits which tend to protect and increase the race tend to survive?

    –Do social institutions lead to biological survival?

      –This is what Darwinism [is].

–Is it progress or retrogression to get more wants than can be satisfied?

  –Huxley – Evolution and Ethics (Romanic lecture)

  –The critical self consciousness seems to be biological suicide.

    –All high civilizations have died off from the "malady of thought."

–The 19th century took economics too seriously.

  –The difference between efficiency and waste is fundamental to the individual and to society.

  –But they were not critical of the rules of the game.

–Did not see the non-economic aspects of all life. – the questioning of the rules of the game.

  –The problem – solving activity.

  –The ecclesiastical game, the military game, the commercial game.

[In near margin, triple vertical lines alongside foregoing line]

  –Neither preaching nor teaching can be sold under an honest label – people won't buy either on the basis of a commodity, as meeting demand.

    –Education can be put over on a slightly more honest level.

    –Knight doesn't know what truth is but is sure he knows what lying is.

      –"It ain't ignorance that does the damage; it's knowin' so darned much that isn't so" – Josh Billings.

Economic Behavior

  1. Generalized physical description of manifestation and empirical order (statistics) (economic history).

  2. Motivated by "wants" (for goods and services). –Classical, neo-classical, mathematical economics.

3. Motivated by "otherwise," ideal principles (Control? [F.T.O. insert]).
4. <u>Ideal</u> manifestations separated from human beings. (As grammar, language) (Historical and institutional economics)
   –An ideal <u>form</u> into which behavior fits but not itself depending on the behavior (as plant fits into soil).
   –Veblen – mostly under this category, but he never definitely stated himself as an idealist, or as a realist.

1 and 4 are incompatible with control.
   –It is the essence of a science that the object studied can not exert self-control.
   –Also, science is static, gives the underlying properties of a subject matter.
   –Predicts only hypothetically, does not predict the original "if."
   –If the scientist's experiments <u>change</u> the subject matter in the process of experimentation, then he is no scientist, and his study no science.
   –Self-control or self-knowledge is transcendental, not within the sphere of knowable cause and effect.
   –The essence of positivism is the denial of <u>activity</u>.
      –The <u>social</u> sciences <u>must</u> talk about activity, it is itself an activity.
         –In order to deny this, one must do the equivalent of saying "I am not talking."
      –The social sciences are an inevitable part of social life.

1 and 2 are confused. If an activity is errorless, it is not an activity.
   1 – Is errorless.

   2 – Has always been considered by classical econ[omics] as errorless (this makes it essentially #1. – but possibility of <u>error</u> <u>inre</u> the <u>means</u>.
   3 – Possibility of <u>error</u> <u>in re</u> <u>ends</u>.
   4 – No error is possible.

4 – Is there an internal contradiction?
   –Can history be regarded as Ideal in content.
   –And at the same time be regarded as positive, with cause and effect relations.

1 – Is history made intelligible by putting it in certain form.
History ≠ science.
Fact ≠ activity.
(cf. Leibnitz – two clocks – control or God. – or Seaton vs. Aquinas – dispute over God and right)
–Bergson seems to have answered this problem most satisfactory [sic].
   –He (Bergson) looks at the problems in terms of the <u>function of thinking</u> – which is an apparatus by which an organism adapts itself to its environment.
   –But intuition, or insight, accomplishes much communication.

–Poetry is the deepest <u>social</u> science. – In social science classes, we try to observe such insights in a strange environment.

–Difference between figurative and literal language – only one of degree.

–No use talking about <u>value judgment</u> <u>except</u> as a part of <u>a</u> culture.

–Likewise, no use talking about <u>fact judgment</u> except as a part of a culture.

–Levy-Bruhl:–no such thing as <u>thinking</u> except as part of a culture.

–The Puritans had the stuff out of which civilization is made.

–If we judge civilization by its <u>peaks</u>, there has been no progress since 1500BC Egypt.

– Most people are <u>poseurs</u>.

–Knight's only theology: A universe of values (and a universe of facts).

–A super-individual, impersonal universe.

–Unless people take themselves seriously there is little to talk about.

–All the individual can "<u>do</u>" is to change configurations; move something from one place to another.

–The significance of anything is symbolic; and <u>only such</u> meaning is significant.

–All science explains the world out of existence, value or physical.

Creation of new demand – is a historical change.

–No change, no history.

–Veblen has no <u>explanation</u> of change, nor Marx, nor Sombart.

–Does Veblen give a real answer to the problem of the effect on human nature of machine technology?

–Arabic notation – 1204 –

–Sombart says in an examination of some number of <u>manuscripts</u>, over half the divisions are wrong – additions good.

–Roman numerals were merely a way of setting down the result – obtained by abacus.

–First book on <u>accounting</u> 1460 or 1490 – not generally in use until 16th century.

–Italian cities first used the new figures in place of Roman, but only to write down the answer. – the Arabic figures replace the abacus, more efficiently.

–When did long division and multiplication become a part of widespread education.

–Could Adam Smith do long division?

–Everyone makes a distinction between fields in which, in an argument, one side must be wrong and one side right, and in which the result is a matter of opinion – <u>but</u>, the classification of <u>any</u> problem into one or the other of these categories is a matter of opinion!

Social Space, Moral Space
  –Historical change; is it movement?
  –Do we have any concept of motion in history?
  –If so we must have a concept of <u>social space</u>.
–We do apply this terminology all the time – is it a mere figure of speech? Or does
  it mean something – Knight doesn't know.
–Is there a "historical <u>field</u>?" Newtonian laws of inertia and force? – There seems
  to be.
–<u>Time</u> is the same in history as in mechanics.
–But space and mass, are they the same?
–In order to talk of inertia we must have a notion of mass. What is social
mass?
      [Preceding eleven lines accompanied for emphasis by double vertical lines in
      near margin]
      –Historiography, Flynn [Fuether, E.]
        Park – <u>Introduction</u> to Park and Burgess
            (Windelband, Rickert point of view)
            –There is something in history that isn't in science.
        Barth, Paul. <u>Die Philosophie der Geschichte als Sociologie</u> [1922].

–Intelligibility
    –Principle of adaptation is fundamental to organic life.
        –Darwin had a very sound <u>rationale</u> in his system – even though now disproved
        for <u>organic</u> evolution.
        –T. H. Morgan

    –Economic principle – satisfying desires, is fundamental to behavior.
        –But not an ultimate explanation.
        –What do we <u>mean</u> by saying that a person gets interested in problem-solving.
–Can you <u>look at social behavior as culture phenomena</u> – Institutionalism.
    –Language the supreme example. (Linguistic change)
      E. Sapir, <u>Language</u>
      Blumfield [Bloomfield] is a behavioristic linguist [Paul Bloomfield,
      Language, 1933].
        –cf. Goldenweiser-Sapir article.
        –Legal history.
            –R. Pound, <u>Interpretations of Legal History</u>.
            –weaver econ[omy], for <u>its</u> institutions do get embodied in law – almost
            a branch of law.
            –cf. also his <u>Spirit of the Common Law</u> and his <u>Law and Morals</u>.

The science of medicine is the tombstone for all social sciences.
  –The doctor does not tell the truth.
  –If he did would he be allowed to continue or not?
  –Most of the time he hopes not to do too much harm.
Social control of medicine raises all the problems of social control.
Social problem of the ethics of educating people.

(Dynamic positivism – a contradiction in terms)

Knowledge of society, if complete, eliminates change, in two ways.
  –Independent change can not occur.
  –Change of society based on knowledge of it can not occur.
History must be continually rewritten – i.e. continual change, activity.
  –Is activity dynamics?
  –There is not any activity in physical dynamics.
    (No change in the total quantity of motion in mechanical dynamics)
  –This is not true of social "dynamics."
Marx – Did Hegel have a philosophy of history that said there was no
end to the process; while Marx said that historical process ended with the
proletariat revolution. cf. pp. 85–89, Vol. III – change of stand on surplus
profit.
  –Got his theory of gluts and crises from Rodbertus (wrote 20 years before –
    crises: wage power, not being the whole of industrial return, can not take all
    products; leads to Hobson's stand)
  –Steps
      1 – Dialectics
      2 – Materialism
      3 – Economic interpretation
      4 – Surplus value
      5 – Concentration of production
      6 – Disappearance of middle classes
      7 – Reserve army of the unemployed increases
      8 – Increasing misery of the mass
      9 – Crises
  –National labor policy in U.S., the government and many business men believe
    in a lag in working-class purchasing power as a cause of depressions.
    –Marx, Foster and Catchings, etc.
    –It is tripe as an argument, except for the case of a lag in the rate of productive
      increase over some parts of industry.
    –Is labor used in production any more homogeneous than the utilities in
      consumption.

–Marx's theory of distribution is no worse than that theory in <u>any</u> classical economist – couldn't be.

  –Assumed that there isn't any competitive bidding for labor (or capital).

  –Thus explained its return on metaphysical grounds.

  –The whole thing goes back to the fool labor theory of value.

   –Adam Smith

    –3rd chapter of Genesis is the root of original sin – "out of the sweat of thy brow shalt thou make bread."

–Hegelian dialectic is the theory that history follows the same course as ideas in the mind.

  –The movement of history conforms to the movement of thought patterns in a mind.

–In Marx's ultimate utopia, what kind of production and distribution would there have been.

  –What will happen to capital? Will it be distributed equally per man? If so, there will be tremendous inequality of net income per man; or else a war between men in small-capital and those in large-capital industries.

  –Same in agriculture (principle of marginality is the principle of efficient production).

–Marx gives a pseudo-intellectual basis for faith.

Veblen

–All drive for accumulation is based on desire to emulate.

–And in the end, <u>all</u> consumption is also based on desire for emulation – a "<u>theory</u>" of consumption.

–Would Veblen have placed himself on the productive or unproductive side of human activity?

–Conspicuous leisure-

  –"Theory of leisure class" (1899) – a psychogenetic rationalization of Marxian theory of class struggle.

   [Alongside the following three questions, in the margin: "F.T.O."]

    1. Was not Veblen a <u>puritan</u> – dislike of <u>consumption</u> and <u>leisure</u>?

    2. Is <u>his</u> kind of deduction just the same as that of classical and neo-clasical economists, as Harris says?

    3. Could he make his statement of the place of science as following invention for <u>modern</u> industry, which is largely chemical?

–What would be Veblen's <u>real values</u>?

–Knight says puritanism is only one possibility.

–And Knight says that Veblen is more closely allied to Carver's point of view – a biological or physical ideal – <u>physical life</u> is the ultimate end. – (Is not Carver also a Puritan?) [F.TO. question?]

–Veblen's work gave him a large degree of "invidious distinction"!

–cf. Laski's review of Veblen in Economica, 1925; <u>Dorfman</u>, Political Science
Review, 1932, "The Satire of the Theory of the Leisure Class"

  –Veblen only understood in terms of satirizing Spencer.

  –Knight objects: Classical economists did not view life as a struggle for
    existence.

–Puritanism is "activist asceticism."

–The Instinct of Workmanship, 1914 [by Veblen]

  –(Knight) Did Veblen ever give any thought to the <u>difference</u> between work
  by/to savages and by/to moderns.

  –Most people overestimate the difference between modern work and
  handicraft work – one was as monotonous as the other.

  –Radcliff-Brown – said the domain of econ[omics] was the domain of <u>work</u>!
  But, aside from the economic naiveté of this, he never came to grips with the
  above problem – the <u>meaning</u> of work to savages. Or the general problem
  of the difference between work and play.

  –Spencer thinks that <u>instinctive</u> and <u>pleasurable</u> are by definition <u>useful</u>.

    –This just isn't true.

    –Read about the instincts of insects.

      –Metchnikoff, "The Nature of Man"

      –Wheeler – The Social Life of the (Ant)

      –Sumner – "Folkways"

    –Insects' interests are incredibly destructive.

–Veblen thinks there is no validity in a <u>preference</u> for <u>future</u> value. – Knight
agrees but individual future value can not extend beyond death.

  –Capital accumulation does, it is never consumed.

–What does Veblen mean by any of his terms? – exploitation, serviceability.

–Some ethical theory, plus some theory of imputation, are both necessary for
a theory of the origin of wealth.

–Veblen's system

  <u>Instincts</u> = give the objective ends of action.

  <u>Instinct of Workmanship</u> = is a means to achieve an end –

    –Most of our "<u>propensities</u>" are in this class.

  [Alongside preceding two lines, in margin: "Contradiction in terms
  here"]

    <u>Intelligence</u>

    <u>Institutions</u>

–How does Veblen work instincts, intelligence, and institutions into a
comprehensive system that has inner consistency? He doesn't!

Max Weber
- –Religion and the Rise of Capitalism.
- –Followed by Sombart and Tawney.
- –Criticized by Brentano and Robertson, Trouch [presumably Ernst Troeltsch, author of *Protestantism and Progress: A Historical Study of the Relation of Protestantism to the Modern World*, 1912; see below]
- –Capitalism was unique, not allied to any other systems of exchange or profit.
- –Was stimulated by Protestantism.
- –A <u>new spirit</u> – closely allied to <u>rationalization</u>, which is the outward sign of the universal spirits.
- –Luther put forth a new interpretation of the word "calling" – Beruf – the pursuit of work for the sake of work. – leading to a worldly asceticism.
  - –Proof: <u>Baxter's Dictionary</u>, <u>PoorRichard'sAlmanac</u> – showing the change in moral concepts.
- –Criticism – Troeltsch (<u>Protestantism and Progress</u>) – says the Protestant Reformation retained large elements of Catholic dogma.
  - –Tawney has proved this for the Anglican Church.
  - –Robertson – looks at capitalism as a borrowed system, taken over from Moslem Society.
  - –Capitalism is the <u>Acquisitive Society</u>, i.e. agrees with Tawney.
  - –Knight says this is wrong – acquisitive society has always existed.
    - –Capitalism is a creative spirit.
- –Robertson sees the reason for the flowering of capitalism in the Western World in the invention of double-entry bookkeeping
  - –Did the Arabs have a bookkeeping system?
  - –Knight – there was more liberalism in Renaissance Italy than in the Protestant Reformation – which was essentially a backwoods, fundamentalist reaction – a move back to former morality.
    - –As the result, the Catholic Church tightened up.
    - –Ashley pointed this out in his Economic History and Theory.
    - –The Jesuits were 100 years ahead of the Calvinists.
    - –How did the rise of Protestantism come, then, to be linked up with liberalism.
  - –<u>Wood</u> – says freedom came in as a by-product of the Reformation – not foreseen by anyone.
  - –<u>Troeltsch</u> – says freedom was a result of the existence of three churches each changing the authority of the other
  - –Knight: Luther was originally an individualist, but he soon turned to the dogma:

cuius regio, eius religio
>   –i.e. the State needed a business man, and Protestantism became a sort of
>   State Catholicism – tied up with Nationalism.
>   –The rulers of the 15th century were more individualistic than the business
>   men of the times.

–Criticism: Lutherism [sic] and Calvinism must be carefully distinguished.

Randall – Making of Modern Mind – says that Puritanism had a close parallel
within the Catholic church.

–An adjustment between Protestantism and Capitalism did not take place
until about the end of the 17th century.

–What is the "Spirit" of capitalism – was it something new in Western society
or had it existed before?

–Weber defines his case much more carefully than Robertson gives him credit
for.

–Remark of Alfred Weber to Knight – 1930 – his lectures on cultural history
worried him, as to their importance and validity, but he said, "You know
we first have to have <u>something</u> to put up against Marxism" [Alongside
beginning of this sentence in margin: double vertical lines].

–Why did these Arabian, Chinese inventions come to flower in Northern
Europe – there was essentially a <u>frontier</u> condition.

>   –Knight thinks it is the marriage of Northern feudalism and Italian arts and
>   culture [Alongside this sentence in margin: double vertical lines].
>
>   >   –Italy was essentially Florence and the Medici.
>   >   –The marriage of two Medici women to the Kings of North Europe and
>   >   France.
>
>   –In Spain under Phillip – the State took over the ultimate authority of the
>   Church – as nationalism within Catholicism.

–What of influence of <u>service life</u> and <u>construction period</u> on trade cycle?
>   cf. Hayek – restate him.

[On top of page 16; nothing else on either side of page]

OSTRANDER'S NOTES ON READINGS

(1) Wesley C. Mitchell, "Commons on the Legal Foundations of Capitalism"
(*American Economic Review*, vol. 14 (June 1924), pp. 240–253)

>   –Substance of capitalism = "production for use of others, acquisition
>   for the use of self" – implying human activity, natural resources,
>   ownership.

–Legal foundations of capitalism were laid by judges, who validated and enforced those ideas relating to property and liberty which are involved in business enterprise.

    a/– What ideas of property and liberty are implicit in business enterprise.

    b/– How have the ideas of property and liberty prevailing under Feudalism been converted into those prevailing under Capitalism?

    –An analysis and history, is a conviction that the decisions of the courts are of paramount importance.

–Economic evolution – new forms of behavior crop up, responsive to new needs or opportunities.

    –Selection among these for social survival is made – tardily – by the courts – on basis of what ideas they think good, and bad.

–The "ultimate unit" in economics, ethics, law, under Capitalism, is not one man balancing sacrifices and satisfactions, nor two men bartering for apples,but is a Transaction – involving a minimum of five persons – two parties directly concerned, two more parties representing the next best alternatives open to the bargainers, and a judge.

–Behavior in all business transactions is molded by a long line of judicial decisions.

(2) Frank H. Knight, "The Development of Economic Institutions and Ideas," mimeographed [See Ross B. Emmett, "Frank H. Knight (1885–1972): A Bibliography of His Writings," *Research in the History of Economic Thought and Methodology, Archival Supplement* 9 (1999), p. 31]

1–Policy assumes freedom of choice.

2–Explanation assumes that causation is applicable – universal law.

    –"Classical economics'' belongs to (1) – its "laws" (what will or would happen under given conditions) assume rational choice.

    –To continue that is the essence of its social policy.

    –Inquiries into the form of organization of activity (wants, resources, technology).

    –Is not interested in what any one of these [arrow pointing to "wants, resources, technology"] is at any particular moment.

–The study of economic activity in the concrete – the whatness calls for a different method of attack: history.

    –Explanation of activity is an exposition of its course of development, which is a detail in a vast social complex; influence of historical forces.

    –Why do people want bread?

        –Because they want to is the answer of economic theory.

–Why do they <u>want to</u>, why are they able to?

–i.e. it is a problem of "<u>conditions</u>" (alternatives to that want and the circumstances under which they present themselves).

  –Conditions are parts of whole <u>patterns</u>.

–i.e. it is an <u>institutional</u> problem.

  –Economic study of the patterning of conditions and the changes in those patterns, is <u>Institutional economics</u>.

 –The primary social institution: <u>language</u>.

  –<u>No</u> individual choice in the matter.

  –Such an institution <u>determines</u> the behavior of humans, and is practically an ultimate cause.

  –It has the <u>characteristics of an Institution</u>.

   –It has a "genius" or "<u>spirit</u>."

   –<u>It grows according to some</u> <u>internal</u> <u>principle of</u> order.

 –Historical   economist   asks,   <u>how far</u>   are   the   conditions   of   choice institutionalized.

   –Immediate conditions of choice: individual's tastes, available articles, their prices, his income.

   –All are historical products.

   –But the most important condition is the <u>whole situation</u>: the fact that an individual gets an income, is confronted by prices and commodities prices, etc.

   –<u>Freedom of choice</u> and <u>Institutionalism</u> are <u>antithetical ideas</u>, but, <u>freedom of choice is itself an institutional fact</u>.

    –"Civilized" men deliberate [and] <u>choose</u> much more than <u>traditionalized</u> <u>primitives</u>, who had <u>narrower possibilities</u>.

–To explain this [arrow pointing to "who had   <u>narrower possibilities</u>"] is the main problem of social history. [In margin, double vertical parallel lines alongside preceding two lines]

 –The great problem arises: <u>what is the role of choice in the process of change</u>?

   –Insofar as there is <u>real</u> choice, not <u>fully</u> explicable in terms of conditions, thus not <u>inevitable</u> in the light of those conditions, there is a <u>limit</u> to the possibility of <u>explanation</u>, and to the validity of scientific categories of thought.

   –Carlyle: "great" man and leadership in history. Tolstoy: social forces all-determining ("War and Peace").

   –Is it possible <u>deliberately</u> to <u>change</u> institutions? – by breaking or defying them?

    –Legislation changing them is itself an institutional product.

–Thus the role of history in explanation is not entirely clear. – <u>However</u>, unquestionably, the mere tracing of continuous evolution does give our minds a <u>feeling</u> of knowing <u>why</u> things have happened.

<u>Economics</u> deals with <u>an aspect</u> of social life; it is not easily defined.
 –Three phases of economics: wants, resources, technology.
<u>Economic History</u>: has three general divisions.
 a – Men's ideas about the ends of economic activity.
  –Means are limited relative to desired ends; "urge" towards increased efficiency of attaining ends: = economy.
  –The <u>ends</u> of activity change.
  –The idea of efficiency changes ["F.T.O." follows; may apply to "<u>and efficiency</u>" – inserted above the line – in next sentence or to preceding two sentences and next sentence].
  –Means and ends <u>and efficiency</u> are characteristic of most human activities; it is not the <u>main</u> characteristic of most activities (artistic creation, etc.) but it is the main characteristic of economic activity.
  [At top of page: "–the evolution of men's view as to the ends of life, comes into economic theory and history, only indirectly."]
 b – <u>Means</u> of achieving ends efficiently: <u>technology</u>.
  –Comes into economic theory only indirectly.
  –Economic <u>history</u> is only indirectly concerned with the details of <u>invention</u>.
 c – <u>Social organization</u> (likewise a part of the means of achieving ends efficiently)
– division of labor, specialization.
  –The field of economic history proper.
  –<u>Political</u> organization and <u>economic</u> organization.
   –Partly external, partly internal.
   –Latter part is mainly economic (both F.T.O.)
  –Organization of modern <u>economic</u> life: an <u>automatic</u> system, production and distribution work through the medium of <u>prices</u>, which result from individuals' activities.
  –Economic theory explains this "automatic" system and tells why it works well or badly.
  –Organization of modern <u>economic</u> life: an <u>automatic</u> system, production and distribution work through the medium of <u>prices</u>, which result from individuals' activities.
  –<u>Political</u> organization supplements and guides the price system through authoritarian action of <u>society</u>, originating in a <u>policy</u> deliberately adopted.
   –What to do about "bad" opportioning of economic activity is a problem shared by <u>economics</u> and <u>politics</u>.

–Political <u>details</u> do not belong in the field of economic theory or history but their general character and relations to economic phenomena do belong to that field.

–Economic history traces the development of modern economic organization – into a social system in which activity is controlled mainly by prices which respond automatically to the self-seeking activities of men.

[In margin, "Development of the price system" and double vertical parallel lines alongside second half of preceding sentence]

–But always various <u>politico-legal forms and institutions</u> are acting on the price structure.

–As are the intellectual and moral develop[ment]s of peoples, and their technological skill.

<u>The Historical Standpoint</u>: Our interest in the past is that of explaining the present; the history "of" <u>something</u>, knowledge of history, as such, is only instrumental in value.

–The <u>meaning</u> of the solution, not the bare fact, is of primary importance in selecting and defining the problems, and as an <u>end</u>.

–Keep this in mind in the <u>exposition</u> of history.

The problem of describing the institutional system of the <u>19th century economic order</u>:

–Any brief characterization is not complete description.

–The theoretical system of "classical-neoclassical-mathematical" economic theory does have priority in discussing the typical organization of 19th century – mechanics of frictionless machine is <u>prior</u> to description of actual machines, regardless of amount of actual friction in <u>any</u> one machine.

–It <u>can not</u> be replaced by <u>any</u> conceivable statistical tabulation of the facts of economic life [Single vertical line alongside in near margin]

–What short phrase can we use to designate the theoretical system referred to above?

[Arrow pointing to preceding line in near margin]

a- "<u>Competitive</u>"

–Monopoly is such an important form of "friction," degree of intelligence used in applying concepts in the realm of social sciences is so low, that monopoly must be explicitly coupled with competition in the general theory.

–Most producers outside agriculture have some monopoly power.

–The element of monopoly does not keep the system from being competitive in principle.

b-"Competition"
  –Implies a <u>sentiment of rivalry</u> or emulation as a motive of activity.
  –A human being acting thus is not acting in accordance with <u>rationality</u> (of 19th century economic theory).
  –Price factors, individual wants in relation to particular marketable services (goods) – are the <u>only</u> motives of 19th century economic behavior.
  –<u>Market</u> competition is very different from other human competition.
  –The word is useful, but not as a general description of the <u>whole</u> economic system of <u>markets</u> – it is probably more a motive, ultimately, than economic rationality.

c-"<u>Exchange</u>"
  –Most typically suggests people living by <u>production</u> and <u>direct</u> exchange; money an unimportant mechanism; a product the end of individual production.
  – i.e. late medieval town life.
  (1) But <u>modern</u> economic life is <u>different</u>; individual does not produce a <u>product</u> nor own even a part of what he does produce. He performs a minute technological operation – one of hundreds by hundreds.
  (2) Laborer or capitalist has no voice in <u>what</u> shall be produced or <u>how</u>; gets his income from <u>sale</u> of personal or property services in a market, for a money return. Expends money income in <u>other</u> markets, to get real income.
  (3) Production carried on by <u>enterprises</u>, to which individuals sell services, from which they buy them: <u>impersonal</u> organization.

d-"<u>Enterprise system</u>"
  –Is the best name for a system having above 1), 2), 3) points; if associated with <u>entrepreneur</u>.
  –However, objections:
    –Ordinary usage calls <u>any</u> project an enterprise.
    –Enterprise is typically an organization, not an individual.
    –Enterprise as a <u>unit</u>, buying and selling productive services rarely exists in a pure example.
      –Popular habits of thought view the enterprise as an "association" of productive interests, of property owners.
      –An association, based on <u>contractual</u> relationship, is <u>very different</u> from the <u>principle</u> of <u>enterprise</u> in which the owners merely buy and sell.
        –In principle and in legal theory, and reality.

e.-"<u>Capitalism</u>"
  –Very bad: is tied up with (and derived from) a definite propaganda attitude.
  –Its <u>main implication is false</u>.

–Marx never grasped the essential mechanism of market relations: e.g. that wages as well as prices are largely determined by impersonal forces, not arbitrary power.

–Implies that "capital" employs and controls (and "exploits") "labor."

  –Of only limited literal truth; re ideal type of modern enterprise is false.

    –Historical fact, the borrowing of capital developed along with and as fast as the hiring of labor [Single vertical line alongside in margin]

    –The enterprise is an entity distinct from both labor and capital, hiring one and borrowing the other.

    –Both labor and capital are in the same relation to the business unit, as "factors of production" (productive services) hired in the market, without voice in control.

    –Under the modern enterprise system – the terms ownership and property are very ambiguous (case of farm with mortgage where the capitalist is the mortgage holder, the owner the farmer; as a business; capitalist holds ("owns") the wealth invested; manager "owns" the enterprise).

    –Most management is labor.

    –Under "competition" the only effective control over production is exercised through the expenditure of income.

    –But ownership of property does confer power.

      a. For ownership is fairly concentrated and thus property owners have large incomes to expend.

      b. And income from property can be capitalized, by sale or use as collateral, whereas our "free" legal system does not permit the individual to capitalize his productive power to any large extent – no way of enforcing a sale of future delivery of labor power.

  –Thus, we conclude, Enterprise Economy is the best title for 19th century economic order.

    –A production and distributive organization, with "enterprise" buying in the market productive services, employ[ing] them to produce consumable goods and services to sell on the market, with the motive a hope of a difference between cost and "profit," and the control being market competition (tends to cut down "profit").

    –Person and property are legal categories.

      –One's control over his person is identical with proprietorship, but, by modern law, only the proprietor can own his productive services, except as he voluntarily sells them from day to day (One can lease, but only for a short time)

Phases or Aspects of Economic Development
- –Comparison of two "societies": one may be more advanced than the other in one respect, less advanced in another.
- –There is some intimate connection between these phases of development. But how far there is a necessary relation between the degrees of advancement is a question.
- –The meaning of the "economic interpretation of history" is that, the evolution of technology goes ahead of other phases, and somehow determines them.
  - –This is also the meaning of "culture lag."
- –Others claim that the "intellectual spiritual" evolution is primary, and determines the other phases.
- –Others could, without unreasonableness, complete the list by suggesting that social relations as such tend to consciousness and into discussion ahead of the other phases.
- –Knight's only remark is to say that this is same presumption in favor of a mutuality of relation, an interaction between the different phases of progress.
  - –At least – only on the basis of careful definition of causality-in-history can such questions be discussed.

Conceptions of the Nature, and of the Dynamic Principle of Development
- –"The phenomena of history are the phenomena of life." (There is some question here. F.T.O.)
- a. – Mechanical causality – does life involve any other principle but this?
- –Even chemistry and physics, to say nothing of biological purposiveness, are not reducible entirely to mechanical cause.
- b. – Biological causality – change is bound up in the individual organism.
  - –Darwinian "natural selection" and "survival of the fittest."
    - –More adaptive of small random variations tend to perpetuate themselves in greater numbers.
    - –Would apply to individual structure; or to changes in the forms of group relations independent of biological base: language, law.
  - –"Mutation theory": large changes, still [in]stantaneous, random, accidental.
    - –Likewise will apply to social-institutional as well as individual-biological.
  - –Directed evolution: changes of whatever size, are not random, but predominantly in same direction – for whatever purpose.
    - –Equally imaginable in social institutional field.
  - –"Lamarckian" theory – struggle of the individual with his environment produces adaptive modifications which are biologically inherited by offspring.
    - –Mechanical influence of environment apart from teleological struggle.

c. – <u>Social-institutional causality</u> (super-individual).

    –Darwinian "natural selection" and "survival of the fittest."

    –Mutation theory.

    –Directed evolution.

    –<u>Individuals</u> consciously think out improvements, teach them to group and they become changes in <u>social habits</u>, except as to origin.

    –<u>Improvements</u> are <u>generated in ideas</u> by <u>conscious discussion</u> among members of a <u>group</u>; <u>social thinking</u>.

    [In margin, alongside preceding two items: "Both must imply <u>real</u> conscious deliberation and choice – not mechanical determinism."]

    –If new ideas are regarded as experiments, not as inherently right, then a non-biological selection, based on social and human values, enters.

d. – cf. Todd, Weatherby for other possible themes.

e. –It seems impossible to talk about history without <u>some</u> principle of continuity and progressive change; in fact, without recognizing the <u>fact</u> of <u>progress</u> in some such sense as "mankind becoming conscious of itself" – not without setbacks – but involving the growth of rationality.

    –The <u>fact of progress</u>, even of progress with respect to the nature and mode (and notion of progress, is an essential social phenomena [sic].

    –In civilized society, the balance shifts to the more conscious kinds of change.

<u>The Schematization of Economic History Stages</u>

    –Any description of the development of a phase of social life takes the form of a series of stages.

    –Such stages are not supposed to be either completely or even very accurately descriptive of life at a dated chronological period.

    –The general scheme is valid and useful, even though different peoples do not follow the same stages, or the order is reversed, or some are left out.

    –The problem is to find out how to formulate concepts that will characterize one period or people as farther along than another – which is itself a <u>matter of fact</u>.

    –There does seem to be some degree of "cyclical" character in the general current of "our" history (from Ancient Near East to Modern Europe).

    –Two breaks are important:

        a/– Europe started at a much more primitive level than the peak of Near East civilization, and its social system developed along other lines.

        b/– The same is true of Northern and Western Europe in comparison with <u>Classical</u> antiquity – and even sharper <u>technological</u> contrast, than social.

    –In both cases, a decline or decadence of <u>older</u> civilization before upswing of the later.

–The "break" – due to a/transition from agricultural basis to that of water-borne trade; from inland to coastal location; b/due to technological revolution – and shift of economic centers to North where slavery was unprofitable.

–Perhaps, some truth in "life cycle" of societies.

–The feature stated as characteristic of a society at any time, <u>need</u> not be exclusively or even predominantly characteristic (Examples: town life, as stage in late Middle Ages; and corporate enterprise in 19th century).

–Stages <u>need</u> not be literally successive – some lag possible between.

<u>The Modern Economic Attitude</u> – Classical tradition of economic theory

  –Maximum satisfaction of individual wants for consumable goods and services.

    a/–Qualified by unequal distribution of incomes.

      –Accepted – partly from political difficulty of doing anything about it; partly because of difficulty of <u>capacity</u> as apart from amount of income in enjoyment of income, partly because income is supposed to depend on effort exerted, which is supposed to depend on getting the full equivalent of what individual contributes to the total.

      –i.e. use of capacity depends on effort, and effort depends on incentive.

      –Some agreement applicable to capacity in form of ownership of external things <u>or</u> of personal powers.

      –Inheritance of money (and personal capacity) is allowed because the good effect on the social function of family life over-weighs the bad effect on distribution of income.

    b/–Qualified by fact that <u>satisfaction</u> is thought of in connection with <u>consumption and with production</u> – with a net balance (algebraic sum) of satisfaction-dissatisfaction.

      –i.e. full individual choice between the value of income and the cost of securing it.

  –Assumption that the values are such as find expression in the <u>economic</u> behavior of individuals in such a system.

    –They (the values) must either be attached to consumption, <u>or</u> inhere in the fact of freedom itself.

    –19th century literature assumed the former, and argued for freedom as a <u>means</u> for maximizing net <u>balance</u> of pleasure in connection with specific items of consumption and production.

    –The fundamental principle was that <u>the</u> individual was the judge of his own good. – any motivation goes.

Stages in the Evolution of the Economic Attitude
   –Doubtful usefulness for history before European – little thought given to
   problems, and <u>no</u> records.
      –Universal tendency to follow religious, metaphysical and esthetic forms –
      rather than our utilitarian view of life.
   –European history.

COURSES FROM HENRY C. SIMONS

F. TAYLOR OSTRANDER'S NOTES ON HENRY CALVERT SIMONS'S COURSE ON PRICE THEORY, ECONOMICS 201, AND ON PUBLIC FINANCE, ECONOMICS 360, UNIVERSITY OF CHICAGO, 1933–1934; AND HELEN HIETT'S NOTES ON PRICE THEORY: INTRODUCTION

Warren J. Samuels

F. Taylor Ostrander had two courses from Henry C. Simons, Economics 201, Price Theory in a Competitive Economy and the Effects of Monopoly, and Economics 360, Public Finance. Ostrander's and one other set of annotations of the Syllabus from Economics 201 and his notes from Economics 360 are presented below.

The 67-page mimeographed syllabus from Economics 201 in 1946 was published by Gordon Tullock in 1983 (Simons, 1983). Tullock actually did not preserve his copy of the syllabus. He was enrolled in the law school at Chicago and all law school students were required to take Simons's course. He reports that it was his first and his last course in economics, and that "Simons's course in a real sense changed [his] whole life." He "began reading the leading journals from

Documents from F. Taylor Ostrander
Research in the History of Economic Thought and Methodology, Volume 23-B, 195–205
Copyright © 2005 by Elsevier Ltd.
All rights of reproduction in any form reserved
ISSN: 0743-4154/doi:10.1016/S0743-4154(05)23104-8

cover to cover." Receiving full credit for the course despite being drafted ten weeks into the course, he went almost immediately into the infantry. When he much later decided to publish the syllabus, he borrowed a copy from a friend, the identity of whom he no longer remembers (Tullock to Samuels, email of January 20, 2004).

Ostrander writes that "It is a mini-textbook for the course, with reading assignments. Despite its formal appearance, it is a very personal document. It is pure Simons; I hear Henry's voice in every carefully crafted sentence." It "reflects his cool passion for teaching his views, as well as his constant struggles to get writing exactly 'right' " (Ostrander, "A Personal Appreciation of Henry C. Simons," December 31, 2002, p. 8).

Ostrander preserved two copies of the syllabus. One was his own copy. It has his notes added to the text of the syllabus. The second copy was given to Ostrander by Helen Hiett and it, too, has her notes added to the text of the syllabus. Given Tullock's publication of the syllabus, it makes no sense to publish its text here. Published below are assignments not included in Tullock's version and the two sets of annotations, one by Ostrander and the other by Helen Hiett, and identifications of the points in the syllabus to which the notes refer. The notes can be considered either or both student annotations and indicators of points and/or expansions made by Simons in his lectures using the text of the syllabus. Also published below is a comparison, by Kirk D. Johnson, of the two editions of Simons' Syllabus.

Apropos of the differences between Hiett's and Ostrander's copies of Simons's *Syllabus*, Ostrander opines as follows:

> I'm not too surprised that even my copy and Helen's copy for the same course taken at the same time are different! . . . Simons was never satisfied with what he had written, and was constantly re-writing everything he wrote, even these mimeographed class notes – despite the obvious fact that their use was in part a device for saving time. I recall his altering the mimeographed text while reading from his mimeographed notes, he would "improve" what he meant to say; sometimes I may have written in these changes on my copy.

> Helen Hiett might have picked up a copy of his class notes for this course a few months before taking it, because she was at the University three years; whereas I had to pick up my copy when I arrived at Chicago in September 1933, or later, when I took the course [Ostrander to Samuels, January 24, 2004, clarified by Ostrander to Samuels, January 25, 2004].

Helen Hiett was born in Chenoa, Illinois, 23 September 1913 and eventually became a pioneer female foreign news correspondent, especially a front-line war correspondent.

Hiett's father, Asa B. Hiett, was superintendent of schools in Pekin. While in high school, she became the protégée of F. F. McNaughton, editor of the Pekin Daily Times. She graduated from the University of Chicago in June 1934, in political science, taking only three years. She likely received permission to enroll in the course as a Senior. Granted a scholarship to study the League of Nations in

Geneva, she embarked for Europe the day after graduation. She later held a number of positions in Geneva and entered the doctoral program at the London School of Economics in the fall of 1937.

Abandoning LSE at the onset of war in 1939, Hiett moved to Paris, briefly broadcasting over Paris Mondial and, in January 1940 starting the *Paris Letter*, covering the war. She engaged in a lecture tour during April and early May in the U.S. She was soon hired by NBC to report from Paris, arriving 21 May 1940, leaving for Bordeaux on 10 June, responding to the French government's departure from Paris. After less than two weeks' shortwave broadcasting, she fled to Geneva, arriving on 4 July.

Hiett scored a scoop reporting on the bombing of British Gibraltar, for which she won the National Headliners Award. She returned to the U.S. in 1941, holding a serious of positions during the war: NBC daily new commentator, lecturer at Stephens College in Columbia, Missouri, and, in 1944, war correspondent in Europe for the Religious News Service. After the war Hiett joined the New York Herald Tribune. In 1945 she accompanied 1,500 displaced persons being involuntarily repatriated by ship back to the Soviet Union (described by her in *Deadline Delayed*, 1947).

Hiett also published *No Matter Where* (1944) and pre-war articles on the causes of European ferment and on public opinion and the Italo-Ethiopian dispute. She served as a forum direct for the New York Herald Tribune for many years and lectured in the Chautauqua Institute.

Marrying publishing executive Theodore Waller in March 1948, she thereafter was known professionally as Helen Hiett Waller. Starting in 1958 she hosted a CBC television series, "Young Worlds."

Ostrander met Hiett again in Westchester, New York, almost 30 years after Chicago, at the home of mutual friends. She was killed on 22 August 1961 by a falling rock while mountain climbing near Chamonix, France.

Ostrander had forgotten that he had her notes and why she gave them to him. He now speculates that she must have given him her class notes from Simons's course as she packed up to go to Geneva (Ostrander e-mail to Samuels, January 12, 2004).

An entry for Helen Hiett Waller is in *Dictionary of American Biography* (volume 4, p. 979); see also McKerns (1989, pp. 339–341).

Henry Calvert Simons (1899–1946) had a BA from the University of Michigan and taught at the University of Iowa (1920–1927) and the University of Chicago (1927–1946). Although Simons wrote on a wide range of topics, typically policy-oriented ones, he was best known for *A Positive Program for Laissez Faire* (1934) and *Economic Policy for a Free Society* (1948), which represented his long-time efforts to combine the general principles and sentiments of libertarianism with the existential reality of social control and both with his view of justice. He was also a recognized specialist in public finance, particularly taxation (Simons, 1938, 1950); his work on taxation had influence on the U.S. federal income tax, especially in its definition of income and the use of progressivity.

After a dispute over his being granted tenure in the Department of Economics, Simons was appointed to the Law School, the first economist to be so treated, where he became, in effect, the founder of the Chicago Law and Economics tradition, succeeded by Aaron Director and, much later, Ronald Coase.

Simons was long associated with Frank H. Knight; both moved to Chicago from Iowa in 1927. They were the founders of the Chicago School – the Older, or Early, Chicago School, to differentiate it from the more doctrinaire and simplistic Later Chicago School of Milton Friedman and George Stigler. Simons shared with his successors in the Later Chicago School – several of whom had been his students – a faith in markets and a general antipathy to government. But he recognized the importance of government for markets and the market economy and also for measures to cease and reverse past inequality-promoting uses of government (Inequality-*correcting* measures of government are typically seen as *government*. But inequality-*causing* measures of government are rarely seen as *government*, in part because of the belief that legally-received income is productively earned by its recipient). He favored strong and effective, pro-competition, antitrust enforcement and progressive taxation of income. His approach to monetary policy favored rules over discretion but rules that performed counter-cyclically. John Maynard Keynes's early emphasis on active monetary management and later emphasis on fiscal policy, as fundamental arrangements, were decidedly not to his liking, though, like others then at Chicago in the early 1930s, he supported deficit finance in economic exigencies – as the Great Depression surely was (see Davis, 1971).

Simons's approach gave effect to the market-plus-framework approach to the theory of economic policy, in which he affirmed the superiority of a market economy and, far from obfuscating and/or finessing the economic role of government, recognized the importance of government in helping to determine the basic economic institutions and relative legal rights. Included in social-control activism were the roles of government in providing a workable monetary system and competitive markets, promoting economic stability and a reasonable distribution of income (in part through progressive taxation of income), and in changing the law in response to changing conditions and tendencies toward the undue concentration of power. He did not believe that markets were inherently competitive, as did the Later Chicago School. He perhaps was overly sanguine that government could act to counter extreme inequality and concentration of power rather than being used by upper hierarchic powers to reinforce and extend their hegemony. He believed that the truly free society did not yet exist and that government had an important role in helping bring it about. He advocated, for example, major reform of the financial structure and in the operation of Federal Reserve monetary policy.

About a quarter-century ago I attended a small conference with the title, as I recall, Classical Liberalism. Members of the group had promised to read in advance of the conference a medley of classic writings by Classical Liberals. The upshot of the conference was that the perfect Classical Liberal did not exist. Every classic work by an ostensible candidate was found to be deficient by most attendees. The presence in the writings of affirmative statements of what government can and should do, was disqualifying.

My qualification for being invited was, presumably, the fact that I was an historian of economic thought who was interested in the treatment of government in the history of the discipline. Among other things, I had written on the Physiocrats and showed that their vaunted laissez-faire was anything but laissez-faire as conventionally understood. I had also written a book on the theory of the economic role of government of the Classical School of Economics. In the latter, somewhat following the lead of Lionel Robbins, I showed the Classical Economists to have had a market-plus-framework approach in which legal and non-legal social control figured prominently alongside their emphasis on market allocation of resources, meaningful private choice, and so on. No wonder they could not get all the way through the front door of the house of Classical Liberalism.

In regard to matters of most pertinent concern to the notes that follow, the Later Chicago School stands for the market, for the idea that markets are by their nature competitive, for monetarism, and against governmental activism in general, thus against activist monetary and fiscal policy, government redistribution of income and wealth, and against all sorts of regulation, including antitrust regulation.

An article in the *Financial Times* of London for March 2, 2003, entitled "A choice between the rich and the market" and authored by Luigi Zingales and Robert McCormack, compared two groups of defenders of the market economy. Their common opponent, decidedly anti-market, "believed that markets did not work and that governments had to step in to substitute for them, as well as to redress the wrongs of the past by redistributing wealth." One pro-market group "believed in the protection of property rights, in the virtues of free markets and in the dangers of government interference." The second pro-market group, or so they claimed to be, was the rich who, "fearing expropriation," sided with the first pro-market group as a matter of convenience.

As the authors see it, "there is a difference between being for free markets and being for the elite that has benefited from them" – and, if the truth be told, through their control of government used its powers to form markets and structure the distributions of opportunity, income and wealth in their favor. Zingales and

McCormack argue that truly free markets not only are competitive but create competition, thereby undermining the positions of established interests. "While this competition is an opportunity for the have-nots, . . . it is a threat for today's elites." Accordingly, many of the rich – "the incumbent super-rich" they call them – "can be against free markets because [the rich see] them only as competition and not as opportunity."

On the other hand, there are those, like Simons, who understand that markets have – must have – a legal foundation, that there "is a distinction between market-supporting and market-suppressing regulation," that "the forms of taxation that benefit markets are not necessarily those that benefit the rich," and so on. "Being pro-market and being pro-rich are not the same."

In his *Personal Income Taxation* (Simons, 1938) affirmed both greater equality and its pursuit through progressive taxation of income. "Taxation," he wrote, "is the proper means for mitigating inequality" (p. v). His concerns about doing so, however, say much about how he comprehended economic politics. Redistribution is often a cover for policies whose actual motivation resides elsewhere; even otherwise it can be overemphasized. The book, Simons acknowledges from the beginning, is both "an academic treatise and a tract for the times" (p. v). Not surprisingly, therefore, he writes,

> When Republicans were the leading disseminators of economic fallacies, it was proper for the academic person to stress the problem of inequality. Nowadays, however, there is no issue as to the need for lesser concentration of wealth and income. Indeed, we tend now toward relative overemphasis . . . (p. vi).
>
> Just as the Republican party consolidated its power by dispensing gigantic subsidies in the form of protective-tariff duties, so now the Democrats have been purchasing allegiance by endless restraints upon internal trade (p. vii).
>
> We need, at the present time, not greater awareness of inequality but greater awareness of the dangers involved in trying to mitigate it by methods which involve restraint of trade (p. viii).

In short, Simons was a sophisticated form of liberal, and also pragmatic. But to those who did not think as deeply as he, he seemed interventionist, hardly the perfect model of a Classical Liberal, a certain type of Classical Liberal. In this, I propose, Simons was a splendid example of another type, one to whom economic theory should not be a monopoly of the reactionaries nor government the monopoly of the already established as they seek to use government both to further entrench their hegemony and to further enrich themselves. This latter type of Classical Liberal did not fare well during the period of the Cold War and afterward or in the minds of those who like ideas and policies ideologically cut and dry – i.e. those who tended to dominate economics during that period. Simons was a dedicated supporter of laissez faire *as he understood it*. Moreover, his support was articulated without the shrillness one associates with the Later Chicago School.

What did laissez faire mean to Simons? Laissez faire meant neither a totally negative view of government nor that changing the law was ipso facto coercion – as it did to so many noninterventionists then and later. Laissez faire to Simons meant an economy that was in fact, and not only in ideology, both competitive and open to further competition; and in which government was an instrument of both promoting competition and negating inherited anti-competitive arrangements and promoting greater equality. Such a laissez faire was a truly radical doctrine. It meant monetary policy no longer in the hands of private commercial bankers. It meant the end of both the corporation and the labor union as institutions of aggressive economic power. It meant a conflict between the two foundations of a free market economy – private property and competition – and an activist government promoting the latter.

Government was important to Simons. It could be a potentially constructive or a potentially destructive tool. Denying "any justification for prevailing inequality in terms of personal dessert" (Simons, 1938, p. 18), Simons asserts,

> Taxation must affect the distribution of income, whether we will it so or not; and it is only sensible to face the question as to what kinds of effects are desirable (p. 18).

The effects themselves are partly a matter of arithmetic and partly of social psychology (p. 19).

John Davenport's obituary article (Davenport, 1946) centers on an apposite quotation from Simons:

> The representation of laissez-faire as a do-nothing policy is unfortunate and misleading. It is an obvious responsibility of the state under this policy to maintain the kind of legal and institutional framework within which competition can function effectively as an agency of control (Davenport, p. 116).

The reasoning behind this position was the following. Government was already, and necessarily, involved willy nilly in the economy. Doing nothing meant continuation of past uses of government for anti-competitive purposes. What is needed is not a quiet government, which only obfuscates its use for anti-competitive purposes, but an activist government that will promote competition in the present and in the future. His *A Positive Program for Laissez-Faire* and *Economic Policy for a Free Society* advance this position. J. Bradford DeLong (1990) presents an analysis in response to those who see Simons's *Positive Program* to be highly interventionist (on Simons in general and his interventionism in particular, see Stigler, 1988, pp. 20–22, 148–149); for example: "Much of his program was almost as harmonious with socialism as with private-enterprise capitalism" (p. 149) – to which, what Stigler says of the argument between Knight and Jacob Viner, over the psychological versus the lost-opportunity theory of cost, also applies: "dangerously close to an argument over words" (p. 22).

Simons wanted monetary policy in the hands of the central bank, a rule-following central bank. He wanted such fundamental reform of the corporation as an instrument of economic power that the changes identified by Berle and Means paled in comparison. He wanted to limit the power of labor unions with a view to both promote capital accumulation and enterprise, and the unorganized. Simons was a member of none of the three groups identified by Zingales and McCormack. He was, alas, a group of one.

In contrast, the Later Chicago School found that the market economy was inherently competitive, that business regulation interfered with markets, and that the only institution other than regulatory agencies requiring fundamental reform was the labor union. An activist program of reform, such as Simons's, was unnecessary and anathema – except for that part dealing with organized labor. It is not too much to say that Simons's program was hijacked and put to use by the defenders of the corporate system – led by Milton Friedman and his colleagues (David Martin (Martin, 1976) contrasts Simons with later Chicago economists). One does not find in Simons's work the proposition that corporate-management decisions are *per se* optimal.

Ostrander writes, in his December 31, 2002 memoir of appreciation, that Economics 360 "was in a way Simons' favorite course." This was "because he saw the income tax as the key instrument to create the Jeffersonian small-scale competitive and egalitarian society he espoused." This view has two points to it. One point is its Jeffersonianism. By the 1930s, many civic-minded thinkers had given up on it, on the grounds that it was doomed by business economies of scale, bigness in business driven by them and by will to power, and the growth of national markets and national problems. Simons was not one of them. The second point is its activist social constructivism. Simons did not feel that government should or could do nothing. Government inevitably provided the legal foundations of markets and business, and these foundations could promote either a plutocratic or a competitive and egalitarian society. The measures to that end included the income tax, rules over authority and, as Ostrander puts it, "100% distribution of corporate earnings (by taxing undistributed earning[s] at 100%) as a means of achieving corporate democracy and ultimate free enterprise." Simons's proposal for controlling business size was similar to Hayek's, which likewise called for a tax on retained earnings and also would have prohibited corporate ownership of the equity securities of other corporations. Little, seemingly, did Simons either appreciate how entrenched was plutocracy or anticipate how politicians would take advantage of, manipulate and reinforce public opinion's anti-tax mentality while – indeed, as a means of – promoting tax policies favorable to the upperpercentiles of income recipients and responding to critics as seeking to play the "class" card. Moreover, the failure of the conditions requisite for Jeffersonianism was matched by the need for jobs in an employee society and thus the power of jobs-provision arguments on behalf of big business.

More specifically, Ostrander, with the Preface to Simons's *Personal Income Taxation* (1938) in mind, says in his "Personal Appreciation" that "it clearly reflects his unfavorable reaction to the New Deal." But Simons was not easily pleased in these matters.

> After 12 years of Republican administrations – Harding, Coolidge, Hoover – Simons had viewed Republicans as only interested in high tariffs and other forms of protection or interference with his highest goal – competition. It took only a few years of Roosevelt's New Deal for Simons to be in full opposition, repudiating its aims and methods.

Nor in other matters was he willing to compromise; e.g. a proposal for less than a 100% tax on undistributed profits left him totally uninterested. Still, his reasoning is weighty: The tax "was not a revenue device; he did not expect any company to pay the tax. Its purpose" was to create a capital market

> *in order that* shareholders, and only shareholders, would determine which companies would grow, by re-attracting new capital investment from satisfied shareholders. In his view, only this would ensure realization of the kind of corporate democracy he sought.

In effect, he didn't trust either Management or Directors to meet the aims and rights of the owner/shareholders of the company.

> ... In the business world of the 1990s, ... Simons's view would have sounded bizarre, or worse. Yet, a world where many Managements made all decisions without even bothering to pay out any portion of earnings as dividends, may have been just as bizarre – and more dangerous.

That is a plutocracy. That people – business managers and politicians – seek to capture and use other people's money independent of the market (the full story is more complicated) should be no mystery. Two mysteries attach, however, to Simons's biography.

First, Blaug (1999, p. 1035), and Stein (1987, p. 333), give his middle name as Christopher; whereas Blaug (1986a, p. 225); Blaug (1986b, p. 786), *Index of Economic Journals*, e.g. Vol. II, p. 429 (American Economic Association, 1961), and many other sources give it as Calvert. Blaug seemingly follows Stein but does not remember why; Stein is now deceased. Simons's middle name is undoubtedly Calvert; but the source of "Christopher" is a mystery.

The second mystery concerns Simons's death in 1946. The obituaries by Davenport, Stigler ("Simons died in Chicago at the age of 46 of an accidental overdose of sleeping pills" (Stigler 1974a, b) and others speak only of death. But the explanation of death due to suicide was widely circulated. Ostrander's "Personal Appreciation" (which notes that he and Stigler entered the Chicago graduate program together in September 1933 and "must have taken both Knight's and Simons' courses together") reports that he learned from Herbert Stein in May 1948 that Simons's death "had been suicide" and queries, "Could Stigler have been protecting Simons' family by giving the best possible interpretation to the tragic event?"

(Simons was not among those remembered in Shils's collection of essays on important Chicago faculty members (Shils, 1991). The economists were Milton Friedman, Harry G. Johnson, Frank H. Knight, Tjalling C. Koopmans, George J. Stigler, and Jacob Viner. Nor were Charles O. Hardy, John U. Nef, or Chester W. Wright – others from whom Ostrander took courses, the notes from which will also be published in this annual).

ACKNOWLEDGMENTS

For help in preparing Simons's biography I am indebted to Taylor Ostrander, Mark Blaug, James Buchanan and Steven Medema for comments in an early draft and to Peter Boettke and Gordon Tullock for graciously replacing my temporarily but frustratingly misplaced copy of *Simons' Syllabus*. For help with Helen Hiett's biography I am indebted to Holly Flynn, Michele R. Giglio, and Kirk D. Johnson.

REFERENCES

American Economic Association (1961). *Index of economic journals* (Vol. II). Homewood, IL: Irwin.

Blaug, M. (1986a). *Great economists before Keynes*. Atlantic Highlands, NJ: Humanities Press.

Blaug, M. (1986b). *Who's who in economics* (2nd ed.). Cambridge, MA: MIT Press.

Blaug, M. (1999). *Who's who in economics* (3rd ed.). Northampton, MA: Edward Elgar.

Davenport, J. (1946). The testament of Henry Simons. *Fortune* (September), 116–119.

Davis, J. R. (1971). *The new economics and the old economists*. Ames, Iowa: Iowa State University Press.

DeLong, J. B. (1990). In defense of Henry Simons' standing as a classical liberal. *Cato Journal*, *9*(3,Winter), 601–618.

Hiett, H. (1944). *No matter where*. New York: E. P. Dutton.

Hiett, H. (1947). Return to the Soviet. In: *Deadline Delayed*. Overseas Press Club of America, New York: E. P. Dutton.

Martin, D. (1976). Industrial organization and reorganization. In: W. J. Samuels (Ed.), *The Chicago School of Political Economy* (pp. 295–310, Reprinted, with a new Introduction, New Brunswick, NJ: Transaction Books, 1993). East Lansing: Division of Research, Graduate School of Business Administration, Michigan State University.

McKerns, J. P. (Ed.) (1989). *Bibliographical dictionary of American journalism*. New York: Greenwood Press.

Shils, E. (Ed.) (1991). *Remembering the University of Chicago*. Chicago, IL: University of Chicago Press.

Simons, H. C. (1934). *A positive program for Laissez-Faire*. Chicago, IL: University of Chicago Press.

Simons, H. C. (1938). *Personal income taxation*. Chicago, IL: University of Chicago Press.

Simons, H. C. (1948). *Economic policy for a free society*. Chicago, IL: University of Chicago Press.

Simons, H. C. (1950). *Federal tax reform*. Chicago, IL: University of Chicago Press.

Simons, H. C. (1983). *The Simons' syllabus*. In: Gordon Tullock (Ed.). Fairfax, VA: Center for the Study of Public Choice, George Mason University.

Stein, H. (1987). Henry Christopher Simons. In: *The New Palgrave* (Vol. 4, pp. 333–335). New York: Stockton Press.

Stigler, G. J. (1974a). Henry Calvert Simons. In: *Dictionary of American Biography, Supplement 4: 1946–1950*. American Council of Learned Societies. New York: Scribner.

Stigler, G. J. (1974b). Henry Calvert Simons. *Journal of Law and Economics*, *17*(1), 1–5.

Stigler, G. J. (1988). *Memoirs of an unregulated economist*. New York: Basic Books.

F. TAYLOR OSTRANDER'S AND HELEN HIETT'S NOTES ON HENRY SIMONS'S COURSE ON PRICE THEORY IN A COMPETITIVE ECONOMY AND THE EFFECTS OF MONOPOLY, ECONOMICS 201, UNIVERSITY OF CHICAGO, 1934

Edited by Warren J. Samuels

As indicated in the introduction to this section, the notes taken by both Ostrander and Hiett are in the form of annotations to the text of the lengthy Syllabus distributed by Simons to his students in Economics 201. The annotations are also presented in such a manner as to require reference to and coordinate with the designated passages in *Simons' Syllabus*, edited by Gordon Tullock in 1983. The annotations were recorded by the two students as they heard Simons work his way through the material. It is impossible to say which annotations were due to their notes taken during Simons's lectures and which were due to the memory and mind of each student.

In the course taken by Ostrander and Hiett, the mimeographed Syllabus commenced with one and one-half pages of Assignments. The pagination system used below is in the form "pp. 3/1," indicating p. 3 of the Syllabus used by Ostrander and Hiett and p. 1 of Tullock's version. The difference is initially due to the assignments section. Only after the assignments section did the Syllabus proper – that

Documents from F. Taylor Ostrander
Research in the History of Economic Thought and Methodology, Volume 23-B, 207–229
Copyright © 2005 by Elsevier Ltd.
ISSN: 0743-4154/doi:10.1016/S0743-4154(05)23105-X

published by Tullock – commence with the "Preliminary" section. After some pages the difference widens, due to differences in coverage.

Ostrander's and Hiett's annotations are not the only source of interest in these materials. The Syllabus published by Tullock in 1983 was 53 pages long and dated from the period he was at Chicago, 1939–1943 and 1946–1947. The Syllabus used by Ostrander and Hiett runs, respectively, about 67 and 69 pages. Adjusting for font size and size of page, the Syllabus used in 1934 covers more ground than that of the later period(s). The substance of what was later excluded is of interest, as is also possibly changed formulations of theory. Kirk D. Johnson has prepared a comparison of the Syllabus used by Hiett and Ostrander with that used by Tullock; this comparison is included below in a separate document. Also provided below are the outlines of the two editions of the Syllabus.

If one considers *Simons' Syllabus* and the annotations as constituting Simons' version of price theory, it is clear that the modern neoclassical treatment of price theory – neoclassicism's fundamental paradigm – was in place by 1933–1934. Profoundly influenced by Alfred Marshall and Frank H. Knight, as well as by two generations of theorists since Marshall's first edition of 1890, this was the first stage of the consolidation and stabilization of price theory. The later contributions of John R. Hicks, Paul A. Samuelson and the post-World War II reformulation of price theory and general economics are largely absent (see E. Roy Weintraub, *Stabilizing Dynamics: Constructing Economic Knowledge*. New York: Cambridge University Press, 1991; and *How Economics Became a Mathematical Science*. Durham, NC: Duke University Press, 2002).

Certain passages are worthy of comment. Ostrander emphasized the following in the Syllabus:

> Every man his own economist vs. intellectual specialization and leadership. Economic theory abounds everywhere – is implicit in almost all arguments and decisions on matters of governmental policy. There is no question of whether we shall use economic theory, but only a problem of selecting and formulating the kind of generalizations which are conducive to sound judgment on, and insight into, practical issues (pp. 5/3).

One wonders, first, as to the basis on which such theory is selected and formulated; and second, whether the "sound judgment" is due to the theory or is the test of the theory itself. This is Simons's language, and it leaves much to be desired – especially since he also writes that positive analysis of the mechanics of a free-enterprise system "inevitably carries some suggestion of apology or justification" (pp. 5/3).

With regard to the assumption "That people as consumers know their own tastes and behave intelligently" in their purchases, one of Ostrander's three added comments reads, "– Problem of 'intelligent behavior' " (pp. 8/7). Here Simons follows Knight in raising the problem of intelligent action, in marked contrast with Late Chicago, which, in assuming "rational expectations," sought to rule out of order certain problems that would have interfered with the orderliness and optimality of market solutions.

As for Hiett's annotations, see below.

The course in Economic Theory at the University of Chicago – taught under varying course numbers – has had legendary status in the economics profession. It was and remains the make-or-break course in the curriculum. The doctoral candidate either mastered the material taught therein – the economic point of view, Chicago style, as a definition of reality, as a catechism of doctrines, as a set of the important questions, as a system of policy – to the satisfaction of the instructor or left the program. The material itself became price theory. The professors who taught the course have been the most eminent of the eminent, and have included Simons, Knight, Mints and Milton Friedman (Ostrander's notes from Knight's course are published in this volume; Glenn Johnson's notes from Friedman's course will be published in a subsequent volume of this annual).

One senses from the notes from Knight's, Simons's and Friedman's courses that the Chicago point of view is an all-encompassing, almost completely self-sufficient view of the economy. It is essentially a deductive system – assumptions and conclusions – and is driven by a distinctive policy position. What the Chicago system contains is what, in its view, the economy is all about; if its system of logic calls for something, that something is how the world is – or would be, if interfering factors, including government interference did not get in the way.

Several epistemological problems do get in the way. First, deduction yields validity, not necessarily truth (descriptive accuracy and/or correct explanation). Second, deduction is not independent of induction, or empiricism: facts are theory laden and theories are derived in part from selective readings of facts. Third, many of the terms used in theory are primitive, in the sense that they lack definition, enabling the reader or auditor to provide their own definition. The fact of these problems does not demonstrate the wrongheadedness of Chicago theory; it does help establish the epistemological limits of meaningfulness of that body of theory (see Warren J. Samuels, "The Roles of Theory in Economics," in Klein, P. A. (Ed.), *The Role of Theory*, Boston, MA: Kluwer, 1994, pp. 21–45; "Some Problems in the Use of Language in Economics," *Review of Political Economy*, 2001, Vol. 13, No. 1, pp. 91–100; and "Thorstein Veblen as Economic Theorist," in Warren J. Samuels, Willie Henderson, Kirk D. Johnson and Marianne Johnson, *Essays on the History of Economics*, London: Routledge, 2004, pp. 271–305).

ACKNOWLEDGMENT

I am indebted to Joyce Christie Trebing for translating Hiett's shorthand.

The following list of assignments is from both Ostrander's and Hiett's copies of the Syllabus.

ECONOMICS 201

MATERIALS AND PROBLEMS FOR CLASS DISCUSSION

ASSIGNMENTS

Indispensable Reading, first five weeks:

Henderson, H. D., <u>Supply and Demand</u> (New York: Harcourt, Brace and Co., 1922. $1.25).

This short treatise should provide a good review of previous work (it is among the materials for Social Science I). It should be read promptly, to renew acquaintance with the terminology and central propositions of economic theory; and the relevant chapters should be re-read later on, in connection with the class discussion of special topics.

Knight, F. H., in Syllabus and Selected Readings for Social Science II, pp. 125–250.

This is also a review assignment; but no other material is likely to prove more valuable in connection with the first part of this course.
The first section (pp. 125–137), on "Social Economic Organization and Its Five Primary Functions," should be read promptly, in connection with the class discussion of the first week.

Ely, R. T. et al., <u>Outlines of Economics</u>, 5th ed. (New York, 1930), Chapters IX, X, XI, XX, and Appendix A (The corresponding chapters in the 4th edition will serve equally well for this course). The first three of these chapters should be read as one assignment. The first part of Chapter XI deals with what are, from the point of view of this course, highly controversial questions. Chapter XX merits very careful study.

Gray, Alexander, <u>The Development of Economic Doctrine</u> (New York, 1931), Chapters III, V, and VI.

The chapters should acquaint students with the main ideas of the mercantilists, and of Hume, Adam Smith, Malthus, and Ricardo. Appendix A of the Ely book should be read in connection with this assignment.

Indispensable Reading, last five weeks:

> Robertson, D. H., Money, new edition revised (New York: Harcourt, Brace and Co., 1929. $1.25).

> This is an excellent, concise treatise by a leading English (Cambridge) economist. It should be studied with care, preferably in advance of class discussion of money and banking.

>> Ely, R. T. et al., Outlines of Economics, 5th ed. (New York, 1930), Chapters XIII to XVIII inclusive.

> These chapters also merit careful, deliberate study.

> Gregory, T. E., The Gold Standard and Its Future, 2nd (or 1st) ed., London (and New York), 1932.

> An unusually fine treatise, excellent for its fundamental analysis, and closely relevant to currently interesting and urgent problems.

> Optional Reading:

> Ely, R. T. et al., Outlines of Economics
> Gray, Alexander, The Development of Economic Doctrine
> Cassel, Gustav, Fundamental Thoughts on Economics
> Cassel, Gustav, The Theory of Social Economy (Barron translation), Book I and Book II
> Marshall, Alfred, Principles of Economics, 8th edition, especially Book V
> Hardy, Charles O., Credit Policies of the Federal Reserve System

OSTRANDER'S SUMMARY OF THE COURSE

Description of the Course, Assignments and Books, Methods.
Purposes – see p. 5, outline.

Definitions of Economics
  (Marshall)
  (Ely)
  Davenport
  Knight

Criticism of definitions
  (Knight)

– Historical origin of the subject

Our approach to the problem of defining
  – By progressive elimination: (see Knight 301 notes and Syllabus)
– Fundamental ingredients:

Traditional Content:
  1 – Analysis of <u>forces</u> governing <u>production, consumption and exchange</u> under <u>given</u> conditions (simplifying assumptions). Economic Statics and Dynamics.
  2 – Study of "<u>historical</u>" <u>changes</u> in the <u>givens</u>, of <u>resources, wants, and technology</u>.
  3 – <u>Methodology</u>: appraisal of the whole system of rules which define particular economic systems. Philosophical problems of its relation to other studies.
  4 – Problems of <u>Policy</u>: applied economics.
    a – Changing the <u>rules of the game</u>.
    b – Probable consequences of such changes.
    c – Economic measures for obtaining <u>specific</u> results (problem of policy).

Relation of Econ[omics] to Policy
  – L. Robbins – criticism
  – <u>Knight</u> – notes [Econ] 301 (302, 303)

Assumptions made in <u>this</u> study of the price system
  – i.e. <u>Institutional framework</u> assumed.
    – Matter of attitude toward, or evaluation of <u>these</u> institutions.
    – Matter of the relevance of the principles based on these assumptions to <u>other</u> institutional frameworks.
      – i.e. problem of the <u>universality</u> of our assumed institutions.
        – Degree of their spread necessary to a system based on them.

Basic Functions of Economic Organization
  – cf. Knight, <u>Syllabus</u>, and Outline summary
  – Transition to.

General Price Theory
  I – a – Methodology
    b – Utility theory
    c – General View; Economic <u>circle</u>

  II – Equilibrium
    a – Pricing process in short period
    b – Pricing process in long period
      (a) One resource, one commodity

  (b) One resource, many commodities
  (c) Many resources, one commodity
  (d) Many resources, many commodities
 c – <u>Conditions</u> of Long-run Equilibrium
 d – Interpretations of Equilibrium Arrangements
  (1) [words indecipherable]
  (2) Taxation added
  (3) Socialism or communism – political bureaucracy
 e – Interrelations

<u>Demand</u>
 1 – <u>Care in Usage</u> – def[inition]
 2 – <u>Relation to</u> Utility
  cf. <u>Risk, Uncertainty and Profit</u> – Chapter III; Syllabus
 3 – Elasticity – Charts
 4 – Substitution

Cost of Production and Price
 A – Competition
  1 – Problem: – charts, exercises; principles.
  2 – Addition of <u>Tax</u> – same problem.
  3 – Problem of <u>Ingredients</u> (productive services) under 1 and 2 (same problem).
   – Increasing cost, constant cost (Decreasing cost).
  4 – Joint Cost and Joint Supply.
  5 – Pricing of the Productive Services, i.e. Distribution: (a) assumptions; (b) table; (c) questions for class discussion; and (d) 3 problems (assumptions and table) for class.

B–Monopoly
 I – <u>Pure Monopoly</u>
  (1) – Assumptions of "pure monopoly."
  (2) – Contrasts with Free Competition.
  (3) – Diagrammatic and algebraic representation of the workings of pure monopoly. Price and production.
  (4) – The place of pure monopoly.

 II – Partial Monopoly – Cartel
  (1) – Assumptions made.
  (2) – Short run adjustment.

(3) – Long run adjustment – algebraic and diagrammatic representation – by stages (i.e. by number of plants).

(4) – Breakdown of cartel.

OSTRANDER'S ANNOTATIONS

pp. 3/1–2: The annotation converts three topics in which students should make special efforts to acquire facility: (a) the language of more rigorous economics, with the main terms and concepts; (b) the assumptions under which particular analytical arguments proceed; and (c) to digest the analysis of particular problems as class proceeds – into three purposes and adds a fourth; (d) "Criticism, in the light of above, of the large output of spurious econ[omics]."

pp. 4/2–3: The annotation notes the juxtaposition of several definitions of economics, Knight's criticism and Knight's own definition by "progressive elimination."

pp. 5/3: The annotation takes one form of emphasis used by Ostrander, namely, double vertical lines in the margin alongside the text. The emphasized material makes the following points: that academic economics is primarily useful "as a prophylactic against popular fallacies;" the need for economic analysis that is "internally coherent, relevant to real problems, and couched in terms which are carefully defined;" "[t]he ultimately practical character of all good theory;" "There is no question of whether we shall use economic theory, but only a problem of selecting and formulating the kind of generalizations which are conducive to sound judgment on, and insight into, practical issues;" "Analysis of the mechanics of a free-enterprise system need imply no invidious comparisons. . . . But mere description . . . and the implication that all is not chaos, inevitably carries some suggestion of apology or justification;" and "The principles applicable to the pricing process under a free-enterprise money economy are definitely relevant to problems under almost every form of economy, and are highly relevant to all systems (socialistic and communistic as well) which employ a money of account, which allow some freedom to [inserted in ink: some] consumers in the use of their purchasing power, and which do not merely assign [inserted in ink: all] people to jobs by conscription and explicit coercion."

pp. 6/4: Under the subsection on Valuation, relating in part to Knight's treatment of the functions of economic organization, Ostrander added that "The Price System = a phase of Society being organized," that "The Price System is the value-scale of economic organization," and that "Individual preference becomes a social choice – via Price System." In the next subsections, he added, "The Price System is the guide of production, of allocation," "Prices of Productive Services – efficiency within industry," and "The Price System is the mechanism of Distribution – at one time [and] over time."

pp. 8/7: With regard to the assumption "That people as consumers know their own tastes and behave intelligently" in their purchases, Ostrander added three comments:

–Consumers not the whole population – cf. MacGregor notes.
–Problem of error in knowing tastes, and in gratifying them.
–Problem of "intelligent behavior."

pp. 9/7: Apropos of the assumption of entrepreneur and factor-owner having "reasonably full knowledge," the annotations raise Knight's issue of risk versus uncertainty. Apropos of the assumption that resource supply per period remains unchanged, the phrase "and are fully employed" is added. Apropos of the assumption that merchandising activities and expenses are ignored, with its implied assumption that "resources are not employed in changing people's tastes (advertising)," the annotation reads, "no uncertainty."

pp. 9/8: A "short digression" on industrial fluctuations and unemployment is given, running almost a full printed page in length. An introductory paragraph in brackets is found on Ostrander's copy but omitted in Tullock's copy:

> One day at least will be devoted to analysis of the cycle phenomenon, – in order to indicate clearly the kind of maladjustments which are to be neglected in the current discussion of relative-price theory – and reserved for special treatment later in the course. This parenthetical discussion of cycles is intended to give students some feeling for the special subject-matter of the different sections or topics which follow, and for the kind of questions which are appropriate to each section. Especial effort will be made, in this digression, to undermine crude notions about "overproduction," and to make clear the nature and justification of the assumption, in price theory, of full employment of resources.

The next paragraph traces fluctuations in production and employment to changes in the relation between the level of product prices and the level of costs (notice the putative influence of Wesley C. Mitchell, among others). Ostrander writes at the beginning, "Cycles result from," in effect presuming a definition of cycles other than the language in Simons's notes.

One of the causal elements is the sensitivity of money circulation to changes in business earnings. The annotation reads: "Abstracting, for the present, from the existence of a banking system."

pp. 11/9: At the beginning of the section "The Pricing Process: EQUILIBRIUM," Ostrander's annotation is, "Discussion of Utility."

pp. 11/10: Alongside several propositions concerning the pricing process for a short period, the annotations indicate, in part, that the topic is one of "fixed supply;" that "all of this is true also of long run pricing;" that utility theory deals with "points in time" and "not over time;" and that "intelligent maximization is a necessary assumption."

pp. 13/11–12 and 51–52/13: Ostrander identifies: (a) the discussion of marginal productivity analysis on p. 13 (and p. 12 of the Tullock edition) as constituting "Distribution Theory;" and (b) the discussion commencing in the Ostrander-Hiett edition on p. 51 and "The Pricing of Productive Services" in the Tullock edition on p. 13 as "Distribution." The discussion on the pricing of productive services, which opens in the Tullock edition on p. 13 with "In an isolated community . . .," is found on p. 52 of the Ostrander-Hiett edition. However, the Ostrander-Hiett edition commences (p. 51) with two short paragraphs omitted from the Tullock edition.

THE PRICING OF PRODUCTIVE SERVICES

Some central principles, to be developed through discussion of a simple problem in an imaginary community where the pricing process has to do only with productive services.

> The problem is introduced at this point to avoid divorcing "price theory" from "distribution theory"; it should also serve to define the terms and clarify the propositions on pages 12–19 above. A more complicated variant of this problem will be introduced later on.

pp. 18/18: The section, taking one-third printed page, entitled "Complexity and Intricacy of the Inter-relations," has the following annotations: "example of disequilibrium and adjustment" and "historical change" alongside the second paragraph and "dynamic change" alongside the third, and final indented, paragraph.

pp. 25/26–27: At the end of page 25 of Ostrander's copy, following the section on demand elasticity and substitution, the following is printed (not an annotation): "THE NEXT PAGE IS NUMBERED 34." Ostrander's copy thus jumps to p. 34. No lacuna is indicated on Tullock's edition. Hiett's copy runs through what is p. 24 in Ostrander's copy but is unnumbered (along with pp. 22–23), lacks pp. 25–27, 29–33. Also, Ostrander's copy stops with p. 67; hers continues through pp. 67a–c, 68 and 69.

pp. 47/38: Ostrander's annotations apropos of the imposition of a tax read as follows:

> Specialization of resources – is a technical matter, character of adjustment
>   Mobility of resources – is a matter of <u>time</u> – <u>rate</u> of adjustment
>   – Are the same in short run.

HIETT'S ANNOTATIONS

Whether these annotations represent Simons's lectures or her understanding thereof, is unknown. At several points, a question mark is placed in the margin alongside text. Whether Hiett is questioning the analysis in the text or indicating a matter for discussion, or something else, is also unknown. The most dramatic annotation is:

> Thus with increase of population and fixity of quantity of land, rent of land constantly increases. Can determine rent at any population.

> We have here determined rent by residual process, of what is left over after wages are paid. This is ridiculous [Facing page 52].

Hiett uses, in a few cases, a version of shorthand which has had to be translated. The omitted annotations relate primarily to number problems.

Hiett's notes help fill a gap: Tullock was unable to find course materials on money, the topic of part of the last third of the course. Hiett's notes include, in her longhand, four pages entitled "Money and Banking" and one and one-half pages entitled "Monetary Theory." Although perhaps more than the usual level of caution needs to be exercised, these presumably reflect what Simons had to say. Also in her longhand are a one-page statement entitled "General Price Theory;" a one and one-half pages on several topics, with "Economics 201/Mints" at the top; a one page statement entitled "Equilibrium in the Long Run," and a one and one-half page document on productivity, dated "4–19–34" [This date suggests Hiett took this course in the Spring quarter of 1934 rather than the Summer quarter].

pp. 8–9/7–8: Apropos of the first assumption – that consumers know their own tastes and behave intelligently, Hiett writes, "Mints thinks about 99% of advertizing [sic] is social waste, i.e. does not increase the quantities of goods and services. Advertizing [sic] that genuinely gives information to the public about amounts and kinds of goods available has a purpose." The notes modify the third assumption, that consumer tastes remain unchanged, saying, "Merely for short-run analysis." Apropos of the tenth assumption, the notes read, "Value of money remains unchanged. Relative prices of things are shifting."

pp. 11/9–10: Just before the section entitled "The Pricing Process: EQUILIBRIUM," an inserted page reads,

> Marginal (incremental) utility.

> Essential concept in old "marginal" definition is that of difference in value. In our definition it is rather, interchangeability.

If any rationality can be assumed on part of consumer he will equilibrate his income in relation to incremental utility.

Current income is sufficient to take surpluses off the market; but other factors enter to keep this from always being the case.

pp. 11/10: Alongside the first two propositions under "The Pricing Process for a Short Period" – propositions about the equation of marginal utilities of last units and marginal utilities proportional to their respective prices – is found, "Equalization margin in consumption." On the back side of the page (like other annotations) is found:

Incremental utility is not the same to all consumers.

By equilibrium we mean the condition from which there is no tendency to change. As it exists we never arrive at a condition of economic equilibrium.

There is the possibility of changing amount of productive resources in long run as compared with short run analysis.

Just below the first condition of equilibrium (annotated thus: "short run) – incremental utility of dollar's work of each commodity must be equal for each consumer – is added: "If this were not true a shift in production would result, more of commodity with greater incremental utility being produced."

pp. 15/13: Alongside discussion of higher prices rendering some owners of productive services unable to sell, Hiett wrote "unemployment." On the back of this page is written:

Long run equilibrium requires full employment of all resources. Is incompatible with unemployment. Requires that equal increments of product must be equal in each industry.

pp. 16/15: On the back of the page presenting most of the discussion of the "many-product system," is the annotation:

In long run prices must equal costs or else resources will be diverted to those fields in which price is greater than cost.

"Viner – if you wiggle one atom in the universe all the others start in motion."

pp. 18/17–18: On the back of the page presenting all of the discussion of "Complexity and Intricacy of the Inter-relations" and the beginning of "Some Supplementary Remarks," are the following annotations:

(1) Demand (amount which consumers will take at a given price).

Other kind is [incomplete]

(2) Demand curve

Result of a change in tastes. Change either in position or shape.
Means that there is a certain relationship between price and quantity.

When price falls you do not have to have a change in tastes to increase the amount taken.

pp. 19/18–19: On the back of first full page of section entitled "Demand, Demand Functions, and Elasticity of Demand," is the following:

Demand from now on will mean the "demand curve."

No way you can pass from subjective element of decreasing utility to decreasing prices.

Can't get at incremental utility of one of a joint product. Not a simple dual relationship between utility and price.

Utility analysis is not valid as an argument for equalization of income in the community. Can't measure ratio of the utilities of the last dollars of the rich and the poor man.

Demand curve could be drawn from the utility curve.

Utility is a subjective rather than an objective notion.

Nature of Demand Curve

In addition to fact that demand curve is negatively inclined, what is the nature of the curve?

Unit Elasticity

If 1% change in price is accompanied by 1% change in amount taken we have unitary elasticity.

pp. 20/20: Apropos of discussion of elasticity of demand, Hiett notes, "Minus sign used because demand curve is negatively inclined."

pp. 21/20ff: Further annotations regarding elasticity:

If you decrease price by 1% and increase quantity by 1% get same product.

1a – violating rules because dealing with [blank]

Get different percentages because amounts of amount demanded is [sic] gradually decreasing.

Elasticity (small change) (small percentage)
$$\Delta q/q/\Delta p/p \quad q = 180 - 10P$$

Derivative of quantity with respect to price is a constant.

$$Dq / dp \cdot p / q = \text{elasticity} -10.\underline{40} / 40 = -1 \text{ (elasticity)}$$

Derivative = rate of change of q with respect to p; slope of curve.

pp. 22ff/21ff [diagrams on elasticity]:

Consider quantity as a function of price. Conventional in math to put independent variable on x axis and dependent variable on y axis.

Curve – small increase in price, up to certain point, make people spend more [Hiett insertion: "total"] when elasticity is less than one; horizontal bar shows unitary elasticity; then curve slopes downward as price increases cause decrease of expenditure.

First page deals only with price times quantity. Figure 2, total expenditure deals with both quantity and price, so curves are different.

[On Figs 3A and 3B, the vertical axis in both versions of the Syllabus is given as Total Revenue (Total Gross Revenue in Tullock's edition) but Hiett changed Total Revenue to Total Expenditure; not so Tullock]

Result of Tax

Amount people consume is reduced and they pay more for what they do consume. With this particular kind of demand curve total amount spent remains the same.

Results are different in short and in long run.

pp. 28/26: On back of p. 28:

Fixed expense – anything for which plant has to pay whether it is operating or not (Capital expenditure). Those which cannot be altered for a short period of time. In a long period of time these can be eradicated.

If can't meet average costs in long run, liquidate to divert part of resources into another line of activity.

In the long run average expense must be met by the price. If not there will be contraction of industry. Thus an average cost curve is not important in short run; it is very important in the long run.

Only way in which cost can influence price is through rate of output. [Repeated]

pp. 34/27: On back of p. 34:

If single firm lowers price, amount taken will increase greatly since industry is highly competitive.

For long run want to fulfill conditions – equilibrium.

[Reiteration of unitary elasticity]

Supply schedule can be obtained from marginal cost curve by determining what output is at each point.

Short run in one industry might be long run in another.

pp. 35/28: On back of p. 35:

Marginal expense

Tax imposed goes down to meet price

Elasticity high
Decline in output rapid. Rise in price slow.

Elasticity low
Small decline in output will bring price up rapidly

Small decline in output Large increase in marginal expence

Price does most of moving

In short run thing that is important is marginal cost.
pp. 36/29: Apropos of U-shaped cost curves: "Decreasing marginal cost curve at beginning is of no significance because marginal cost curves must increase.
pp. 37/30: On back of p. 37:
Result of tax per unit on commodity [shorthand unclear]. Industry's [shorthand unclear] costs of resources does not change as different quantities of resources are used. Would be true the less specialized the resources are to this one industry.

Result of tax:

1 – reduction in output of each plant left in existence
2 – reduction in number of plants

New curve will be flatter because each rise in price will bring less rise in output [diagram intending to illustrate same]

Imposition of tax on commodity produced of perfectly specialized ingredients will not effect [sic] the price of that product at all, because the ingredients could not be turn to any other use. Where unspecialized ingredients are used, tax will effect [sic] price to full extent of the tax.

Influences on an industry are said to be <u>forward shifting</u> so far as they affect the price of the produce and <u>backward shifting</u> so far as the[y] effect [sic] price of ingredients.

pp. 38/30: The eighth point emphasized is that "From the data on average expense, it is <u>impossible</u> to obtain a long-run supply curve." For whatever reason or purpose, Hiett wrote, "Why? Will be no fixed expenses."

pp. 42/33: Apropos of Syllabus problem regarding temporary stability of price, on back of p. 42, annotations make the following points: that "Consistency of this [levels of price and output] depends on whether or not product can be sold at that price," and that when "quantity which people are willing to take . . . is less than quantity offered so price will fall. Firms will not be meeting minimum average expense and more firms will bee eliminated from the industry."

pp. 43–44/35–36: Apropos of discussion of results of tax, on back of page 43, the annotations read:

> Can get stable equilibrium conditions by dividing amount that will be taken . . . by output of individual firm and thus get equilibrium price.

> In long run then in this case price will rise by full amount of tax and industry will be contracted accordingly.

> (10) Short run supply curve flattens as number of plants diminishes because each rise in price brings less rise in output.

With regard to question 12a, about equilibrium with demand elasticity greater than unity, on page 44/36, Hiett wrote:

> Industry would contract more quickly. Rise in price would be accomplished more quickly.

On back of page 44 she wrote:

> In our assumption only way you could get a change in price was through change in output. Now consider also change in cost of resources.

> Resources – any industry[,] are likely to be of all kinds.

> Thus you can't make general statements about the effect of a tax.

> Marginal costs increase because of diminishing returns.

pp. 49/39–40: On the back of page 49, the annotation is:

In competitive economy entrepreneur has to assume:
1 Price is fixed (elasticity of demand is infinite)
2 Price of resources is fixed

pp. 51–53/41–42: At this point the section entitled "The Pricing of Productive Services" is found in the Ostrander and Hiett copies of the Syllabus and not in the Tullock edition. Hiett's annotations include the following.
On back of page 51:

Thus with increase of population and fixity of quantity of land, rent of land constantly increases. Can determine rent at any population.

We have here determined rent by residual process, of what is left over after wages are paid. This is ridiculous.
On back of page 53:

If monopolist increases output price will change. He can't calculate any price remaining the same and bring his output up to it.

Must procede on entirely different principle. Monopolist achieves maximum profit by making cost equal additional revenue.

pp. 54/42: at the beginning of the section, "Monopoly and Monopoly Price," which is abbreviated in Tullock's edition, the annotation on the back of page 54 reads:

How find maximum for monopoly revenue. Largest when slopes of the two curves are sloping in same direction. [Insertion: "Calculus"] Will become maximum where addition to total revenue is equal to cost. [Insertion: "Algebra"]

p. 55: The annotation on the back of page 55, relating to an example:

Understand complete contrast between monopoly and competitive conditions. Under latter price would be $30. Under monopoly price more than doubles, quantity cut in two, total expenditure is increased.

p. 57 (back): Still within the expanded section on monopoly price and preceding the section on cartels, on the back of page 57, the annotation reads:

Monopolist in contrast to cartel, controls new investment in industry.

Cartel would be more wasteful than monopolies. Thus worth [sic: worst] form of organization.

If new investment in cartel brings no change in output, existing firms must control – wasteful. A monopoly can contract scale of operations. Cartel must accept number of plants as given.

p. 62 (back):

Gap in syllabus is transition to monopoly.

Suppose monopolist acquires all 1000 plants. He has to accept number of plants as given. He tries to establish largest differential between total cost and selling price. Must maximize his monopoly revenue. Must equate marginal revenue to cost.

[Numerical illustration]

Will probably alter the number of plants in time.

Will maximize profit by restricting output to point where marginal revenue is equal to incremental cost.

p. 64 (back):

No distinction in short run between monopoly and cartel. Both try to maximize monopoly revenue.

p. 65 (back):

Price will not fall to what it was under competitive conditions because there is much more investment in the industry.

This cartel kind of monopoly worse than real kind from social point of view, because monopoly can liquidate 500 of the 1000 plants while 500 more plants are added under cartel.

If cartel should break down, marginal cost would have to equal price.

This ends the annotations by Hiett on the mimeographed Syllabus used in the course in early 1934. The following notes by her can be assumed to derive from lectures.

Money and Banking

1 – Medium of exchange – efficient substitute for barter

2 – Measure of value

   Could have the two separately

3 – Standard of deferred payments

Federal Reserve System

   (Get organization from reading. Amendments of last year).

1913 – That they wanted an elastic bank note currency. Introduced Federal Reserve as central bank to hold ultimate reserves. Non-profit seeking institution probably will not loan up to the limit of its reserves. To extent that Federal Reserve bank loan [indecipherable] competing [indecipherable] member banks themselves.

How could gold standard operate with central banking system. Entire system would expand to extent that their cash permitted them, to limit of their reserves. This could preclude further expansion.

Before 1913 could count on banks lending up to full amount. Additional gold supplies immediately brought expansion.

Involves discretionary power in a large degree. Granting absence of decent monetary theories [indecipherable] part [indecipherable] present authorities, Federal Reserve system seems to be worse than the automatic system.

Reasons for Federal Reserve System

1 – elastic element
   – centralized control of reserves
      a – not necessarily meritorious in allowing for discretionary power

Don't need to reconstruct economic system to get a good monetary system [Placed within large parentheses for emphasis because it is one of Simons's typical comments, per Ostrander].

Federal Reserve – non-profit seeking.
Manner in which they operate to permit expansion on the part of member banks.
1 – Not make loan at all
2 – Increase reserves of member bank by rediscounting operation, i.e. so much money is deposited in reserve bank by member banks. ... This is selling commercial paper to Federal Reserve bank, thus increasing its own reserve and will thus be able to expand loans.

Only limit is
1 – exhaustion of Federal surplus reserves, or
2 – change of policy

Relation of hand to hand currency to deposit
   Open market negotiations
     1 – government bonds
     2 – bills of exchange [within brace: bankers acceptances, trade acceptance]

Open market purchase increases reserves of member bank, decreases reserve ratio with Federal Reserve Bank. Difference between this and rediscounting. Re[serve] d[eposits] can only be carried on between Federal Reserve Bank and member bank at set rate. Open market negotiations can be carried on with anybody.

In depression, buy in open market to increase reserves – member banks, permit them to lend more.

Selling in open market increases reserve ratio of Federal Reserve Bank. Restrictive policy.
Differences between England and America. Banking.
One central bank versus 12 connected ones.
Bank of England owned by private stockholders; Federal Reserve by member banks.
Bank of England will be permitted to issue notes only on deposit of one pound of bullion for every note issued. Is no such thing as issue without banking [sic: backing?]
Are 260,000,000 already out based on nothing but government security. Are no legally prescribed reserve[s].
Joint stock banks never borrow from Bank of England. Yet Bank of England is source of funds. Gives money to bill brokers when joint stock banks call in their loans to bill brokers.
Bank of England established 1694.

Monetary Theory
Important to distinguish two points of view:
   1 – short run
   2 – long run
Have not always been kept separate. Issues in one and the other are distinctly different.
Long Run
What determines the value of money in the long run?

Changing　　　　　　rising(orfalling)
Changed (different) high (or low)

We are here interested in factors that determine where the price level is fixed. Is it high or low?
Read Robertson on Index numbers.

10 billion of money in "circulation." Government decides to double this over night by doubling amount each person and business corporation has so each person will be in same position as it was before. Possible that price would double over night. If not, people who had more money could begin to buy more goods and price could go up.

Assume we have no paper money but only gold coins in circulation. ISOLATED ECONOMY – no trade connections. Free coinage of gold. Over period of four to five years discover new mines of gold that in five years doubles amount of gold available for monetary system.

Miners would take gold to mint for coinage and then buy more goods. Would demand both consumers and producers goods. Prices would go up. Producers would demand more ingredients, making their prices rise. More labor demanded so wages of labor would rise.

In gold standard country with rising prices, gold is peculiar because price of producing rise[s] but what you can get for [an] ounce of gold when taking it to the mint remains the same.

General Price Theory

Adjustments that must take place through the price system. Not only are equilibrations made in two particular industries by supply and demand, but price system also operates to bring about readjustments of total amount of resources put into all industries.

Adjustments do not take place always quickly and without friction. Free enterprise system is disrupted by:

1 – Monopoly
2 – Widespread depression
3 – Changes in value of money.

Free enterprise does not mean a laissez-faire system.

To have a free enterprise system we need some kinds of government interference. We should have fluidity of prices (system in which prices fluctuate rapidly with changes in conditions); yet when government steps in it attempts to fix prices.

Business depression results when competitive system fails to make adjustments; fails because of sticky prices of monopoly.

Price fixing could only operate to advantage in a completely static system.

Economics 201
Mints

Concept of capacity is an elastic one.

Distribution of income

Major problem of economics. Very small group in Egypt determined what should be done. Large group [indecipherable] wouldn't have opted for pyramids. Particular way each individual gets his income is through wages.

Provision for the Future

Conservation of resources.
Provision, technological improvement in the future.
Distinction between capital and capital goods.
Continuously increasing amount of capital available for productive purposes.

Conservation

Not generally provided for in system of free enterprise? Perhaps not, but when exhaustion approaches, price rises, and substitute is sought; wastes are reduced; more careful in use of commodity.

Economic system does not adequately do this. We are not apologizing for the present system. Question of adequacy causes difficulty, but economic system does conserve to a certain extent.

NRA [National Recovery Act, or Administration] good in so far as idea is used to spread work, but is no good in trying to get us out of the depression; reduction of hours of work and increase of hourly wage increases cost to producer and thus prices of commodities rise and consumer is in no better position than he was before. Have been many ways of getting around it.

Spreading work might conceivably increase purchasing, but total purchasing power isn't increased by giving one man's $40 salary to two $20 a week men. $40 man would have excess income spent on charity, taxes, etc.

Equilibrium in the Long Run

1 – Will assume fixity of individual tastes
2 – Assume the resources will remain fixed. Assume ownership of resources remains fixed.
    a – Makes no inquiry into changing population.

[3 –] Assume consumer spends his whole income.

Propositions

If you have fixed resources by adding to one [indecipherable] increase [indecipherable] value or product but by repeating the process the value added diminishes. For practical purposes [indecipherable] law of diminishing returns must set in very early.

Increment in Value to the Entrepreneur

To determine value of particular resource in an industry, keeping all other resources fixed, either add or take away an increment of the particular resource and then determine how much product has been increased or decreased.

This is procedure entrepreneurs follow.

Single Page, Dated 4–19–34

Productivity of a resource – Value of product increment-Amount by which the output could be increased or decreased by adding or subtracting a unit of resources.

How allocate for different productive resources? Product increment of equal expenditure in each resource brings equal results.

Product increment of one resource is to product increment of another as price of 1st is to price of 2nd.

If there are changes in the relative prices of resources, entrepreneur will take more of low priced than high priced; would alter his resources.

Entrepreneur will continue to add units of particular resource until the value of product increment equals cost. He will cut down use of any resource until point, reached where productivity value increment, equal to the price.

Will carry output until value of added unit equals cost of making it. If he stopped short of this point, he would be losing chances for profit and would not be a good business man.

NOTES ON THE 1933–1934 AND 1946 VERSIONS OF *SIMONS' SYLLABUS*

Kirk D. Johnson

The following notes are intended to identify differences existing between two versions of Henry Calvert Simons' Syllabus of lecture notes distributed to students in Economics 201: the edition used by Helen Hiett and F. Taylor Ostrander in 1933–1934 (hereafter Hiett, inasmuch as her copy was used) and the edition used by Gordon Tullock in 1946 and published by him in 1983 (hereafter Tullock).

Warren Samuels' initial charge to this author was to identify the changes rather than pointing to the similarities. This approach might have resulted in a document that led the reader to believe that the changes are greater than they were. The similarities between the 1933 and 1946 notes are quite striking. The larger significance of the two syllabi is that Simons gave practically the identical course, though not in all details. In considering the possible reasons for this degree of similarity, different plausible explanations arise. These include: (1) Simons delivered a course he intended to be nearly exactly the same. Through the process of developing the 1983 reprint, hand-written emphasis not directly attributable to Simons may have appeared with emphasis in this reprint. Not having direct access to the original with which Gordon Tullock worked, I do not know the answer to this issue; (2) The course was part of the department's curriculum and was a required course for students. The content may not have been entirely of Simons' choosing. It may have reflected general department expectations based on the course's role in the curriculum as worked out, possibly, by understandings among Simon, Knight and Viner. If emphasis on particular points was the predominant

Documents from F. Taylor Ostrander
Research in the History of Economic Thought and Methodology, Volume 23-B, 231–239
Copyright © 2005 by Elsevier Ltd.
ISSN: 0743-4154/doi:10.1016/S0743-4154(05)23106-1

form of editorial expression available to Simons, then, as we shall see below, some significant alterations appear, but still only alterations. Some wider set of decision-makers may be partly responsible for both the overall continuity and the changing details of these syllabi; and (3)The underlying course content remained strikingly similar: Introduction and definitions, General Price Theory, Demand, Cost of Production, and Monopoly and Monopoly Price remain from 1933 to 1946. They are, bluntly, the traditional stuff of which Microeconomics is made and still a large part of the curriculum 70 years later in 2003. The emphasis, de-emphasis, and choice of language underwent changes in the 13-year period between editions of the syllabus. Combine these with the changes in the supplementary appendices, and while the topics may have remained the same, the economics and its exposition did change somewhat.

The foregoing explanations of the relative continuity of the contents of the Syllabus and presumably the course can be summarized thusly: Simons did not intend to change very much, the Department (or group of theorists) did not intend to change very much, or the profession did not change very much. These are all arguments which can be turned to explain the changes that did occur; Simons did change, though perhaps unintentionally; or he changed while confronted with the inertia of a department and/or profession. In any case, the purpose of this paper is to examine the changes between the 1933 and 1946 (1983) editions of Simons' syllabus.

Several layers of comparison are necessitated by such differentiation. For example, Tullock contains two appendices absent in Hiett, and Hiett contains one missing from Tullock ("I have not been able to find the paper on money which he distributed" (p. vi)).

Some changes are explicit and unequivocal. Other changes are more or less objective, some being clearly subjective.

CHANGES IN SUBSTANTIVE CONTENT

Significant passages involve substantive changes in content. This is especially the case in the discussion of monopolies (Tullock, pp. 42–50), and the choice of problems for students to complete in preparation for class discussions. Many short segments appear in Tullock. Short sections of handwritten notes, especially for clarifying assumptions and recording short versions of answers to in-class questions, are found in Hiett. Finally, there are many differences in the placement of emphasis. All of these suggest several questions:

(1) The missing section on money is described by Tullock as "straightforward" (p. vi) and "Simons' position on money is well known" (p. vi). Is there another source for this paper? As we shall see, these notes display significant differences that might also exist in the discussion of money.

(2) The interpretive process of editing and reprinting invariably leads to alterations in the object subjected to individual filtering. Does Tullock's version reflect the exact choice of emphasis (through underlining, italics, and bold-face font) of Simons, or is there editorial interpretation occurring here? Hiett's mimeograph of the typed document reflects a very different set of demarcated concepts.

(3) Hiett's hand notes indicate a significant discussion of elasticity, both in application and as a modeling concept. Several problems are worked through, and equations are notated as explanations of the meaning of the relevant equations. Tullock's notes have a series of questions on *The Pricing of Productive Services* (p. 13), and make a different series of points regarding *Contrasts between Complete Monopoly and Perfect Competition* (pp. 42–47). Are these differences the result of students' notes regarding sections with which they had particular difficulties in their studies, or has Simons changed the course content during the intervening years?

In addition to editorial changes in spelling and grammar (Tullock, pp. 19, 20 and 40), the numerical example of pages 27–29 is presented later in Hiett. The following identifies changes in common content. Initial pagination is again from Tullock.

pp. 13–15: The pricing of productive services exercise through class discussion questions is absent in Hiett.

p. 16: The parenthetical explanation of the equimarginal concept is absent in Hiett. This paragraph does two things: it asserts that productive resources will be used on the margin in the most productive activities as measured by pecuniary reward, and individuals can violate this assumption if their preferences are to avoid undesirable industries and are willing to accept the resultant equilibrium. This is an interesting contradiction. Simons writes, "seek always to maximize their pecuniary incomes" and "are indifferent to any consideration other than price." This is followed by workers choosing "employment in industry *a* at lower wages to employment in industry *b*, then equilibrium will be reached . . ." This section will be discussed again later as Hiett's hand notes emphasize the impact of disequilibrating activities, while Tullock's emphasize the tendency toward the equimarginal conditions.

p. 19: Spelling change of "infinitessimal" (Hiett to "infinitesimal" (Tullock).

p. 20: Spelling change of "rotation" (Hiett) to "notation" (Tullock); and the addition of a qualifier ("roughly") to the numerical examples of the definitions of elastic and inelastic.

p. 31: *Preliminary exercises* question *b*: reads "What is the lowest price at which it will be more profitable to operate a plant than to suspend operation entirely?" In Hiett (1933–1934, p. 38) it reads: "What is the lowest price at which it will be profitable to operate one of these plants at all?" Although the problems are substantively identical, the appearance of a series of hand notes by Hiett, both in the lecture content and the problems section, make this change a curiosity. Why would Simons change the wording of this particular problem while leaving the rest of the discussion unchanged? Do Tullock's personal notes, if his edition reflects only the content of the mimeograph packet distributed by Simons, display the same series of comments and concerns regarding the presence of sunk costs and loss minimization? Is there a piece on sunk costs appearing about this time that might have garnered the extra attention during Hiett's time in the course that Tullock's class would not have experienced?

p. 32: In problems 2–5 (Tullock), $19 and the 14 and 15th units are $20 and the 16 and 17th units (Hiett).

p. 40: "removal of" is added to Tullock's version. Hiett has the correction made by hand. The omission was apparently intended to be corrected for later notes and was done by the time Tullock took the course.

pp. 42–47: The section *Monopoly and Monopoly Price* is significantly altered from Hiett's course notes:

The opening assumptions in Hiett include the uniqueness of the monopoly to only the market under examination. Tullock does not.

First discussion paragraph: Tullock assumes that firms in perfect competition recognize that price is always equal to marginal revenue with extra emphasis on marginal revenue. Additionally, the price elasticity of demand for a monopolist is less that infinity implying any change in quantity results in a change in price, therefore, price must exceed marginal revenue. In this discussion, firms will want to reduce quantities of output to increase the price. In Hiett, marginal revenue is generally not used; rather such terms as *increment of product value*, *value of product increment*, or *value of incremental output*. Hiett's hand notes indicate an application of the total revenue test as a means of demonstrating the effects of increased output on price, marginal, and total revenues.

The second discussion point in Tullock, and not in Hiett, the smallness of firms in perfect competition, results in potential differentiation in performance. Firms may have an identical output, but different production methods are made possible in this discussion. These small firms have no impact on input prices, while the monopolist does have price impact. This last point appears elsewhere in Hiett's notes.

The third discussion point in Tullock, and not in Hiett, is the *free* entry/exit of firms in perfectly competitive markets; but a monopolist is defined as preventing any transfer of capital into more productive uses.

p. 47: Emphasis in Tullock, and not Hiett, on the impregnable position of monopoly in the pure theory of the market structure.

p. 48: Perhaps the most interesting alteration in this section of the notes, Tullock has additional emphasis on the social losses from any restriction on the movement of resources. The restrictions are not specified as particular to monopolies, the sectional title of the course, but rather to any institution preventing the movement of capital.

p. 48: The text is changed from Hiett to include more usage of the marginal terminology. Additionally, in Tullock, the role of the group's profits is the point of emphasis in the very first sentence in the cartel discussion. In Hiett, these profits are incidentally mentioned later in the section.

p. 49: Tullock has cartel equilibrium divided into short run and long run components. In the short run, quantity is increased to the point where marginal revenues and marginal costs are equalized, and allocated to equalize marginal expenses between production facilities. In the long run, the cartel has no investment control implying a result similar to perfectly competitive markets with the presence of a tariff as a barrier to entry. Hiett also contains the reference to the tariff situation as analogous to the formation of a cartel.

Questions for Discussion changes: While the problems are different, they have the same fundamental characteristics, comparing and contrasting the differences in the market prices and quantity of output under the three market structures. One interesting difference of note is the language of the expectations from students' solutions. In Hiett, students are to approximate the particular numerical solutions. In Tullock, they are to solve for the exact numerical solutions, with an increased usage of the marginal terminology.

Two sets of explicit changes occurred between Hiett and Tullock. The first set is the increased usage of the marginalist language and the predilection for equilibrium solutions with determinacy. This is reflected both in the content of the notes, and in the expectations from students in their discussion questions for which they were to prepare responses. The second set is the increased formality of the definitions of theoretical constructs – the definition of perfect competition, pure monopoly, and the increased level of abstraction – through the removal of parenthetical discussions. No single change would have significantly altered the content of the notes, but the series of small changes in language and examples has changed the tone and content noticeably.

CHANGES OF EMPHASIS

Additional emphasis appearing in Tullock

p. 3: Increases emphasis on the role of explanation as justification.

p. 10: Reinforces individual(ism) and the uniqueness of outcomes from the consumer perspective. This will be quite different when the producer perspective is described.

p. 11: Inputs are variable and interchangeable, but outcomes are singular. Firms have some choice in the method by which they produce goods, but the level of output is exogenous.

p. 12: Producers in a market have no control over input prices, availability, mutability, output prices, competitive presence, or consumer choices.

p. 16: Lack of producer control is reinforced again.

p. 17: Individual consumers have choices and the word *tastes* highlights the ability to, and changing nature of, consumers' choices. The emphasis here reinforces the sovereignty of the consumer over markets.

p. 18: Emphasis appears to be used to differentiate a *key-term* from the remaining text.

The foregoing changes in emphasis appear to increase the role of consumer choice and the lack of choice and/or control confronting businesses. As part of the "general" material on pricing, this would indicate a change from a system with less determinate outcomes (as appearing in Hiett's notes) toward one in which producers have few (if any) choices. They are reduced to the "black-box" system in which the consumer has all the power in the market with their ability to direct personal resources into alternative goods.

p. 27: Lack of control in business; assumes away fixed costs and implies the predominance of marginal/variable importance; restates the smallness in the range of choices that confront businesses.

p. 33: Emphasis on the role of marginal costs; rewrites production decisions into marginal terms.

p. 38: "Will the rent of this land fall more or less, if *demand for the product of this second industry* is elastic or inelastic?" was "Will the rent of this land fall more or less, if demand for the product of this second industry is elastic or inelastic?" in Hiett, p. 48.

p. 39: Role of constancy confronting businesses, both fixed costs and their inability to change outcomes.

p. 40: Emphasis appears to be used to differentiate a definition of *joint-cost* from the remaining text.

p. 41: Emphasis appears to be used to differentiate a *key term* from the remaining text.

p. 42: Atypical nature of joint cost separates firms' decision making into individual commodity units at the margin.

p. 47: Emphasis appears to be used to differentiate a *key term* from the remaining text.

p. 48: Asserts the need for the lack of any form of control, the emphasis being careful to include social controls (this is discussed further in the section, Changes in Substantive Content, *supra*).

ADDITIONAL EMPHASIS APPEARING IN HIETT

p. 12: Mechanics of the marginal productivity equation when mathematical functions are italicized.

p. 17: emphasis on the lack of value in saying costs of production = price, or using prices as = costs of production "*and each proposition is, by itself, inadequate and misleading*"; the indeterminacy of their relationship is stressed "*Everything depends upon everything else.*"

p. 26: Question is emphasized, are commodity demand curves elastic because they are luxuries? Or is this coincidental? "*Is the demand for luxuries elastic?*"

p. 27: Hiett emphasis is on the individual firm being able to ignore their impact on the market while Tullock is on the smallness of the individual producer. Hiett has "*his*" emphasized, implying that individual producers do understand that if sufficient numbers of their peers in an industry also make changes to their output that the summation of their individual choices will have an effect on the market price. The fallacy of composition problem of moving from individual activity to market activity is emphasized in Hiett.

p. 28: Emphasis appears to be used to differentiate a *key term* from the remaining text.

p. 31: Hiett provides handwritten underlining for the profit maximization rule assumed in equilibrium "*Price must be equal to marginal cost in every plant.*"

p. 33: Handwritten emphasis by Hiett on the key term *capacity* and its suggested dependence on a very large number of factors, making it immeasurable, "*dependent both on the price of the product, and on the level of prices of everything (especially labor) which goes into production of that product?*"

p. 34: Shut-down rule emphasis by hand, accompanied by a significant amount of handwritten notes taken by Hiett on this concept and the conditions under which it works.

p. 35: Handwritten emphasis on the long-run move toward equilibrium as new production facilities will replace/augment old ones; this emphasis appears just prior to a series of class discussion questions on related material. It would be impossible to differentiate the underlining as an act of reminder for use in the problems needing to be answered, or as a representation of Hiett's perception of important concepts presented in Simons' lectures.

p. 37: *unspecialized* was emphasized in the description of factor inputs assumptions used to that point in the course.

SOME DISCUSSION POINTS REGARDING THE USE OF EMPHASIS THROUGH UNDERLINING, ITALICS, OR BOLD-FACE FONT

Emphasis in Tullock differentiates itself from that in Hiett in some very consistent ways. During the discussion of general pricing theory, the consumer's role as a chooser, capable of saying "yes" or "no" to all alternatives is increased. Simultaneously, businesses are without choices as they are trapped in an industry with prices and goods as exogenously determined variables. This results in an appearance of sympathy with the plight of zero-profit firms that can be replaced *ad infinitum*. During the production section of the course, greater emphasis on the role of marginalism is achieved by defining away fixed costs. Additionally, the role of constancy of operating conditions facing an industry makes individual firms unable to make choices. In an unusual inconsistency, the joint-cost definition is emphasized, but then its discarding as useless is also italicized. Emphasis in the monopoly section is on the definition of monopoly and the need to eliminating *all* constraints on business activity rather than simply monopoly power. All of the additional, non-definitional emphasis in Tullock increases the perception of the power of individuals to make choices and reduces the power of firms to influence market outcomes.

Additional emphasis in Hiett, in the notes on general pricing theory, is on the indeterminacy of the relationship between "objective" measures of costs and prices. In the production sections, her additional emphasis, missing in Tullock, is mostly geared to the definition and explanation of the shut-down rule. The additional appearance of emphasis in her copy of this section of the course is the relaxing of the unrealistic assumption of the uniformity of all inputs. The additional use of underlining in Hiett appears directed toward increasing the appearance of qualifiers (hedging), and the choice of identification of key terms and definitions for class discussion problems.

In their entirety, while the wording of the sections may be nearly identical, the specific emphasis within Tullock, relative to Hiett, gives an appearance of a system that is much more deterministic. Consumers are applying the equimarginal principle with greater control over their personal economic outcomes; while producers have less control through fewer choices and less market impact. Hiett makes less use emphasis. The presence of emphasis in Hiett, absent in Tullock, is largely due to *key terms*. Most places of content where Hiett has additional emphasis are on passages increasing the indeterminacy of markets and particularistic solutions (pp. 17, 27, 33 and 37). The sole exception to this is her handwritten emphasis of page 31.

CONCLUDING REMARKS

These two sets of notes have brought up several interesting questions. The few offered in the opening remarks are geared toward the general problems of interpretation in the history of economics. As documents pass through different sets of hands, the original author loses significant control over the ongoing dialogue. Here, the sets of hands are the same (Simons), but the period over which the content was delivered was long enough for changes in the perspective on the content to have occurred. This shows up in the choice of identifying different *key*, *important*, and/or *significant* concepts to the different groups of students. This could be an intentional movement toward a more deterministic view of economics, but it could also be a pedagogical device. His use of emphasis could merely be an instructor's way of alerting students to concepts overlooked by their predecessors in this course. If this latter is the case, then the perceived change in viewpoint would be the result of an instructor's pedagogical decision, or interpretive problem; the passage of time distances the reader from the author. Changes in the use of terminology are striking. The later notes make much more use of the language of marginalism. While the general concepts remain the same – firms maximize profits at the point where marginal costs and marginal revenues are equalized, and consumers maximize utility by equating the relative marginal benefits through incremental changes in their consumption bundle – the language used to describe these was changed. With this change, most appearances of hedging are lost, and students are expected to find particular solutions for discussion rather than approximating solutions.

As an interpretive exercise, studying the alterations between two sets of notes for the same course, delivered by the same professor, provides an interesting opportunity to examine the subtle ways in which the choice of language can fundamentally alter the presentation of what seems, or is, superficially the same. But if the initial appearance of a phenomenon exactly coincided with its explanation, the role of interpretation and the interpreter (historian) would be superfluous.

ACKNOWLEDGMENT

The author and the editor are grateful to F. Taylor Ostrander for comments on an earlier draft of this paper.

F. TAYLOR OSTRANDER'S NOTES ON HENRY SIMONS'S COURSE ON PUBLIC FINANCE, ECONOMICS 360, UNIVERSITY OF CHICAGO, 1933–1934

Edited by Warren J. Samuels

This is the second set of lecture notes from courses in public finance published in an archival volume in this series. Volume 19-C (2001) was entirely devoted to notes from lectures by E. R. A. Seligman at Columbia University. Two differences mark Seligman's lectures and the lectures by Henry C. Simons at Chicago, as reported below. Seligman seems to have been lecturing primarily to students in tax administration, hence he presented very little economic theory; whereas Simons was lecturing to graduate students in economics, and presented relatively more theory. Seligman did not refrain from some passing of judgment but his lectures were largely descriptive and non-judgmental; whereas Simons has no hesitation in presenting his own normative approach on various issues. These issues tended strongly to focus on inequality, tax justice, and progressivity.

Simons begins his lectures with a statement that is hardly credible 70 years later. He says that public expenditure is "not of much interest to the economist." Then we read the question, posed by Ostrander: "what of economic effects of government spending as against private spending? Simons, however, was correct: Public finance specialists in the U.S., until well after World War Two, paid negligible attention to the causes and consequences, the mechanisms, and the significance of pubic expenditure (interestingly, in Simons's discussion of sacrifice theories of taxation,

Documents from F. Taylor Ostrander
Research in the History of Economic Thought and Methodology, Volume 23-B, 241–301
Copyright © 2005 by Elsevier Ltd.
All rights of reproduction in any form reserved
ISSN: 0743-4154/doi:10.1016/S0743-4154(05)23107-3

incongruously, is found the statement, "Maximum utility of expenditure is the other part").

But Simons does have some material things to say about public expenditure. His first point is that the question of types of public expenditure should not be tied to the question of types of financing expenditures. Rather, he holds that "public expenditure from taxation vs from borrowing is entirely a matter of monetary policy and trade cycle policy." If one did not know better one would think that this was uttered after John Maynard Keynes's theory of compensatory fiscal policy or, for that matter, Abba Lerner's theory of functional finance, that is, for taxing and spending by government to be used to counter adverse macroeconomic developments. But this is 1933–1934 and the University of Chicago, where, contrary to likely expectations, contra-cyclical deficit financing was applauded (see J. Ronnie Davis, *The New Economics and the Old Economists*, Ames, Iowa: Iowa State University Press, 1971). Thus Simons says, some pages later, "As Keynes says, the only hope and weapon against cumulative deflation is the government deficit." The next recorded statement after this last one, we now know, is an exaggeration in one respect and wrong in another. The statement is, "The only thing which prevents universal bankruptcy and unemployment is the bankruptcy of the government." It is an exaggeration to say or imply that deficits – and, by implication, national debt – necessarily leads to bankruptcy; the evidence is that it does not. It is wrong to say that government bankruptcy will prevent "universal bankruptcy and unemployment" (in the private sector); most likely the failure of the government – whatever that means – would destroy the psychological, monetary and legal foundations of the economy.

The remainder of his initial recorded exposition (on the main point of the preceding paragraph) deals with monetary policy. That "monetary policy is inseparable from banking policy" is unsurprising in retrospect; but for Simons to apparently explicitly affirm both a "political element" in monetary policy and that "there are no accepted rules of monetary policy; rather, chaos," is remarkable. Seventy years later, notwithstanding the efforts of Milton Friedman's monetarism, the two statements remain true.

At the same point where Simons is recorded as seconding Keynes, that "the only hope and weapon against cumulative deflation is the government deficit," he makes two other noteworthy points. The first has to do with the cause of the business cycle: "Cycle is a problem of cost and price relationships as intensified by short-term monetary changes." It is not too much to say that this is the general Veblen-Mitchell theory. The second, also echoing Keynes, has to do with the locus of monetary-policy decision making:

–Governments have let the free, private banking system usurp their powers.

–Took note issue out of private hands after Civil War.

–Have not yet taken deposit-loan issue out of private hands.

In time, the Federal Reserve System will effectively do something like that through its control of bank reserves and interest rates.

The reader should be alert to the possibility, even likelihood, that Simons is using terms quite differently from their present-day definitions. "Depression" today refers to a particularly deep and long-lasting contraction; "recession," to a contraction that is of conventional depth and duration (9% and greater unemployment versus c. 6–9% unemployment); whereas he uses "depression" to signify the downward, contractionary phase of the business cycle. He uses deflation to refer to the same period (and inflation the reverse), whereas today deflation generally means a more or less sustained decrease (inflation: increase) in the general price level.

When Simons says, therefore, "Taxation is deflationary," he means, in today's language, that it is contractionary. Two points: First, a contraction may also be deflationary; i.e. a sustained decline in income, output, and employment may also be accompanied by a sustained fall in prices – but it need not. Second, while taxation *per se* is a leakage from spending, or from the funds available for spending, tax revenue may be spent by the government; the net fiscal effect depends on the relative magnitudes of taxation and government spending (also the relative private-sector and public-sector marginal propensity to spend – terms in use much later). Simons is recorded as saying that government "borrowing tends to absorb money that would be hoarded otherwise." That depends on how the debt is financed. Also that there is "no difference between expenditure based on borrowing, and printing paper." That, too, depends on how the debt is financed. And as for government borrowing, "at zero interest it is terrible!" but borrowing at $\frac{1}{2}$% interest is "all right!" Several economists (from across the ideological spectrum, including James M. Buchanan and A. Allan Schmid) later argued that government is paying interest to the banking system to whom it has, in effect, delegated the power to create money in the form of bank credit (deposits created in the process of lending). Whereas it could require commercial banks, as a condition of doing business as such, to hold stipulated levels of interest-free (zero-interest) government bonds. Simons recognized the delegation of power; to my knowledge the stipulation condition did not come up during his lifetime – but if it did, these statements suggest he would have disagreed, perhaps for the reasons given in his lecture, but, as seen in a later discussion, he did recognize non-interest bearing legal tender ("inflationary borrowing").

But Simons was not as complete a precursor of Keynes or Lerner as might seem from the foregoing:

–The ultimate absurdity is to resort to fiscal inflation in an attempt to bring prices
and costs in alignment again.
  –When we raise costs as well as prices – we have the same depression at an
  ever-increasing price level.
  –i.e. the absurdity of the debt-deflation theory of the cycle – the rise of prices
  would pay off old debt but we would still have the depression.

The political influences on the treasury are vitally opposed to the correct economic
course – usually the exact opposite course to the sensible one is taken.
  "Balancing for the cycle" – to have expenditure equal income without borrowing,
for the whole cycle phase is absurd – depends on exact prediction of the whole
course of the next cycle.
  Simons compares fiscal and monetary policies, one the product of legislative
bargaining and the other of a group of experts (as for monetary policy, it is hard to
envision Simons being concerned that it is "extremely expensive – buying bonds
at high prices" given the money-creating power of the central bank. He next says,
however, "The greater the 'loss', the more effective the policy"). His attention
then shifts to monetary policy and the coordination of monetary and fiscal policy.

–A satisfactory system would <u>have</u> to imply a close connection of fiscal and
monetary policy.
  –At present the other extreme is reached, with fiscal policy and monetary policy
  at opposite poles.
  –And a preposterous division of power.
  –Expenditure and taxation in control of legislature.
  –Monetary control in hands of Central Bank.

He does not specify how coordination is to be institutionalized, though, since one
could infer monetary policy to be the senior partner, he likely would have objected
to the Accord, a decade later, obligating the Federal Reserve System to support
Treasury low-interest financing of the war. He does say, however,

–The rules of fiscal policy are the rules of monetary policy.
  –Changes in excise taxes are most important to business conditions; income
  taxes least.

The meaning of the first sentence, however, is unclear: which set of rules is the
senior partner, to be adopted by the other? Which is to be constrained by the other?
His answer may be inferred from the next recorded sentence:

–We shall deal with the subject of sound fiscal policy under: (a) conditions of free
enterprise; and (b) <u>present</u> monetary policy.

Already striking is the degree of attention to monetary policy in a course in public finance. Not surprising is the limited attention to public expenditures as such: Simons was a specialist in taxation and public expenditure theory in the U.S. was largely undeveloped – as he says at the start of the course, government expenditure is "not of much interest to the economist." Nor was fiscal policy – the manipulation of government spending and taxation in pursuance of macroeconomic objectives – anything but rudimentary and idiosyncratic in both understanding and practice.

When Simons turns to the specific goals of monetary policy, we find them to be all the components of the equation of exchange, $MV = PT$, and thus the Quantity Theory, $P = (MV)/T$, with the exception of T (presumably tracked by output (O)). If successful, all adjustments will be through income, output and employment. Two further points: When he has the stabilization of M carried out by fiscal policy, he limits policy to debt-management policy. When he says, "Has greater tendency to stop tinkering with money," he engages in wishful thinking; in any event, one person's "tinkering" is another's effort at "sound money." Suggestive of the perceived potential for major systemic change or political instability if not revolution is Simons's introduction – apropos, one thinks, of the "conditions of free enterprise" – of the importance of keeping "the present system alive'' into the policy calculus.

One interesting aspect of Simons's diagnosis is his emphasis on the stickiness of wages and transport costs as a cause of what we now call the Great Depression. One of the reasons that is interesting is that among those who also emphasized sticky prices – including Keynes and, later, John R. Hicks – was Gardiner C. Means. Means later became a target of George J. Stigler, in part for that reason.

An interesting aspect of Simons's self-perceived role is suggested by an earlier statement, to wit: "Difference: – a limit in one case, not in other – so people think" but is more explicit in this one:

The Gold Standard was a religion and worked fairly well.

The problem now is to build up a new system and especially a new religion.

–The price–level stabilization religion ought to be more attractive than a neutral-money religion.

–It capitalizes a movement which is already well under way, and an opposition admirably suited to defeat! (Kemmerer, etc.).

This is sometimes called the "High Priest" role, the performance of latent function over manifest function – sociological terms for theorizing and acting on a systemic level and pretending to others that something else entirely is going one. Nonetheless, Simons seems to have had doubts about the pretense of "is" when the topic is one of "ought." Speaking of the ends of fiscal policy, he is recorded as arguing,

–Most writers on the problem keep up a <u>pretense</u> of complete objectivity, never admitting a discussion of ends.
–A completely hollow formula – it is impossible to discuss "sound" policy without seeing that the <u>whole</u> problem is really one of ends.

And ends are matters of "ought."

In these notes, Simons affirms progressive taxation:

–Under any <u>reasonable</u> assumption as to slope of these curves – this leads to <u>progressive</u> taxation.
  –Not only now – but <u>increasingly progressive</u> for all time to come. Few utility curves, like indignation, will continuously recede.

By comparison, Ostrander's notes for Economics 303, Current Tendencies, shows Frank Knight opposed to progressivity. What Simons finds a substantial base – ethics – Knight considers ambiguous, inconclusive and not dispositive of his objections. On the other hand, Knight would agree with Simons's statements,

–The long-run utility function is itself a function of the distribution of income.
  –e.g., man, utility of a $10,000 income will be high or low depending, in part, on how many other people there are at that income.
  –The short and long run marginal utility curves would be functions, likewise, of what happened to the rest of people.

I.e. relative, not absolute, income matters – in one word or another, position or rank.

Whatever one's view of equality and redistribution and the various bases thereof, one has to be impressed with Simons's neutral use of the "no free lunch" principle. Simons acknowledges that "progressive taxation will <u>of course</u> affect capital production and accumulation," adding that, "To say progressive taxation should stop before it touches production and accumulation is to say progressive taxation should stop before it becomes <u>effective</u>." His point is that "equality of distribution <u>does cost</u> something. Ethics and justice (and progress) are all costly, enormously costly." This is very different from those economists, and others, who act as if the mere mention of a cost is ipso facto sufficient to negate a policy proposal (see Warren J. Samuels and A. Allan Schmid, "The Concept of Cost in Economics," in Warren J. Samuels, Steven G. Medema and A. Allan Schmid, *The Economy as a Process of Valuation*. Lyme, NH: Edward Elgar, 1997, pp. 208–298).

As Knight, and likely Simons, might say, policy is the choice of which cost, whose cost, and whose immunity from which cost.

Simons's position seems close to Knight's on two points – the economy as a game [in the sense of play, not game theoretic] and the importance of status emulation and rank (on which both join Adam Smith). The two points are made, both stunningly, the latter inventively, with regard to entrepreneurial effort:

–Entrepreneurial effort:
  –Is mainly a game – based on prestige symbols.
  –Some prestige now attached to expenditure on glamorous goods, <u>could</u> be attached to publication of income tax lists, and place on <u>it</u> and based on mainly psychological benefits.

This is a very unconventional understanding of entrepreneurship. Seventy years-' later, after the explosion in managerial salaries and other access to private wealth (e.g. stock options), especially after noted corporate failures, Simons's lamentation is quaint:

–Community is squandering large sums on high salaries – which are not necessary to keep the system working – it might as well get some amount back in taxation.

Simons's heretical analysis does not stop there. Conventional economic analysis has had a complex view of capital accumulation. It has technically held that whatever is the equilibrium level of saving = investment, whatever the determining factors (the demand and supply of money, the supply of savings and the investment demand, the marginal efficiency of capital and the marginal propensity to save, the warranted rate of growth, etc.), is socially optimal. It also has often been understood to argue that more investment is better than less, and that investment can be financed through credit creation (all with ceteris paribus qualifications). Both those who intentionally or otherwise promote a savings-investment elite and those who are concerned with unemployment consequent to inadequate investment, tend to applaud "greater investment." The meaning, therefore, of the optimal level of investment/rate of accumulation varies.

Simons – as was common at the time – ties accumulation to the supply of savings but also says that the "rate of savings in a community depends on a great many <u>more factors</u> than the rate of interest." One of those factors is the "stability of [the] social system," and it "is <u>aided</u> by redistributive taxation." He goes on to say, therefore, that the "cause of justice may slow up the increases in accumulation some," adding, "though the government may become a saver to some extent and make up for this." If Keynes had in mind the partial "socialization of investment," Simons contemplated the partial "socialization" of saving – meaning by "socialization," in both cases, compensatory government action to achieve macroeconomic goals, but likely through different means.

That, however, is not all. Simons heretically moves, first, in the direction of the Keynesian revolution – that, given the level of investment, the level of saving may be too large – doing so by the argument that "saving may be a severe affliction – in times of depression [business-cycle downswing]." He then argues that the [progressive] "income tax may do a lot for ending this problem," even though "this does not yield a reliable and stable tax."

Secondly, Simons questions another orthodox doctrine: "Should society save so much for a distant future?" What is the obligation of the present generation? "Is it not enough to provide a decent ratio between population and resources – for the future? – and no need to work entirely on the resource side[?]."

These notes suggest that Henry Simons was much more radical in his economic theory and doctrine than we hitherto have thought. Not only that, but his radicalism took place within a largely otherwise conventional body of mainstream economics. If one combines the early Knight, dating from at least the early 1920s, with not only the Simons we knew from his published work but the Simons of these notes, the suggestion comes readily to hand that the first Chicago School was not only different and somewhat more radical than its famous successor, but *much more* radical than we have been led to believe; e.g. it involves much more than strong and effective antitrust enforcement. The more complete and precise dimensions of this radicalism remain to be studied, as do the questions of what brought about the change in what "Chicago" stood for, and why our knowledge of the change has taken so long to be revised. As for the latter two questions, was it mere change in personnel, was it the Cold-War-generated needs of social control and psychic balm coupled with a desire for economics to be "safe" in such a world, status emulation, and/or other factors, more likely in some changing combination? Is the "safety" status-emulation story to be supplemented by a status-emulation account of the genesis of mathematical formalism? Here the prestige of mathematics as a mode of doing science is such that mathematization is not a secondary feature but the basic reason for the creation and development of new forms of economic theory, such as general equilibrium theory (see Bruna Ingrao and Giorgio Israel, *The Invisible Hand: Economic Equilibrium in the History of Science*, Cambridge, MA: MIT Press, 1990, and various works by Philip Mirowski).

How is it possible that the group of brilliant students who attended Chicago during the mid-1930s and who became famous themselves after the war – their names are legion – have been so silent? It is especially important that the first Chicago School be studied as neither precursor to nor faint anticipation of the later "true" Chicago School but on its own terms as an independent phenomenon as if the later Chicago School did not exist.

Simons continues his discussion along the same heretical lines, taking on the system of property and the conventional understanding of what a market, capitalist, free enterprise system entails:

–Each generation inherits a certain system of property rights and certain mortgages.
–The retirement of the public debt out of progressive taxation diminishes the problem of its results on accumulation.
–Or the government might invest in private industry.
–Or provide free consumption service.
–Or go into business itself.

No Milton Friedman or George Stigler here.

Simons rejects any tax on land value to capture the unearned increment in value due to the growth of society and population pressure, as proposed by Henry George and others. He would, however, tax some land differentially under the aegis of having the owners pay for benefits received from government. In his view, "The whole property tax may be looked at as a tax of special benefit assessment – for the municipal expenditures benefit the whole municipality against another municipality."

Simons's analysis of the burden of a tax might seem incomplete. He argues first that "the real burden to the community of the tax is the less efficient use of resources in the new position," and "they will be less efficient uses, since we have assumed the community to have been in the most efficient position of production." The incompleteness lies in his seeming neglect of two things, first, that the efficiency in question is not productive efficiency – efficiency in the production of particular goods – but social or economic efficiency as between the production of different goods, and, second, that the tax is levied, presumably, to provide government with the purchasing power to acquire a particular but different good. The result is no different allocatively from the case where consumers change their preferences.

Another way of making the point refers to Simons's version of the argument that points to the adverse impact on someone: "Agricultural land is not a matter of capitalization, but of the return the farmer is able to get from the maintenance of his farm. The government takes a part of the return; he is no longer willing to put into the farm so much money, because of the low return." The decision to tax is presumably consequent to a decision to spend, i.e. to transfer resources from private spending to government spending. The logical necessity, at full employment, is for some area of private spending to be reduced. Simons's formulation tends to give effect to the common view that taxes are bad (he twice is recorded as speaking of "government confiscation" of some part of the equity) and to the neoclassical economist's view privileging (in this case) the pre-tax optimal allocation. The irony of the latter is that it gives effect to all pre-existing taxes and other actions of government – just as in the future when a new tax is added, the present new tax will recede from view and be a part of the given institutional background (*Vide* the old saw that the best tax is an old tax).

Simons also argues, a few pages later, "the burden of excise taxes is to the degree that people change their demand programs–diverting their expenditures from the lines which they really want." Three points: (1) To repeat, the purpose of taxation is to transfer purchasing power to government for its purchases, the necessary effect of which, in full employment conditions, is the reduction of consumption and investment; (2) People also change expenditures in response to many other variables, not least the prices of other goods; and (3) Saying "really want" creates a misleading if not wrong picture. On neoclassical reasoning, one would expect them to divert expenditures first from those *least* "really wanted," whatever "really want" may mean.

Immediately thereafter, Simons repeats his faulty earlier argument, saying both "less productive use of resources" and "altered allocation of resources is the ultimate real burden."

But, his linguistic formulations notwithstanding, Simons goes on in the earlier discussion to acknowledge the other half of the matter, the government expenditure. He queries, "what do we assume as to the effect of the kind of expenditure the government chooses to make with its tax-collected money?" But he seems only to have raised the point to make another, the "Justification of talking about taxation incidence without talking about expenditure incidence." His justification is simple and straightforward: "We are interested in the advantage of one tax as against another." But logical thinking requires the question be asked, "If that is your interest, then why use the ideological language of confiscation when it applies to both taxes being compared?" (It is, strictly speaking, clear that he is comparing two taxes).

The explanation of (excise) tax regressivity recorded in the notes is potentially misleading. The text starts out clearly and correctly. The lower income is taxed on its entirety because all of it is spent on taxed goods and the higher income is taxed on only the half spent on taxed goods. The total amount of tax increases but is regressive with regard to total income. The notes proceed to make regressivity a function of the generality of the tax. If everyone consumed the same proportion of their income in taxed goods, the fact that not all objects of expenditure are taxed would make no difference; the proportion of taxes paid to income would be proportional. But, as in the example given in the notes, items of purchase of personal service, more typically undertaken by higher income groups, are not taxed, so the proportion of income spent by them falls and with it the effective rate of tax with respect to total income. The omission is not inadvertent.

In the midst of an informative, partly administration oriented, discussion of the property tax, Simons is recorded as insisting on "Rules instead of judgment" as the basis of appraisal-assessment systems. The analytical and practical problem is that judgment nonetheless exists on two levels, or in two respects. First, there are multiple theories on critical aspects of real estate appraisal, e.g. valuation of corner parcels. Judgment, i.e. choice/discretion, exists in the adoption of a theory, in the instrumentation of a theory, and in the particular rules to be applied. Second, judgment, i.e. choice-discretion, exists in the application of the particular rules adopted. The appraisal situation is analogous to that of criminal law. The legislature determines what behavior is to be designated criminal, a decision(s) that often involves gradations and therefore subtleties (e.g. as to "intent" or degree) and therefore judgment; and the prosecutor exercises discretion in filing charges, a decision that often involves judgments as to the strength of the evidence. The similar situation occurs with rule-based monetary policy. The meaning of any law must include how it is enforced or administered, and therein resides judgment. Administration means choice and judgment.

One part of the recorded discussion of special assessments makes three strange points: (1) "The essential policy in the problem of special assessments is to avoid any change in anything that is well established." The specific objective of this dictum relevant to special assessments is unclear. Does it mean no new areas of special assessment may be added?; (2) "Many types of improvements should be taxed by special assessments but there is no theory that is really tenable for doing it." The theory, or rationale, is both well known and considered tenable – appropriate, just, and administrable – by many: the existence of a reasonably clearly identifiable group – class, in the legal sense – receiving a particular benefit from public expenditure. The argument against proliferation of special assessment districts may involve sufficiently high administrative costs as to outweigh the argument and maintain reliance on the general property tax (The idea of "a reasonably clearly identifiable group" is elastic but the practice is nonetheless tenable. Consider a new special assessment district for street lighting. It makes sense for a residential subdivision; transient drivers likely comprise a small percentage of the traffic, and there are non-traffic benefits accruing to residents. It likely does not make sense for the owners of businesses along a major thoroughfare; transient drivers likely comprise a large percentage of the traffic); and (3) "They are the only taxes that can save the world from the realtors." The motivation behind this dictum is not obvious.

Early in his discussion of the income tax, Simons is reported as noting, "Tax a response to demand for more equitable distribution of income." Several points: (1) That is correct, in part, especially when combined with a progressive rate structure and/or exclusion of low incomes from taxation by one device or another; (2) Taxation of income was also supported by those interests who wanted to shift the tax burden away from property; (3) A problem common to both taxes is that under the property tax some property owners have little income, and under the income tax some income receivers (of labor income) have little property; both are problems of ability to pay, the former also a matter of illiquidity of assets and the latter, of unequal distribution of wealth. A similar problem within the domain of income taxation is that of income in kind and income in money; and (4) Where transactions are conducted in literal barter terms – goods for goods – both income and sales taxation is almost impossible.

The notes define accumulation as "change of ownership within a period." This is mistaken: Accumulation is an increase in capital stock – the implicit definition in what follows; not a change of ownership.

Simons was a specialist in both monetary theory and income taxation. The notes record him saying, "Impossible to have a good income tax until there is a good monetary system; given a bad system, very little by way of recompense can be made by income tax." It is not clear what he means by this statement, beyond the contribution of "sound" monetary policy. Two very different institutions are involved. A monetary economy can do without an income tax but not a monetary system. One would not expect to remedy faults with the latter by tax provisions. Interest is hardly a fault, making it tax-deductible solves no monetary problem – that is his point, but who would have thought otherwise?

Simons knew – the 1930s was a major period in its development – that national income accounting involved definitions that were the results of numerous judgments as to inclusion and exclusion. No wonder he remarks, "Thus every bit of measurement of income is a <u>construction</u>."

Simons manifests considerable insight in his recorded statement, "But such rules-of-thumb are soon elevated into principles!" In a socially constructed world, once the construction is finished, the result tends to be treated as part of the natural order of things (this phenomenon is given the name, naturalistic fallacy, or reification), and the rules-of-thumb, etc. used in construction are given the status of principles (or natural laws). Similarly, particular definitions of income and other terms depend upon, embody, and give effect to particular theories – of income, etc.

Simons treats certain problems as either non-economic, arising outside of the economy and therefore of economics, and/or as beyond the ability of economists to handle. Because, for example, the rate of saving is grounded in people's personal philosophy and psychology, he assumes it to be fixed – in effect, parametric to what transpires in the economy. That "some people save, instead of consuming," he says, is something about which "the economist is hardly capable of deciding why? . . . it is an institutional, philosophical, psychological problem." Moreover, "using the utility principle to explain it, is only a confusion of the issue." But this discussion takes place with regard to the interest rate, and Simons's treatment of that, both in retrospect and as of the state of thought at the time, leaves much to be desired.

Simons is also recorded as saying that "interest theory is merely superimposed on price theory, to bring out certain important relations." To some extent this is true: given the elements of a theory, the analyst can seek to determine what those elements further imply. But it is also, and perhaps predominantly, the case that a theory is constructed on the basis of certain assumed relations. Thus, given alternative models of interest-rate determination – the interest rate as a function of the demand and supply of money (or of liquid assets), or of investment demand and saving supply, or of liquidity preference *vis-à-vis* the marginal efficiency of capital, and so on – , the implications of each can be worked out. But the more important step is the antecedent one, namely, the choice of model of interest-rate determination.

One of the techniques of defending a theory – formulated in part by Imre Lakatos – is to create defenses for its "hard core" propositions. One of those defenses is to assume away all the factors or variables that interfere with what one wants the theory to "prove." Thus, for example, during the 19th and 20th centuries believers in Say's Law erected a series of assumptions the effect of which was to insulate the theory from negative developments. The so-called Coase Theorem, actually the formulation of George J. Stigler, invoked zero transaction costs in order to generate the proposition that the assignment of rights is allocatively neutral. The irony here is that Coase eventually insisted that he had been trying to show the opposite, namely, that the world does not have zero transaction costs and that (the assignment of) rights and other institutions matter. Interestingly, Simons gets involved in the same type of model building:

Income, in the sense we have outlined it, would be almost impossible to use as a basis of a tax.

–Except in a world with costless transportation, indestructible factors, heterogeneous properties, and a price on everything.
–If no one owned any property, and everyone rented or borrowed, it would be easier.

Apropos of Stigler and Simons before him – Stigler having been a student of Simons's – there seems to be something about Chicago Economics-Department thinking (Coase is in the Law School) that leads its protagonists to such an approach or situation.

Actually more is involved than the defense of particular theories. It is one aspect of theory to abstract from some variables in order to concentrate on others and/or to keep the analysis manageable. But as with the choice of model of interest-rate determination, which is but one example, the choice of included versus excluded variables is key; one can prove – as a matter of logic, not necessarily a matter of truth – *any* proposition, through the careful choice of included variables.

Simons, no less than Frank Knight, was prepared to be candid in his criticism, unrestrained in his desire to say something like, "The whole thing was absurd," as he is quoted, for example, in his discussion of the income tax treatment of capital gains and losses – the *ninth* time he is recorded using the term. Simons was, therefore, like most other professional economic specialists, willing and able to pass judgment on the work of the legislature, if not also the courts. This represents the tendency for tension between professional opinion and representative democracy – two sides neither of which are of one voice on any issue (A half-century later some economists hold that government decision making could not yield an optimal solution to a problem, whereas others think it could).

On the other hand, Simons, disappointingly, let certain matters drop. Several times he says that "a very good case" for something can be made – and then does not even summarize the case he has in mind. Undoubtedly the readings contain relevant material and even more certainly Simons did not contemplate our listening in to his lectures through a student's notes, but surely this was an opportunity to show students how his mind worked and to engage in discussion.

On two other, different points: Simons exhibits much implicit belief in government and politicians as benevolently seeking the public interest. Simons perhaps exhibits not enough belief in Harold M. Groves's dictum, which he expressed in Quaker terms, that people's motivation is to shift taxes "from me to thee." In these respects, Knight had less faith in the applicability and adoptability of rational proposals made by professional economists.

One of Simons's proposals has been made by several people, not least by Friedrich von Hayek. The idea is to replace the capital market internal to the firm with the capital market encompassing putatively all firms.

It might be possible to pass a law that all corporations distribute 100% of earnings as dividends.

–Finance all new capital out of new issues.

Like Hayek, Simons also supported limits on corporate holdings of the securities of other corporations.

These lectures touch on a great many topics. The point is not that they are not public finance but that public finance is inextricably enmeshed in the other institutions of society.

F. TAYLOR OSTRANDER'S NOTES ON HENRY SIMONS'S COURSE ON PUBLIC FINANCE, ECONOMICS 360, UNIVERSITY OF CHICAGO, 1933–1934

ECON 360   Public Finance   Simons

Expenditure – not of much interest to the economist (? [F.T.O. QUESTION] – what of economic effects of <u>government</u> spending as against private spending?)

Justice

Fiscal policy and the Monetary and Trade Cycle Policy

  –Old question – what kinds of expenditure should be financed out of taxation, what <u>kinds</u> out of borrowing.

    –Simons finds no sense in the question, put thus.

    –Although most treatises do draw such a distinction.

  –New formulation: public expenditure from taxation <u>vs</u> from borrowing is entirely a matter of monetary policy and trade cycle policy.

    –But monetary policy is inseparable from banking policy (political element).

    –And there <u>are</u> <u>no</u> accepted rules of monetary policy; rather, chaos.

[Bibliography:]

  Standard Texts:

    <u>Schultz</u>, (best) [William J. Shultz, *American Public Finance and Taxation*, 1931]

    Lutz

      –Useful for information, not inspiring, on intellectual[ly] <u>low</u> level.

      –Omit <u>Incidence</u>, skim over <u>Distribution</u>.

    Pigou, <u>Public Finance–Distribution</u> – intellectually respectable.

    Bastable – Public Finance, last edition, 1903

    <u>The Cost of Government in the United States</u>, N. I. C. B. [National Industrial Conference Board] – yearly

    <u>Recent Social Trends–Taxation</u>

    Bulloch [Charles Jesse Bullock], <u>Readings in Public Finance</u>, Chaps. 1–3.

    –Rate of increase of public expenditure, proportions within distribution, and proportionate increases within distribution.

    –Most things dealt with under heading "Expenditures" should be dealt with later under Taxation.

    –Read <u>data</u> on expenditures with suspicion, e.g. government expenditure, 1918, based on borrowings – borrowings repaid out of taxes, 192(2) – same item.

– In 1918, a <u>real</u> expenditure; goods and services were actually consumed.

– In 192(2) borrowings repaid out of taxes, meant only the <u>transfer</u> of funds from taxpayers' pockets to bond holders' pockets.
– Most estimates of "war cost" make the double count!
– Thus it is better to look at data for taxation, than for expenditure (? [F.T.O.]).
[Bibliography:]
<u>Commerce Clearing House</u>, Annual publication on taxation
Simons's <u>Memorandum on Banking Reform</u>

The Cycle – Simons's interpretation views in terms of <u>banking</u> and <u>monopoly</u>.

(1) – <u>Cumulative</u> maladjustments in the system.
(2) – Stickiness, inflexibility of the price structure due to
  (a) Plain lag
  (b) Monopoly
    – Trade unions – stigma attached to lowering of wages, no stigma attached to laying off men.
    – Where competition is moribund or over[ly] "polite."
    – Public utilities – inertia
      "This is a depression of wage rates and freight rates" – C[harles] O. Hardy.
        – Prices which largely govern cost are <u>sticky</u>.
        – (a) Thus only income <u>over</u> costs tends to continue or (b) any loss of income below costs tends to continue.
          – a – Induces <u>dishoarding, increased velocity, increased investment</u>
            – until costs catch up.
          – b – Then a minor jar to confidence and costs catch up with income
            – thus they pass income – hoarding, falling off in investment, drop in velocity.
(3) – In modern <u>banking system</u> – cumulative changes are likely to be extremely severe.
  – In addition to hoarding and velocity change.
  – There is a change in the quantity of "effective money."
  a – Larger earnings bring increased loans, no bank in the whole expanding system <u>loses</u> reserves.
    – Dishoarding brings increased deposits.
    – Much profitable investment – banks unload government low-return bonds, getting more cash, which they put in bank – increasing deposits.

- When Federal Reserve Banks <u>buy</u> government bonds, from private holders – who regard them as nearly money – not circulating but sound – putting money into their hands, which they put in banks, increasing deposits again.
- Is the whole thing in the hands of Central Bank Board and its policy?
  - Simons says, so long as their action is <u>discretionary</u>, no Central Bank Board will oppose expansion – or either its powers or its make-up would be changed.
  - The check to expansion must be exercised by the Central Bank Board as a result of <u>legislative mandate</u>.
- Thus there is a substantial opportunity for increased loan circulation.

b – In the downswing, no one bank can gain at the expense of others unless it deflates faster – swift deflation inevitable.
- In a country like this – with sticky <u>wage</u> and <u>freight</u> rates and expansive bank credit – there is no <u>necessary</u> limit to inflation once it is begun.
- The only correction to keep income and rates at a par.
- One of the most important steps in this direction is to abolish bank credit [Single vertical line alongside in margin].
- b – In downswing, banks become insolvent, many close – government insures.
- Cycle is a problem of cost and price relationships as intensified by short-term monetary changes.
- Governments have let the free, private banking system usurp their powers.
  - Took note issue out of private hands after Civil War.
  - Have not yet taken deposit-loan issue out of private hands.
- As Keynes says, the only hope and weapon against <u>cumulative</u> deflation is the <u>government</u> <u>deficit</u>.
  - The <u>only</u> thing which <u>prevents universal</u> bankruptcy and unemployment is the bankruptcy of the government.

Taxation is deflationary.

Expenditure offset by taxation is an offset, neither inflationary nor deflationary.

Expenditure without taxation is inflationary.

- Borrowing tends to absorb money that would be hoarded otherwise.
- No difference between expenditure based on borrowing, and printing paper.
  - So long as government borrows at zero interest it is terrible!
  - If it borrows at 1/2% interest it is all right!
  [Double vertical lines in margin alongside preceding two lines].
  - 1– Difference: – a <u>limit</u> in one case, not in other – so people think.

- 2 – Difference: – what of effects of the alternative systems on redistribution of income – possibly different.
  - Simons says this is important in wartime (high interest), not when government can borrow at 1/2%.
  - Hard money people say: you can't stop inflation when you start it; you can't start it – because business men get scared.

The behavior of agriculture in depression is <u>ideal</u> – the behavior of industry and transport is wrong, and results in agriculture's bearing the brunt.

  - ! – The A. A. A. [Agricultural Adjustment Act] thus decides to make agriculture behave like other industry!!
  - Rigidity <u>does</u> break down in depression.
  - The ultimate absurdity is to resort to fiscal inflation in an attempt to bring prices and costs in alignment again.
    - When we raise costs as well as prices – we have the same depression at an ever-increasing price level.
    - i.e. the absurdity of the debt-deflation theory of the cycle. – The rise of prices would pay off old debt but we would still have the depression.

The political influences on the Treasury are vitally opposed to the correct economic course – usually the exact opposite course to the sensible one is taken.

"Balancing for the cycle" – to have expenditure equal income without borrowing, for the whole cycle phase is absurd – depends on exact prediction of the whole course of the next cycle.

  - An example of the common fallacy of formulating policy in terms of the objective.
  - e.g. [Rexford Guy] Tugwell – having determined the objective, thinks the problem of formulating the means is completed.
  - We try to think of fiscal systems and policy, without having defined for ourselves <u>any monetary policy</u> – result: chaos.
    - When government controls <u>all</u> money – then <u>budget balancing</u> <u>is</u> the stable money policy – M is constant.
    - Stabilizing MV means a changing price level.

Monetary policy is long-run planning.
  - Fiscal policy is usually short-run and the product of legislative halls.
  - Monetary policy as the product of a Central Bank.
    - Is extremely expensive – buying bonds at high prices.
    - The greater the "loss," the more effective the policy.
  - A <u>satisfactory</u> system would <u>have</u> to imply a close connection of fiscal and monetary policy [Single vertical line in margin alongside this point].

– At present the other extreme is reached, with fiscal policy and monetary policy at opposite poles.
  – And a preposterous division of power.
    – Expenditure and taxation in control of legislature.
    – Monetary control in hands of Central Bank.
  – The rules of fiscal policy are the rules of monetary policy.
    – Changes in excise taxes are most important to business conditions; income taxes least.
  – We shall deal with the subject of sound fiscal policy under: (a) conditions of free enterprise; and (b) present monetary policy.

Goals of Monetary Policy
  (– Stabilization of quantity of money – carried out by fiscal policy.
  (– Stabilization of turnover of money (MV).
  (– Stabilization of price level (index).
    – Gives reconciliation of short run and long run policy.
    – On the basis of some price index.
    – Has greater tendency to stop tinkering with money.
  "Price level" doesn't mean much over a period of more than four or five years.
    – Long-run policy isn't of much use in present-day conditions.
    – It is more important to keep the present system alive.
    – So long as the index is highly inclusive, it won't be so bad.

The Gold Standard was a religion and worked fairly well.
The problem now is to build up a new system and especially a new religion.

– The price level stabilization religion ought to be more attractive than a neutral-money religion.
– It capitalizes [on] a movement which is already well under way, and an opposition admirably suited to defeat! (Kemmerer, etc.).

Monetary policy is only secondary to the main problem of eliminating stickiness of prices.

– Which is primarily a problem of altering corporation laws, trust busting, a problem of monopoly.
– Rigid distinction between ordinary corporations and investment trusts.
– Limit the size of corporations.
– i.e. enforce competition everywhere it can be done, – and put monopolies under government control.
– Labor monopolies might be dealt with by the establishment of employment exchanges.

Money and banking proposals – even if you accept some necessary reforms – they are not inclusive (problem of saving banking, short-term borrowing, problem of maturities) and there remains the problem of getting them in.

- And the problem of <u>relation of monetary reforms to Treasury</u> – i.e. to government finance; borrowing.
    - Short-term debt – next thing to currency.
    - Long-term debt – least inflationary.
    - Non-interest-bearing legal tender – inflationary.
- Tax structure:
    - Increase of excise taxes – deflationary.
    - Decrease of excise taxes – inflationary.
- i.e. necessity of cooperation between Treasury and Monetary Authority – so that their policies do not conflict, and so that their means of carrying out policy do not conflict with policy.
- Any discussion of public finance except in these terms is <u>very</u> shallow [Double vertical lines alongside this point].

<u>Problem of Justice</u> – Any discussion of <u>sound</u> fiscal policy is a kind of propaganda – and depends on a discussion of <u>ends</u> of policy.

- This is a subject never rigorously studied.
- But means of achieving ends – i.e. fiscal policy, <u>are going</u> to favor <u>some</u> ends
- it is best to discuss the problem.
    - Most writers on the problem keep up a <u>pretense</u> of complete objectivity, never admitting a discussion of ends.
    - A completely hollow formula – it is impossible to discuss "sound" policy without seeing that the <u>whole</u> problem is really one of ends.
    - We discuss Justice – a confession of faith.
        - Once stable government and freedom appear – justice in fiscal policy comes to the fore.
    - Main problem – <u>how should the burden be allocated</u>? Taxation [Double vertical lines in margin alongside this line].
- a – <u>According to benefit</u> – to what he receives.
    - Is now of only historical interest.
    - 18th century – nobles and clergy escaped taxation, "benefit taxation" became the cry of liberal reform.
    - 19th century – the standby of a reactionary policy of "laissez-faire" econ[omics],
        - Is sometimes upheld by courts.
    - Is an emotional appeal – demanding the reform of escapes and holes in the tax system, or demanding an end to democracy in taxation.

– It <u>defines</u> <u>no</u> rule of allocation.

– As to national defense – what is criterion of benefit received?

– As to subsidies.

– It says nothing to the objection that different forms of taxation and expenditure <u>do</u> affect different groups unequally.

<u>Personal Taxation</u>

– <u>Ad rem</u> taxation should form <u>some</u> part of total system.

   – Land taxes; as a result of their long history, there is a presumption against <u>change</u> – all buyers and all sellers have acted on assumption that present system will last.

   – Gasoline tax – <u>special benefit</u> <u>is</u> given to some groups by government, and they are made to pay for it by gasoline tax.

   Special assessment.

– Only after you decide how much should be taxes in this way, can you go on to talk about the balance (Point of view of questioning traditional utility satisfaction concepts).

   (1) – Taxation of a kind to <u>equalize</u> sacrifice.

      – <u>Mill</u>, says that community with equal sacrifice for each, has a total <u>minimum</u> sacrifice.

      [In margin: Loss of utility due to less income, the same for every person].

      – Edgeworth disproved this – showed that minimum sacrifice of the whole depended on equal sacrifice of the marginal group.

   (2) – Taxation according to <u>proportional sacrifice</u> – Seligman.

      – Sacrifice shall be proportional to total utility of income. [In margin: Loss of utility shall be proportional to the total utility of income].

[Four diagrams, each with income on horizontal axis; and the first with marginal utility per unit of income on vertical axis, the other three with marginal utility. Each has downward sloping (unlabelled) marginal utility curve. First two diagrams each has pair of lines from curve perpendicular to horizontal axis, the areas between them darkened. Alongside first diagram: "<u>Equal sacrifice</u> – utility areas." Third and fourth diagrams lack darkening but each has third line perpendicular to horizontal axis, to left of initial pair of lines; new lines labeled a and a', respectively]

      – Curves may not cut axes.

      – Thus they said give him a certain amount, a, a', then mark the proportion sacrifice.

   (3)– <u>Minimum</u> sacrifice

- Assumes all individuals are about the same in their capacity to get marginal utility out of income, i.e. equal pleasure machines.
- Taxes so allocated that the sacrifice imposed by the last dollar of taxation would be equal over the whole community.
- Taxing the dollars of least utility.

[Diagram similarly constructed, now with four lines drawn perpendicular to horizontal axis, the 2nd labeled 2,000 and the 4th, 3,000.

Thus regressive taxation would give equal sacrifice
[Same diagram, except that marginal utility curve is now a flat line].

- If curve were [diagram with utility on vertical axis and income on horizontal axis, with horizontal line][,] equal sacrifice would be obtained by a poll tax.
- A 10% tax on $3,000 man would cut off 300 from marginal utility area, and a larger proportional tax would be needed to cut off the same area of the utility area of the $2,000 man.
- Thus equal taxation would give a rough proportional sacrifice.
- We have no notion of what these utility functions are.
- In reality, the kind of taxation necessary to provide same criteria of taxation, depends on the shape of the income utility curve at every point.
- All that is necessary in order to get minimum sacrifice, is that the curve be negative in slope.
- The simplicity of this minimum sacrifice doctrine disappears as soon as the different standard of living is introduced.
- Maximum utility of expenditure is the other part.

[Same diagram, again with downward sloping marginal utility function, now with steeper downward marginal utility curves drawn through points where 2nd and 3rd vertical lines cross less steep, original marginal utility function. 1st original vertical line labeled at $1000 and 3rd at $2000 on horizontal axis].

- Man accustomed to 2,000, has steep utility curve for changes, but soon gets accustomed to 1,000, his curve slips back.

[Diagram, reproducing double steep marginal utility lines at 1,000, and repeating same at 2,000]

– As people got adjusted to new situation, you would have to add new taxes – one would have continually to change taxes, in order continually to minimize sacrifice.
– Finally, all tax would be on upper man.

– This only tells us what the short-run man's utility of income will be after he has changed his whole standard of living and budget.
– Under any reasonable assumption as to shape of these curves – this leads to progressive taxation.
 – Not only now – but increasingly progressive for all time to come. Few utility curves, like indignation, will continuously recede.
 [Single vertical line alongside preceding sentences, starting with "Under any . . ."]
– The long-run utility function is itself a function of the distribution of income.
 – e.g. mar[ginal] utility of a $10,000 income will be high or low depending, in part, on how many other people there are at that income.
– The short and long run marginal utility curves would be functions, likewise, of what happened to the rest of people.
– Bosanquet – it doesn't matter what one's total income in a community is, it only matters what relative incomes are.
 – If taxation were general progressive taxation [it] might be even more long lasting.
– Minimum sacrifice principle has perpetuated the doctrine of hedonism far beyond its day.
 – What really is attractive in the argument is the conclusion – it is a pseudo-scientific case against inequality.
 – The doctrine derives all its practical force as well as all of its ethical uprightness from the introduction of an assumption that all individuals are equal, parallel, similar pleasure machines.
  – Pigou admits his ignorance of individual differences – thus takes equality as a working basis – his conclusions are thus based on ignorance, in the ultimate.
 – There is no basis for assuming that there is a positive correlation between utility and income [Double vertical line alongside].
 – "Productivity ethics" – John Bates Clark.
  – "Strong backs" says Simons.
  – Should be rewarded, said Clark – i.e. taxation would not remove inequalities.

– Should be penalized for an initial advantage – one might say.
b-Suppose people equal in all respects <u>but</u> as efficiency machines. It is possible to determine good and poor machines.
   – Hedonistic minimum-sacrifice taxation would <u>subsidize efficient</u> machines [Double vertical lines alongside].
– Simons thinks the case <u>against</u> inequality is enormously stronger than any hedonistic platform.
      – The discussion of inequality of <u>income</u> is sensible.
      – A discussion of equality of pleasure machines is absurd [Single vertical line alongside in margin].
      – Why accept the <u>weakest</u> case for redistribution?
– This has not been so much a refutation, as an exposé that the position is untenable.
– Based on Viner's] utility articles.
– Is a refutation of the <u>basis</u> of all welfare economics.
   – As soon as you drop that basic assumption, or make another one, the results of "welfare" economics are unacceptable.

   [In margin at top of page: Bulloch [Charles Jesse Bullock], "Justice" – Selections 37 – Wagner, 38 – Seligman, 39 – Seligman]

   – So that what you accept, in following welfare economics, is a hedonistic assumption as to individual psychology – uniform pleasure ability.
      – Hedonistic economics – aims at the greatest good.
      – But it makes quite a difference to add "greatest good for greatest number."
<u>Wagner</u> – taxation is an instrument for redistribution of income.
   – If we support <u>existing</u> distribution, there is no use going any further, into sacrifice and utility excuses.
   – cf. Selection 38, Bulloch [Charles Jesse Bullock] – <u>Seligman</u>: (a) gets mad ("communism"); and (b) gets profound – worse.
   – "Legal justice = legal equality."
      – "To redistribute income (based on inherently different abilities) is a travesty on justice" – i.e. redistribution of <u>income</u> is a creation of legal <u>inequality</u>!
      – Pure productivity ethics; as his theory of taxation.
         – Taxation according to "faculty" or "ability."
            – Both completely incapable of quantitative measurement!
            – Yet taxation is to be a proportion of an (unmeasurable) quantity!
   – Simons starts with a presumption in favor of equality of income.
      – Any taxation favoring that is sound taxation.

– Usually progressive taxation of income will achieve this.
– Starts with <u>denying</u> the existing <u>inequality</u> of distribution of income as being a "good thing" – at least the ethics is stated plainly.
– Case <u>for</u> progressive taxation bases itself squarely on the case <u>for</u> equality of income.
  – This rests on an entirely <u>ethical</u> base – as Simons thinks it should.
  – I [F.T.O.] ask – possible to rest it on economic bases?
    – Simons: All economic arguments are slightly in favor of <u>in</u>equality of income.
      – Especially the argument of the effect of inequality on business cycles is bad.
      – He likes it left securely and openly on the ethical (and political) base.
    – Progressive taxation will <u>of course</u> affect capital production and accumulation.
      – To say progressive taxation should stop before it touches production and accumulation is to say progressive taxation should stop before it becomes <u>effective</u>.
      – Equality of distribution <u>does cost</u> something.
        – Ethics and justice (and progress) are <u>all</u> costly, enormously costly.

– Equal Taxation affects:
  – Supply of highly productive personal services.
    – Jobs have a good correlation with salary.
    – But a progressive tax on high salaried technicians would not change their [the tax's] distribution, or their efforts.

  – Entrepreneurial effort:
    – Is mainly a game – based on prestige symbols.
      – Some prestige now attached to expenditure on glamorous goods, <u>could</u> be attached to publication of income tax lists, and place on <u>it</u> and based on mainly psychological benefits.
      – Community is squandering large sums on high salaries – which are not necessary to keep the system working – it might as well get some amount back in taxation.
    – On accumulation – supply of savings.
  – Rate of savings in a community depends on a great many <u>more factors</u> than the rate of interest.
    – Partially on stability of social system, which is <u>aided</u> by redistributive taxation.

– Cause of justice may slow up the increase of accumulation some,
  – Though the government may become a saver to some extent and make up for this [In margin: – repay borrowings].

[Double vertical lines alongside principal text].
– Saving <u>may</u> be a severe affliction – in times of depression – income tax may do a lot for ending this problem – but this does not yield a <u>reliable</u> and stable tax.
– <u>Should</u> society save so much for a distant future?
  – Is it not enough to provide a decent ratio between population and resources – for the future?
    – And no need to work entirely on the resource side.
  – Each generation inherits a certain system of property rights and certain mortgages.
  – The retirement of the public debt out of progressive taxation dismisses the problem of its results on accumulation.
  – Or the government might invest in private industry.
  – Or provide free consumption service.
  – Or go into business itself.

   [In margin at top of page, two references:
   Spahr and others – N. Y. U. – chapter by <u>Studenski</u>

   [Walter E. Spahr, ed., *Economic Principles and Problems*, 2 vols., New York: R. Long & R. R. Smith, 1932. Paul Studenski has chapters on public expenditures, public revenues and public credit].
   Brown, H. G., Chapters 3 and 4 (<u>Incidence</u>) "Economics of Taxation" [Harry G. Brown, *The Economics of Taxation*, New York: Holt, 1924. Chapters 2–5, 7–8, 10 deal with incidence of taxation].

Justice – Taxation and the distribution of personal income.
  – Tax burden should bear equally on persons in the same situation (Sidgwick).
  – No discrimination; at least a minimum of obvious inequalities should be found in the good income tax.
  – "Reasonableness" as between individuals leads back to the "sacrifice" wilderness.
  – The purpose of the income tax is to reduce obvious inequalities.
    – Not to worry too much about the way people spend their income.
    – Be careful of "concessions" – they greatly reduce effectiveness of a tax.

– It is very unfortunate to make a separation between <u>taxes for revenue</u> and <u>taxes for adjusting the distribution of income</u>.

– Taxation for "non-fiscal" purposes (Lutz) – seems to be a way of hanging a <u>name</u> on forms of taxation conservatives don't like.

– <u>Single Tax</u> – peculiar form of insanity;

   – A form of medieval reforming, applied to modern conditions.

   – Is based on a <u>Ricardian</u> concept of land.

      – Might apply to urban land.

      – Not to anything else.

   – Overlooks the impossibility of readjusting vested interests in land.

   – There is no such thing as the State's inalienable right to land or <u>in</u> land.

      – How far back are we to go before we find the original value of land.

   – Assumes the State knows more about land speculation than anyone else.

– All taxes have <u>both</u> functions.

– Proper distribution of revenue burden can not be adequately discussed without knowing <u>how</u> the revenue is going to be spent – and how it has <u>formerly</u> been obtained.

   a/– For it is inequitable to change <u>existing</u> burdens of taxation.

      – All change will create windfalls of gains or losses.

      – At least some <u>long period</u> of readjustment should be granted.

         – A tax once existent has an effect on the flow of investment into the field taxed.

         – And a tax changed changes the forms of investment.

         – Property is bought and sold with the tax in mind – tax affects the equity.

            – Any change in tax changes the equity – should be done <u>slowly</u>.

         – Property tax was in effect a <u>mortgage</u> on all land.

        – i.e. property tax is a tax on the <u>capital value</u> of the land, in which value the effects of the tax have already been capitalized.

   b/– Benefit levies.

– When a commodity is state produced and is enjoyed by only a <u>part</u> of the citizenry – it should be sold at at least cost – i.e. a benefit levy.

   – Construction of a municipal electric railroad into a new suburban area.

– People will pay for this: (a) in fares; and (b) in higher rents –
payment for access – goes to real estate.

   – Property owners should pay by <u>a benefit levy</u> in order to make
up for payments made to them which are due to the railroad
– also this enables the railroad to be built and paid for earlier.

      – What of a subway in Chicago (22nd [Street] to Chicago).

        – What formula to use?

          – Build first, tax what property values <u>do</u> [to] price later?

            – But what property values will do, will <u>depend on</u> the
taxation that may be levied.

            – You would have to keep it secret.

            – U.S. has been very zealous in assessing special
assessments – very lax in paying compensation for
worsening of value.

            – England has been zealous in paying compensation
– very lax in making special assessment for
improvement.

          – Studies of the effect of placement of <u>station</u> on the New
York subway lines.

            – American Telephone and Telegraph studies of the
size of switchboards – extrapolation from rates of
growth.

          – The whole property tax may be looked at as a tax
of special benefit assessment – for the municipal
expenditures benefit the whole municipality as against
another municipality.

          – It is better to have <u>bases</u> of assessment and have them
understood by everyone in the real estate world, than
to worry too much about their nature.

          – Any inequalities are taken account of by capitalization.

        – Then the worst thing to do is to <u>change</u> [Double vertical
lines alongside].

          – Only in the case of <u>entirely new bases</u> of assessment is
there any chance of inequality.

      – In many cases application of benefit principle is
impossible – schools, arterial highways, etc.

        – For its application requires an estimate of what property
values <u>would have been</u> without the benefit.

    – Extrapolation is usually weak, or wrong.

  – Gasoline tax may be considered a form of <u>ad rem</u> taxation.

– For highways provide <u>one</u> part of the population with an
  expensive luxury.
  – The gas tax is an ideal way of taxing that part of the
    population.
  – It is a problem of charging for a government service.
  – One of the few commodity taxes than can be justified.
  – Most <u>ad rem</u> taxes are not defensible.
<u>Income tax</u> on <u>personal</u> income.
  – <u>Capital gains</u> is hard to get at – but so far as it is done, the U.S. systems
    gets at it better than any other tax system.
  – The only distinguished feature of the U.S. tax.
  – As a matter of good monetary policy the present system is excellent
    – it brings in lots of money in boom, and by falling off in yield in
    depression forces the Treasury to borrow.
  – One good tax – yields about 10% of total revenue.
Ad rem taxes – <u>gasoline</u>, etc. <u>Property</u> taxes.

Taxes that are poor in effect – <u>corporation income</u> taxes.
Taxes that are <u>invidious</u> in effect – tariffs, commodity taxes like cigarette
tax, etc.
  – Even most liberals support this.
    – A "luxury" tax.
      – "Poor people do not need to consume it" – does this <u>reduce</u> the
        <u>burden</u> of taxation?
        – They <u>won't</u> stop consuming it, and the tax was levied with
          this in mind. [Vertical line alongside "and the tax . . ."]

Any progress in rationalizing our tax system must depend on
<u>an extension</u> of <u>the personal income tax</u> – until it yields at least 50% of
total income.
  – <u>Primarily</u> to effect a <u>redistribution of income</u>.
  – <u>Secondarily</u> to <u>get rid of taxes poor</u> in their theoretical backing, or
    in their means of collection, or in their effects.
  – <u>Offsets</u> – is only sensible <u>within</u> the system of personal taxation.
  – Means nothing as <u>between</u> systems, as between income, and excise,
    e.g.
  – Argument that <u>all</u> citizens should pay <u>some</u> tax – for the good of their
    souls.
    – But if it is a <u>moral</u> problem – how is it that this argument is always
      used to support an <u>excise tax</u>, which draws from everyone, yes! and

makes the whole population contribute to government, <u>but</u>, in a way that <u>no one</u> recognizes.
- How can it be good for a man's soul to pay a tax when he does not realize that he is paying it? [Single vertical line alongside in margin].
- Corporation tax – is useless.
    - With a good personal income tax there is no need of this.
    - It is an <u>easy</u> tax to get passed in legislatures, and easy to administer.
    - There is <u>no</u> excuse for applying the ideas of justice, burden, etc. to inanimate corporations.
        - What does it mean to redistribute income among corporations?
        - Same excuse for a tax on <u>undivided</u> income.
        - This started out as a form of personal income tax, 1913 – 1% on personal income, and 1% on corporation income – and dividends were exempt from personal income tax.
            - Later the English system of collecting personal income tax at the source.

<u>Perfect competition</u>

- Price is the independent variable – is such in mathematical analysis.
- Any individual firm can not aalter the price by its output variations.

[Diagram in margin with notation, "201 Syllabus." Diagram has quantity on vertical axis and price on horizontal axis, with unlabelled vertical line. Also arrow to the following]

- The output curve of the individual firm – i.e. elasticity of demand is treated as infinite. – Don't use the word "demand" – say <u>demand function</u> (of price), <u>demand schedule</u>, <u>demand elasticity</u>, etc.

- Cost of production affects producers only through the <u>rate</u> of production.
    - Most textbook discussion speaks of what the producers think they can get; this applies only to a monopoly situation.
- <u>No</u> connection of cost curves and <u>any</u> supply curve.
    - Lowest price at which any output will be forthcoming is $7
    $15 price = 17 units output
        = 17,000 units total output
    Price times quantity = 17,000 × 15 = $255,000
- Tax imposed, variable, total, average, marginal expenses are all raised by the amount of the tax ($4).
    - Don't talk about what producers <u>will</u> do or think, but ask, what of it?

– If price remains at $15 and cost is now $19, producers will try to reduce output.
– 255,000/P, if price is $18, there will be some of output not taken, price will fall – at price of $16, the price will have to be bid up as more is demanded than available – at a price of $17, the amount produced (15) is equal to the amount demanded.

[In margin alongside the above, in form of short division:
18 divided into 255,000 yields 14,100 demanded, 16 is produced.
16 divided into 255,000 yields 15,900 demanded, 14 is produced]

– Marginal price will do most of the shifting – but the final position will be dependent on the relative elasticities of demand and of supply [Double vertical lines alongside].
– Assumption is made that the price of productive services does not rise.
– Simons doesn't see much sense in drawing long–run supply curve.
    – They are curves for the industry as a whole.
    – The average expenses curve is for a firm.
    – Consumer's surplus means the utility a commodity would have if it were enormously more scarce [Double vertical lines alongside].
    – This never enters into one's budgeting.
    – Decreasing costs – more palpably absurd in the tax-problem case than usually.

Monopoly:

[Two diagrams, incomprehensible. In A, output is on vertical axis and price is on horizontal axis. Bumpy downward sloping curve, unlabelled; arbitrary points, starting from left: competitive price, monopoly price, monopoly price with $4.00 tax, alternative monopoly price with tax; also $4.00 tax indicated below horizontal axis. In B, presumably labeled like A, similar curve, with arbitrary points, starting from left: competitive price, monopoly price, monopoly price plus tax. Darkly shaded area of rectangle formed by horizontal length between competitive price and monopoly price with tax and height at monopoly price with tax. Lightly shaded area is to the right of darkly shaded area, and is formed by horizontal length between monopoly price and monopoly price with tax and height at monopoly price with tax. Below horizontal axis are several numbers – 1.00, [0.]80, 0.30].

– Question of incidence of taxation on pure monopoly.
– For partial monopoly, there is not much use in talking about incidence.

– It is <u>likely</u> that the price is increased by about the amount of the tax [Single vertical line alongside].

In A and B above – the shaded area is the amount of the "tax burden" on the monopolist – his monopoly profit is <u>cut</u>.

   – The lightly shaded area is the amount of "burden" on the consumer – rise in price by about the amount of the tax – but <u>all</u> the "burden" is not on the consumer.

   – The only <u>ultimate</u> burden of an excise tax (like the same for a tariff) can't be found anywhere.

   – Its meaning is that the community, with a higher price of a formerly most efficiently used product, now buys less of the product and more of other products,

   – And the real <u>burden</u> to the <u>community</u> of the tax is the less efficient use of resources in the new position.

   – For they <u>will be less efficient uses</u>, since we have assumed the community to have been in the most efficient position of production.

   – A <u>general</u> excise tax is in a different position.

   – In the cases of <u>some</u> <u>luxuries</u> (diamonds), if the price is doubled by taxation, it represents no loss, but only shifts of expenditure from an invidious direction to a better direction.

     – A one-carat diamond will then serve as the former 2-carat diamond, and everything will be better – except for hangovers of sentiment.

     – If we turn away from what people want to what <u>costs</u> people least, we have dropped our assumption of free competitive <u>allocation</u>.

cf. Brown, Harry Gunnison – Chapter on "Incidence of Workman's Compensation" – one of the best chapters to read on the problem of incidence.

<u>Sales taxes</u> are usually a collection of excise taxes.

<u>Incidence of</u>:

  – <u>Tax on real estate (land, sites) in an urban center</u>.

    – Will alter the capital value of the land.

    – The same effect as a government confiscation of some of the equity.

    – No effect on amount of land so used.

[In margin alongside: Selling price falls, with the lower capital value, thus there is <u>no</u> effect on the density of buildings there].

    – <u>Tax on buildings also</u>:

    – Immediate effect the same – confiscation of a part of the equity.

- But long run effect would probably be some change in the amount of building in that center (as against another).
- Tax on agricultural land (higher in one place than in another).
- Immediate decrease of capital values.
- Continuous reinvestment is necessary in agricultural land (not in urban).
- Thus the long run effect would be a slow, continuous disinvestment in land – increase in "mining" of the soil.
- The difference between the case of urban and of agricultural land is that the upkeep can be entirely dissociated from the urban site, but not at all from the agricultural land.
- Agricultural land is not a matter of capitalization, but of the return the farmer is able to get from the maintenance of his farm. The government takes a part of the return; he is no longer willing to put into the farm so much money, because of the low return.
- General tax on all tangible property – isolated community – on land, buildings, machinery, farm land, improvements, etc.
- How could such a tax be capitalized?
- It would lower the rate of return on investment generally.
  $1,000 return [@] 5% $20,000 capitalized
  $\underline{-200}$
  $\underline{\$800}$ return [@] 5% $16,000 capitalized
  $-\underline{But}$, the return is reduced on all real property, lowering the rate of interest, to, say, 4% in which case the result is
  $1,000 return [@] 5% $20,000 capitalized -200 tax
  $\underline{\$800}$ return [@] 4% $20,000 capitalized
- Assuming no change in the rate of saving, and realizing that nothing can change the $1000 return – what do we assume as to the effects of the kind of expenditure the government chooses to make with its tax-collected money.

Justification of talking about taxation incidence without talking about expenditure incidence.

Excise Tax

- We are interested in the advantage of one tax as against another.
- Statement that general excise taxes are regressive – how prove this?
  - $1,000-a-year man spends his whole income; it is all taxed.
  - $10,000-a-year man saves $5,000, spends and is taxed on half of his income
- but there is an effect on savings and dividends.

– Very few of our (general) excise taxes reach expenditure for purchase of
personal service.
  – This is a large item in the expenditure of higher income groups.
  – i.e. is regressive due to a lack of generality of the tax.
– Specific excise taxes are regressive
– Ad valorem excise taxes usually regressive because not general enough.

– Defense of excise taxes?

1a- Ease of collection – Federal figures show excise and customs taxes cost
twice as much per dollar of revenue, as income taxes.
b- But, efficiency of administration and collection is very high – for every degree
of efficiency, the cost is less – it would cost heavily to bring income tax collection
up to the same degree.
  – Criticism of excise taxes.
    – Extremely regressive.
    – Pump government revenues out of the lowest income group.
  – Defense.
2a – Everybody should pay something for taxes.
  B – Excise taxes are easier to collect from lower income groups than income
taxes.
    – Criticism:
      – ? – German post-war experience shows high degree of efficiency in
        collecting income taxes on all wages.
      – Poll tax is better than excise tax which is generally unknown to the
        people.
      – How do they get moral benefit out of paying taxes whose burden they
        do not realize.
      – In effect – an unrealized excise tax is the best way of incurring political
        immorality.
      – We are coming to realize them more, however (Cigarette tax, $3.00
        per 1,000, 0.06 per pack).

3 – Generally, they are assessed on luxuries – luxuries, "are commodities which
    poor people ought to do without"!
  – Tobacco – in the U.S. in normal times, yields one-half as much as personal
    income tax; now yields more than personal income tax!
  Criticism: – It is absurd to say that people might do without these if they do
  not actually do without them [Single vertical line alongside].
  – Also – the government has to tax commodities of an inelastic demand in
    order to get stable revenue.

– It makes the tacit assumption that the demand will <u>not</u> vary, i.e. that they are <u>not</u> luxuries.

– <u>Administrative</u> and <u>equity</u> considerations contradict.

– <u>Are</u> luxuries commodities of <u>e</u>lastic demand?

    – Aside from "<u>Sunday School</u>" connotations it is best to <u>define</u> luxury as commodities that <u>do</u> have elastic demand – i.e. shoes, clothes, food, etc.!

– Revenue considerations are to get more money, <u>not</u> to cut down the consumption of the goods – except in a few cases – as Swiss tax on Brandy (50% or 100%).

Conclusion: – <u>no</u> merit in excise taxes except as revenue producers – as an alternative to a poll tax.

    – The <u>burden of excise taxes</u> is to the degree:

      a/ That people <u>change</u> their demand programs.

        – Diverting their expenditure from the lines which they <u>really</u> want.

        – Less productive use of resources.

        – Altered allocation of resources is the ultimate real burden.

          – Which can not be found.

          – Exists even if the tax yields <u>no</u> revenue.

      b/ That they pay more for things they do buy – a double burden, for the allocation of resources exists too.

Should excise taxes be levied on articles of elastic or inelastic demand?

    – Elastic – not revenue producing.

    – Inelastic – not just.

i.e. answer is that they should not be levied.

Qualifications:

    a/ Gasoline tax – is a form of special assessment – to prevent the government from subsidizing [a] class which does not <u>need it</u>.

    b/ Excise tax on diamonds – on a product whose demand is <u>entirely</u> due to <u>invidious competition</u>.

– A large literature on classification of taxes, or names – taxonomic questions – are worthless for the most part.

Shirras: best on excise taxes.

Shoup: "The Sales Tax in France" – best for general sales tax.

– For the <u>States</u>, the only practical excise tax is the general retailers tax – administratively difficult to an extreme.

– Manufacturers' excise is fairly impossible as a <u>State</u> measure – because of the inequalities in inter-state trade – works well over <u>all</u> Canada.

– Retailing sales tax is practical in a crisis – for a time before evasion is learned and practiced widely.

   – In France, they had to compromise; stopped trying to assess the tax on <u>small</u> retailers; put a license charge on them to come to <u>about</u> the same rate.

– History: early "modern" Spanish excise tax.

([George Frederick] Warren – has said: that all his schemes were intended mainly to improve the position of agriculture as against industry – wheat and cotton – their prices moved with devaluation)

Tariffs – have all the shortcomings of excise duties, and then some.

– <u>Differential</u> in <u>two</u> ways, a/ as to <u>kinds</u> of goods; b/ as to <u>sources</u> of some of those kinds.

(Shirras – best chapter on this subject)

– As attempts to control trade, are simply <u>subsidies</u>.

– A commodity that <u>would</u> be largely imported costing $1.00 a unit.

– A tariff puts the domestic rate up to $1.50.

   – The government collects an excise tax from <u>its</u> nationals, <u>uses</u> that tax to subsidize domestic industry.

   – Duty 0.50 = tax of 0.50 plus subsidy of 0.50.

   – The same as a tax on the sale of the product plus a subsidy given to the producer by the government.

– All uniform subsidies – are absurd.

– All <u>differential</u> subsidies are unjust, and demoralizing.

[Vertical line in margin alongside previous two lines]

– In addition to the most undesirable form of tax – excise.

– It is necessary to add to the usual statistical summary of taxes and expenditures, about: <u>$5</u> bill, as <u>excise tax</u>, <u>$5 bill</u>, as <u>subsidy</u>.

– If there are going to <u>be</u> subsidies, they should stand out and not be backhand.

– Subsidy might be given by keeping price at $1.00, but giving a subsidy direct to producers to the amount necessary to keep out foreign goods.

   – Without the non-revenue producing aspects of a rise in price.

Assume: (1) [? Empty line]

(2) Imposition of tariff has no effect on cyclical movement.

(3) Inconvertible paper money.

– Tariff put on, falling off in important of good, rise of price of domestic product, then an increased return on investment in the field of that product; finally, an increase of investment in that field, will reduce <u>somewhat</u> the price.

  – Suppose the exchange rates did <u>not</u> change, some of old volume of exports could not be paid for in foreign exchange and U.S. bank balances with foreign correspondents would rise. Then there would <u>have</u> to be either: (a) loaning abroad by U.S. banks; or (b) a lowering of the rate.

    – <u>Some</u> limit on proportion of their reserves any bank would like to see invested abroad.

    – Even if rates of interest were the same in the two countries, there would have to be severe <u>limit</u> to loaning abroad.

    – But, as investment outlook is improving in first country, there would be some <u>shift in comparative discount rates.</u>

    – And thus a decline in the dollar value of exchange.

      – Exporters will get lower dollar prices.

      – Lower dollar prices of imports.

      – Decline in internal domestic prices.

      – Increase in <u>domestic</u> consumption of <u>exportable</u> products.

      – Shift in production away from exportable goods towards non-traded goods.

    – i.e. a reduction of <u>volume</u> of international trade.

    – And, lower rates of exchanges, lower volume of international trade.

  – With inconvertible paper, the burdens and benefits of tariffs are easily apparent, and there is the possibility of articulate demand for tariff reductions.

  – What of effect of tariff on comparative <u>efficiencies</u> of investment in the two countries? Does it affect the rise of rate of return of the first country?

    – So many other factors that this is not very important.

  – Suppose <u>gold standard</u>:

    – Why should rates get to the gold points?

      – The banks would want to get their foreign balances back into <u>this</u> country – new investment opportunities [in this country].

    – Volume of imports curtailed, volume of exports goes on (?)

      – Exchanges move to the gold points at once.

      – We assume that a flow of gold, then, is the last thing that will happen.

      – There is more credit expansion then credit contraction, due to excess of exports over imports.

      – Thus some rise in level of domestic prices.

        – <u>Prices of non–traded goods rise</u>.

        – Prices of export products do not rise so much.

- Import prices do not rise so much.
- Shift of commodities to non-traded.
- Expansion of bank credit, upward tendency of good[s] traded internally.
- Prices of internationally traded goods are held down by the influence of the world markets.
- Expansion of domestic consumption of internationally traded goods.
- Contraction of domestic consumption of domestic goods.
- Expansion of production of goods not traded internationally.
- Contraction of production of goods internationally traded.
- i.e. a new equilibrium at a new point – with a reduction in the community's real income.
- Unless the former equilibrium was a false one with too high prices, wages, rents.
- i.e. a change in the level of wages, of rents, etc.
- Or, it is an argument against a common international standard.
  - For it [a common international standard] causes changes in the income structure of a country.
  - Wherever prices are hooked to the internationally common standard and the income structure (rents, wages, etc.) is rigid, sticky, then there is an argument against this common standard.
- Trade unions are against devaluation, employers are against increase of par.
- Wages are almost always too high, necessitating the continuous presence of a "reserve army" if unemployed: problem of elasticity of demand for labor.
- Any meticulous statement of theory of international trade adjustment is impossible.
  - It brings in the problem of what banking theory, what theory of prices, etc., the [sic] which this is based on.
  - Ricardo and Mill – vs Ohlin – the latter is only a more intelligible version of the former.
- After War – the International Gold Standard was shifting to a Gold Exchange Standard based on the dollar.
  - Necessity of revision of usual theory of gold movements.
  - Every country acting as a pivot of a Gold Exchange Standard system would keep larger reserves than under old conditions.
  - Gold Exchange Standard is the ultimate corruption of the old Gold Standard.
  - Nothing is left but the evils.

- No automaticism [sic]; what automatic elements there are are overcome by rigid wage levels, etc.
- i.e. another argument for interdependent currencies – perhaps in blocks of countries.
- The requirements of Gold Standard management are often – especially in depression – out of line with the requirements of domestic currency management.
    - Thus leading to tariffs, quotas, etc.
    - Which are to be regarded as a very <u>crude</u> form of monetary policy – an erosion of the ultimate necessity of a Gold Standard system:– lack of restriction.
    - Emphasis changes to bear on short-term capital movements; instead of on commodity movements.

    - cf. <u>Comstock</u> – tables, for comparative importance of excise and tariff duties. [Alzada Peckham Constock, *Taxation in the Modern State*, New York: Longmans, Green, 1929]
    - Excises and tariffs are of necessity the main types of taxation in backward countries.
    - Comstock gives figures for <u>central</u> government – not for all taxing bodies; but the role of the central government varies widely from country to country.

Property Taxes

cf. Cannan – best [Presumably Edwin Cannan, *The History of Local Rates in England*, New York: Longmans, Greeen, 1896].
  Seligman, Chapter – for historical background.
- The main revenue is obtained from <u>real</u> property – a hangover of the time when political and economic and social position was dependent on landed property.
- Far predominant in the U.S.
    - 90% of municipal and small community taxation.
    - Largest part of State taxation.
    - More important than State plus Federal income tax.
- Property tax – on all real property.
    - Tangible property.
    - Intangible property – a difference of mobility – is usually representative of property.
    - Originally a tax on land; has been legally extended to cover all property, but is still predominantly on land.

– Most countries (except U.S.) have gone back to a limitation of property
  tax to land (real).
– The U.S. States levy property tax – Federal government would have to
  tax a State proportionally to population (Nevada vs Rhode Island).
– Property tax as an ad rem levy on real estate.
– Property tax as a personal levy, on persons according to their net ownership.
  – In England a differentiation, higher taxes on persons whose income is
    derived from land than persons of the same income but which is derived
    from "labor" – i.e. heavier taxation of funded, unearned, income.
  – Such differentiation can be done via discrimination in the income tax.
  – Germany's Ergäntzungsteuer – an addition to the income tax according
    to the person's ownership of property.
– Property tax as an ad rem levy must be based on real estate without reference
  to the ability of the person to bear the tax.
– Most literature is unduly critical of the property tax, naïve.
  – Condemn it because it is not ad rem and because it is not personal – i.e.
    do not make this separation.
– In U.S. most of the property tax is an ad rem tax, i.e. we dismiss a discusion
  of ability to pay, etc.
  – Would our property tax be a good one if it were well administered?
    a – A good case for taxation of property on the basis of its location – not
        on basis of the location of the person.
    b – Property has some merit as a basis of ability to pay – such a tax does
        reach people according to their circumstances.
    c – There is a considerable amount of property which does not come
        under the reach of the U.S. or Canadian income tax.
        – i.e. income in kind from consumer's capital.
        – This is taxed, to some extent, in European income taxes.
        – But it is quite impossible to deal adequately with this problem of
          funded and unfunded income.
        – German idea of a progressive tax on net ownership of property, as
          an addition to the income tax – is an excellent idea.
    d – The property tax is a very good way of checking up on the income
        tax, and vice vera.
    – These arguments do not support the property tax as a main source of
      revenue.
      – Because it is that in the U.S., our system is called backward.
    – The main argument is that we have had the tax for a long time and are
      accustomed to it; it is discounted in [blank space].

Administration

- Real estate is most easily taxable, more easily than cows and wagons.
- Person who holds property is taxable for the whole value, even if it is heavily mortgaged.
  - This is not inequitable so long as the tax is considered an <u>ad rem</u> tax, as <u>real</u> property.
  - Criticism that a <u>corporation</u> is taxable for all the property it owns, and each stockholder is taxed according to his ownership of stock.
  - Problem of multiple taxation.
  - Inequalities of assessment of property tax (cf. Leland).
    - There <u>always</u> are inequalities between owners of property of different size and value.
      - <u>A consistent regression in the property tax.</u>
      - Virginia study.
      - Wisconsin study: under $1000, 60% of value taxed.
      1,000–10,000 – 49%, 10,000–50,000, 39%,
      50,000–500,00, – 25%, 500,000 & over – 17%.
    - i.e. uniform actual levy, but the proportion of whole value assessed for taxation varies regressively.
    - This seems to be universal, even for farm land.
      - People of large parcels are the ones who go before Boards of Appeal
        – who usually give <u>some</u> reduction just for coming around – they are the ones who <u>can</u> go and can hire good lawyers.
      - Zangerle – Real Estate Assessment [John Adam Zingerle, *Principles of Real Estate Appraisal*, 2nd ed., 1927]
      - Buck – Municipal Finance [Arthur Eugene Buck, *Municipal Finance*, 1926]

  - The important thing is not the degree of <u>regression</u>, but the degree of <u>dispersion</u> – where the latter is high, the former is likely to be high.
- Other inequalities of assessment between jurisdictions:
  - Townships, county, state, etc.
  - <u>Apportioned</u> rather than <u>proportional</u> taxes.
  - State levies are distributed among minor jurisdictions according to <u>assessed</u> value of properties.
  - No excuse for a property tax unless it is on <u>real</u> property (real estate).
    - i.e. get rid of tax on intangibles (representative property), and of tax on <u>personal</u> property.

- In most communities there hasn't been any real administration of the property tax – it is set up, a local barber put in charge, and it is expected to run itself.
- Vile practice: set up a progressive income tax, then subtract the income already taxed (by way of capital) in other (property) taxes – i.e. the income tax is on salaries and wages.

Program of Segregation and Separation
  - A grand scheme for getting rid of the inequalities of territorial taxation.
  - And for segregating off certain forms of assets for State taxation only (or utilities) – in California, this represented an attack on sites as the basis.

Classification of intangible property for taxation
  - Any result on the total revenue is hard to distinguish.
  - Under old system most tax officials realized that a correct payment on a full holding of intangibles would be exceedingly heavy far more so than tax on tangible property – demoralization of tax officials.
  - Classification allows for two rates, a low percentage rate (say, 4 mills) on intangibles, and a higher rate (say, 2%) on real property.
    - Accompanied by much higher registration of intangibles,
    - And the tax officials can count evaders as fair game – under old system, an extra-legal rule grew up as to the proportion of intangible value to be entered on tax returns.
      - Mr. Fairweather suggested, for Chicago, 10% of value of bank deposits, 20% of value of securities – it worked.
    - Minnesota has a noteworthy system.
      - 4 mill registration fee on securities and intangibles.
      - Four classes of real property (same rate on all four).
        - Mining property, 50% (U.S. Steel)
        - Subdivided property, 43% (city)
        - Unsubdivided property, 33% (farm) – farmers elect legislature.
        - Other tangible properties, 25%
  - Classification can become a mere system of political conflict between groups.
  - Once one gets away from the uniformity rule, the way is open for all kinds of discrimination.
  - On the whole, intangibles should be excluded from a property tax; for a tax on them is neither one thing or another, and is bad in administration, unless a classification system with a low (4 mill) registration fee is charged.

- Property taxation should be uniform, but cannot be so as long as intangibles are included [Double vertical lines in margin alongside these six lines].
- Problem of assessment of corporations:
  - How distinguish real and intangible property?
    - With respect to corporate surpluses.
    - How value a plant? What is real, what is intangible in that value?
  - Corporate excess – difference between property assessment and total value of an enterprise.
  - (How does Massachusetts assess this excess?)
- Retain county as the best administrative unit.
  - Perhaps election of county assessor.
  - If only men who have passed an exam are eligible to run for office.
  - If state committee can remove assessors from office, for a cause, until new election.
  - Central Tax Commission (State) should prescribe all administrative forms, paper, maps, procedure.
    - It should educate the county assessors.
    - It should make all the difficult assessments – railroads, forests, mines, etc.
    - It should equalize assessment ratios among counties.
    - It should have means of testing the results of county assessments.
- Rules of assessment
  - Good tax maps.
  - Sales data is the best criterion.
    - Not sheriff sales, nor condemnation sales, sales to owners of adjoining property, to relatives, etc.
    - Necessity of full statement of "other considerations" on deeds of sale.
  - Real estate appraisal (Zangerle [John Adam Zingerle, *Principles of Real Estate Appraisal*, 2nd ed., 1927], Buck [Arthur Eugene Buck, *Municipal Finance*, 1926]).
  - Most assessment systems start with the appraisal of a front foot of property, of standard depth, at the middle of a block.
    - Sales records, consultation of real estate experts.
    - Publication of tentative appraisals.
    - Public meetings, assessors meet owners.
    - Revision of tentative lists and publication of final front foot values to be used.

- Correction for undue depth, etc.; for corner bock, for alley, for non-rectangular parcels.
  - These things follow mechanically from the unit values. Rules instead of judgment.
  - Such a system is far better than assessment by an individual tax official on the basis of a particular judgment.
  - It has no question of personal inefficiency when a complaint is made; there is plenty of room for protest.
  - Unit values are something really comparable.
  - The merits of the system are with reference to urbanland.
- The comparable system for urban buildings is more complex.
    (– cf. Buck: Buffalo system is best).
  - Classification of all improvements to land – 20– 100 classifications.
  - First value obtained is reproduction cost, if new.
  - Then comes matter of depreciation, etc.
  - Is particular formula more valuable than accurate description? Probably not.
  - Buffalo has a complete record of every house, a picture, description, date of building, etc.

[Digression]

A monopolist "dumps" abroad in order to lower unit costs with a larger output.
  - This, of course, depends on an assumption of decreasing cost, which is allowable only if the first assumption of pure monopoly also holds.
  - Most dumping abroad, actually, is an attempt to break in on a foreign monopoly, is aimed at power at creating a position in a market, etc.
    - i.e. profit motives are not primary.
    - Decreasing costs – possible under partial or only pure monopoly?
      - No sense to question: partial monopoly is an unstable position, an on-the-way position toward pure monopoly.
      - i.e. of what use is discussion of duopoly?
      - Most of it leaves you where you started plus a few equations dependent on strange assumptions.

[Back to Taxation]

- Necessity of year-round appointment of assessment staff.
- Possibility of using insurance on a house as a criterion.
- New York – provides for grants-in-aid on the basis of land assessment.
  - An incentive towards pushing up the assessment level.

– Tax-rate limits are supposed to pull up the general level of assessment – a particularly <u>weak</u> argument.
– Movement is strong to relieve <u>residences</u> from <u>all</u> assessment.
– Dangerous: creates a hopeless financial situation in most communities (no police or fire, etc.) – is bad anyway.
– The case for central assessment of <u>public utility</u> property is evident – it is impossible to assess a <u>part</u> of a utility concern in a particular locality.

[Digression]

(There is <u>nothing</u> that is <u>coherent</u> as a <u>principle</u> of <u>valuation</u> of <u>utilities</u>.
– There is no sense of regulation anyway.
– Where competition (if protected) will regulate and value, that is enough.
– Where it won't, then there is a plain case for government ownership and operation.
– Regulation is a bastard; if it is without principle, it is bad; but there can be no principle.
– The N. R. A. [National Recovery Act] is an attempt to apply to <u>all</u> of industry the bastard form of regulation.
– If the U.S. railroad conditions are typical of all industry – an assumption behind the N. R. A. – then there is no science of economics.
– The Federal Trade Commission represents the highest level of regulation ever reached in U.S. – scrupulous, honest, efficient.
– But look at the freight-rate classification!!
– Plain case for government ownership and a nearly complete abolition of the freight-rate classification.

[Return to Main Topic]

– State-wide assessments of public utilities do work fairly well – with a property tax.
– Cooperation with utility commissions.
– Case for taxation of utilities leads to opinion that [sentence incomplete]

<u>Taxation of National Bank Stock</u> (Section 5219) – Tax allowed to be collected from National Bank real estate.
– Might be taxed by States, but not higher than State Banks <u>in</u> competition.
– Banks paid, down to 1921 – gentlemanly.
– 1921 Richmond case – changed this. Tax on National Bank stock can be collected <u>only</u> in the jurisdiction in which the bank is.

– Court held that the provision above meant that the State could not tax National Bank stock <u>higher</u> than local rates on <u>intangible</u> property – all of which was to be conceived of as in competition.

– The <u>problem</u> is: why <u>should</u> bank stock be taxed? – this question is always taken as axiomatic.

Taxation of Insurance Companies

– Theory of taxation of fire insurance companies in order to finance fire protection.
  – What "theory" is this?
  – A tax easily levied; but no reason for levying it.
  – What is the "corporate excess" of an insurance company?

The essential policy in the problem of special assessments is to avoid any change in anything that is well established.

  – Many types of improvements should be taxed by special assessments but there is no theory that is really tenable for doing it.
  – They are the only taxes that can save the world from the realtors.

IncomeTax

– [Harrison Bray] Spaulding, "Income Tax in Great Britain and United States," [1927] chapters 2, 4, 5.
– Hardy, "Tax Exempt Securities and the Surtax," Chap. 4. [Charles O. Hardy, *Tax-Exempt Securities and the Surtax*, 1926.
– Federal Income Tax, Columbia University Lectures, Robert Murray Haig.
– Schultz, "Taxation of Inheritance," Parts III and IV. [William J. Shultz, *The Taxation of Inheritance*, 1926]
– Schanz, G., Finanz Archiv, 1896. [Georg von Schanz]
– Bruno Mall, "Finanzwissenschaft."
– Handworterbuch der Staatswissenschaft, article on income tax.
– What is income? – very little discussion of this, of any value (cf. Haig and Germans).
– [Income] Tax a response to demand for more equitable distribution of income.
  – Follows the introduction of finance and trade all over the world.
  – First approved in Italy and France, then disappeared.
  – Came to England in late 18th century.
  – Germany, mid-19th century.
– Is the outstanding example of democratic and liberal government to fiscal problems.

- It is an instrument of control and justice.
- "Income" is widely used in the discussion of justice – not badly used when not defined – but a consistent badness in its theoretical definition.
  - What definition provides the basis for the most equitable distribution? – least ambiguous.
  - Must define something that is measurable and must imply a technique of taxation in the definition–. "Meaning of Income," W. W. Huett [William Wallace Hewett, possibly "The Concept of Income in Federal Taxation," *Journal of Political Economy*, vol. 33 (April, 1925), pp. 155– 178, or "The Definition of Income," *American Economic Review*, vol. 15 (June, 1925), pp. 239–246] – especially court decisions.
  - Income from things – productivity.
    - No reference to an individual, the income is from source.
  - Income from transactions.
  - National or social income.
  - Personal income.
[Bracket in left margin of these four types of income]
  - All different, no relation.

Income

- Consumption – destruction of economic goods.
- Accumulation – change of ownership within a period
- Income
  - Must be income per period (some specified time interval).
  - To say that income can not be assigned per period is to say it can not be measured.
  - Consumption and Accumulation per specified period.
    - No necessity of qualification as to past or future.
    - Everything that is relevant is to be found within the period.
    - Market rights exercised in consumption algebraically plus or minus the property rights accumulated or disaccumulated.
    - i.e. add consumption to wealth at end of period, subtract wealth at beginning of period.
    - i.e. Income = Gain to someone.
  - Both consumption and accumulation are to be measured by market prices and values.
    "Income has only an accounting reality."
    - Income is not goods and services, it is not a flow.
    - Income is an arithmetic result.

– Society – a giant partnership.
  – A member's income is his <u>withdrawal</u> plus the change in value of his share in the enterprise.
– Don't forget: perpetual maintenance must be provided for before there is <u>income</u> (or <u>interest</u>).
  – It is not an assumption, but a <u>sine quo non</u> of the existence of income or interest.
– Where draw a line between economic or non-economic income? (Flowers, housewives, self-shaving?) (Home education, doctors' home services)
  – <u>Leisure</u> itself is a major item of income.
  – These elements of income vary with considerable exactitude at different income levels.
    – The poor family with a large income of personal services loses a large element of leisure.
  – Some income in kind ought to be brought in, but to do so opens up a morass – insolvable.
  – But even articles that go through the market are ambiguous.
    – Articles bought for business by one person are bought for play by others (paint brushes). – What of schooling – investment in self.
    – i.e. even the distinction between consumption and accumulation is arbitrary.
    – [Frank H.] Knight tries to explain the <u>economic</u> and <u>non</u>-economic choice in terms of orthodox economics – margin of indifference.
    – Simons says this is impossible – gives up.
    – How set down consumption or accumulation in regard to personal services.
      – Only solution is to recognize the limitations, make some arbitrary concessions.
  – Distribution of total consumption over a period is important, for it may change things.
  – Impossible to have a good income tax until there is a good monetary system; given a bad system, very little by way of recompense can be made by income tax.
  – Even where market prices are present, they are not final.
  – Thus <u>every</u> bit of measurement of income is a <u>construction</u>.
    – Problem of gifts and bequests.
      – Why make a special problem of it? For it requires a consideration of <u>intention</u> in the mind of a second person.
      – Gifts are (not expense, but) consumption to the donor, and are income to the donee.

– Controversy over our definition of income – In <u>Germany</u> (a large literature):
  – G. Schanz, 1896: "net accretion of property of a specified time period, including uses of consumption goods and valuable services of third parties."
  – Der Begriff erweist sich als <u>Reinvemögenszergang</u> eines bestimmten Zeitabschittes [The concept turns out to be the pure incremental value in a certain time period].
  – What was independent economic <u>power</u> of a person in a period; what means were available for his disposition without his resorting to consumption of his capital (property).
  – We calculate as income, all net revenues and services, all services of third parties having a money value, all gifts, insurance indemnities, gambling and speculative gains; and we deduct all interest on investment and capital losses [Single vertical line in margin alongside this point].
  – Robert Murray Haig – "Income is the net accretion to one's economic power between two points in time."
    – If literally construed, this excludes all consumption; is synonymous with accumulation.
    – Does imply that income is a <u>value factor</u>, and the thing to be valued is <u>right, gain</u>.
(German controversy went on mostly during early years of income tax. Schanz got his ideas into the national tax in 1921; it was carried out during the inflation and discarded by 1925).
Herrmann, b. 1820 [F. B. W. von Hermann, 1795–1868?] – Income is the sum of exchangeable goods which accrue during a period, without reduction of capital stock.
– Was focusing attention on social income.
– Consumption fund theory.
– Fisher – income <u>is consumption</u>.
  – German writers say this, but at least add the proviso that <u>something</u> should not be exceeded.
– Prussian tax became model for all German taxes.
  – Broke income up into that obtained from land, from industry, etc. – the Neumann [?] position.
  – Like the British <u>schedules</u>.
– Plane's [?] presidential address – is the German position without quotation marks.
  – Income is "current consumption receipts."
  – Every word is wrong – Receipts is awful. "Consumed" does not imply the necessity of maintained capital. Income is not receipts, it is <u>receipts less</u> something, <u>at least</u> less depreciation.

– This is the Herrmann position; but it has been somewhat out of date in Germany since the 1880s!
– cf. [name indecipherable, possibly [John Ironside] Falconer] – "The Concept of Private Income."
– Criteria of income:
  a – Regularity – can not be strictly construed.
  b – Should be a <u>source</u> – no advocate ever defined a source.
  c – <u>Personal</u>.
– Not "what <u>items</u> are income?" but "how should income be calculated?" is the question.
  – Income is gain – and can not be defined aside from the individual.
  – Nor can it be determined <u>merely</u> by looking at sources.
  – The distinction between expenditure and consumption leads to the personal.
– Yields and receipts <u>can</u> be characterized as recurrent or non-recurrent.
– But <u>gain can not</u> be so characterized – "Income is the net result of arithmetic calculations with value factors."
– Accountant never aims to find what income is, but only to find means of calculation that can find an answer acceptable to naive directors – worship of conservatism.
  – A professional warfare against truth.
  – A denunciation of judgment's place in the world.
    – By setting up rule-of-thumb procedure.
    – But such rules-of-thumb are soon elevated into principles!
      – "Income not realized is not income!"
– Constitution allows the federal government to tax <u>income</u> without apportionment; but the courts have set it up as <u>their</u> province to decide what is income and what is not!
– Advocates of regularity concept, have attempted to create a personal income out of concepts of recurrent yield on things – based on <u>transactions</u>.
– "One may grow richer indefinitely without ever increasing his income." (Th. Reid Powell)
  – This <u>is</u> the usual concept.
– One never does know what the final result of a business venture is until it's over, but it is never over <u>within</u> the income period.
d – Realization criterion
  – Stock bought January 1st at $100, and another share bought December 31st at $150 – they must both be listed at $100 (or any lower figure the stock <u>has ever</u> reached).

– Bond bought January 1st at $90, and goes up to $100 on December 31st. If he <u>does not</u> sell it, there is no income except interest. If he sells it on December 31st (and puts the money in a worthless stock) he has received a realized income of $10 plus interest.

  – One can realize without gaining, one can gain without realizing (<u>Payton</u> – sees this problem). [Likely William A. Paton].

  – If there is going to be a criterion of one, <u>or</u> the other; the alternative must be thrown away.

– <u>Seligman</u> defines income so as to exclude stock dividends – then finds that stock dividends are not income.

  – Finds that income must be <u>realized</u>, and separated.

  – [These] conclusions aren't worth anything, and have no weight except author's name.

  – Income is satisfaction – <u>benefit income</u>.

  – Next page, income is satisfaction attained by money.

  – Then savings slips in.

  – Income becomes <u>yield</u>, productivity.

  – But it is not that because depreciation must be taken account of.

– Suddenly, the quality of periodicity is essential, only to be discarded in a few pages.

– Income from [a] herd – new births is income, is realized and separated.

– Income from forest – new growth – is <u>not</u> income, for it is not separated and <u>thus</u> (sic) not realized.

– Income depends on the number of trees cut – so long as capital is not impaired. But how <u>separate</u> capital in trees from income?

– Then, realization occurs <u>when</u> separation has been effected.

  – i.e. <u>income is defined differently on every page</u>.

  – Cause is [the] shifting from the point of view of income (things, sources) to that of the individual enjoyment of income.

  – Gain and realization can not usually be found simultaneously.

  – What Seligman says is that all <u>positive</u> elements of income must be realized, but he is not at all exacting as to the things that go into the negative-expense side.

<u>There is</u> no case for taxing stock dividends as income, but Seligman's case for proving this is absurd.

  – How is income to be <u>realized</u> unless it is turned into <u>cash</u>; yet most advocates of it do not go this far.

  – In effect – realization is attained <u>only</u> in consumption, which is by definition an <u>end</u>, a using up.

- Only [Irving] Fisher supports this.
- Fisher recognizes clearly the income concept Simons has defined – but would call it earnings – a verbal issue only; except that Fisher thinks there is more to it.
  - Simons says "Income is determined by value."
    - The "time preference" group claims that value is determined by income.
  - In effect – this is not a paradox, for income means a different thing in each case.
    - Fisher uses income to mean realization in consumption, but when he comes to capital goods, he uses income to mean gain – says so in a footnote.
  - In the latter case he uses income to mean discounted future yield; this is not consumption.
- Interest – classical theory gives a cost theory of interest.
  - Such is absolute, we need a relative theory: cost explanation of wages is as uninteresting.
  - Interest is really a problem of saving, not of supply.
  - Assumption – new investment funds will pour into most profitable channels of production.
    - A given structure of prices.
    - Every instrument of production has a certain productivity, and a certain cost.
  - Knowing the productivity of an instrument for its service life, and knowing how that productivity is spread through that life, and knowing the cost –
    - We can calculate the interest rate which gives a present value of those future rents.
  - The rate of interest must be the same for all instruments.
    - And the costs of all must be related, in such a way that the interest rates are all the same.
    - Thus cost governs yield.
    - This is then an equilibrium position.
    - Rate of interest is determined by the available opportunities for profitable productivity.
    - Constant cost is adequate.
    - Varying cost is possible with only a few more complications.
    - Most people will try to treat the rate of saving as a function of cost.
    - But if the resources are treated as variables, then how treat their supplies in terms of their costs? [Double vertical line in margin alongside].

– What is being economized? if the fund of resources is not a fixed one.
– We have treated the problem of interest without giving consideration to the problem of supply – that is a problem of philosophy – we assume the rate of saving (philosophy, psychology, etc.) to be fixed.

– To get back to Fisher – to say that yield or income causes value, is only a half-truth.
  – On our definition, income can not determine value, but is determined by value.
– Most confusion could be avoided if people would leave out the problem of trying to discount future income.
  – Necessity of distinguishing between income to persons, and income to things.
– Some people save, instead of consuming – others smoke instead of drinking; [they] are the same.
  – The economist is hardly capable of deciding why? – it is an institutional, philosophical, psychological problem.
    – Using the utility principle to explain it, is only a confusion of the issue.
– From the point of view of the community, saving increases social wealth, consuming decreases it.
  – Accumulation is as much a matter of invidious consumption.
    – In the 20th century – there is something lacking in the 18th or 19th [century] concept that saving is a postponement of income; it is, in fact, a form of consumption.
  – It is a property right both to eat eggs and to clip coupons. Both represent income.
  – Saving is accomplished by expending purchasing power to buy savings deposits.
  – Savings is a mixture of accumulation and consumption.
  – To say that saving is not income is to say that income is only consumption [Double vertical lines alongside].
    – Fisher does say both these things, trying to change the whole course of historical definition.
  – The value of an instrument is the value of its products – any physical good on which the market will set a price.
  – Interest theory is merely superimposed on price theory, to bring out certain important relations [Double vertical lines alongside].
    – Interest is a cost – probably the most important cost – Fisher can not say this.

– When does income accrue?
  – The notion of accrual should be reserved for assets and liabilities.
– Interest accrues as one or the other.
– When income accrues, is better stated as, when is value imputed.
– Income is a description of the <u>results</u> of an individual's activity during <u>any</u> period.
– All assets and liabilities should be revalued at the end of <u>every</u> period.

<u>Income tax</u>

– Income, in the sense we have outlined it, would be almost impossible to use as a basis of a tax.
  – Except in a world with costless transportation, indestructible factors, heterogeneous properties, and a price on everything.
  – If no one <u>owned</u> any property, and everyone rented or borrowed, it would be easier.
– Yet there is a long history of the tax over the world, especially a helpful history in England and Germany.
  – No tax has ever been levied on personal income.
  – All taxes are attempts to get a simple method of calculation of <u>some</u> elements of personal income.
  – Most tax laws do not really <u>define</u> income, but merely ways of description and prohibitions of evasion.
  – Even such regulations do not <u>define</u> income.
  – That the tax worls are proof of a scarcity of ingenious liars.
  – We must have a <u>broad</u> and <u>admittedly impractical</u> standard of income in order to be able to judge practical measures.
– In terms of <u>discrimination</u>, we shall form a concept of income.
  – The best tax is the one that bears equally on persons in a similar income position.
    – Consumption and business incomes.
    – Earned and unearned incomes.
    – Family is the unit, usually, and it is not a homogeneous quantity.
    [Brace in margin connects preceding three lines, indicating that they:]
  – Are all things that destroy the possibility of a just tax.
    – Also, the problem of those who inherit money and consume income <u>and capital</u>.
      – German system is inconsistent but a good idea: when consumption exceeds income, then consumption is the basis of the tax.
      – Still, we must allow for capital losses, etc.

– We must be on the lookout for evasion.
  – Especially when the system allows people to order their affairs in order to evade, and this is common knowledge.
  – When evasion is widespread, there is an unfortunate effect on community morale.
    – From the point of view of morale and of evasion, the equitableness is much more important than the severity.

Problem of income in kind: measuring consumption.
  – How account for the value of goods and services produced in the household.
  – If such value is not included, there is an obvious inequality, and an over-taxation of those who do less in the household.
  – But there must be some line drawn, and there is <u>no</u> place to draw that line, satisfactorily.
    – Leisure is only a form of consumption.
  1– Perhaps we may separate "<u>earned income in kind</u>" and say that it is impossible to tax.
    – Except for a wife (English system allows this exemption).
    – And in some cases for <u>farmers</u>.
      – In England, tenants are taxed according to the rental value of the farm – a mere <u>presumption</u> as to income.
      – The problem is not a great one; farmers reap large earned income in kind
        – but, they make tremendous <u>sacrifices</u> of <u>leisure</u>.
    – Where <u>property</u> is concerned, there is some excuse for an exemption, as between homeowners and lessees.
      – This is done in all tax systems except the U.S. and Canada.
    – On the whole, earned income in kind is, as in the farmers' case, largely offset by negative leisure income; and it is large only in the cases of those people who would not pay any tax anyway; and, <u>within</u> income groups, the earned income in kind is about the same. Also, it decreases as we go up the income scale – so that the effect of leaving it out is only to increase somewhat the slope of the rate of progressiveness, [thus it is] all right [Single vertical line alongside text from "the effect of" to end of sentence].
      – Likewise, such exemption is gratuitous, and does not lead to evasion.
    – A tax which does not take into account consumers' capital and the income from it, will discriminate against the lessee, against the business worker, etc.
  – How deduct rents – as between homes rented bare, apartments rented furnished, and apartment-hotel service?
    – A discrimination against the latter groups unless different classes.

- If deductions for renters were made, then homeowners might demand deduction for repairs, etc.
- Is it not better to get income from those who rent houses out, than to deduct the rent paid.
  - In England the lessee pays the tax and deducts it from the rent – i.e. it links up with collection at the source.
  - This would not work well here.
    - Our whole system does not collect at the source, and we are unfamiliar with annual value – we think of whole value, Englishman thinks of annual rent.
  - Australia – takes 5% of capital value [of] all real property used for consumption income and adds it to tax base. This taxes homeowners, etc.
  - That a tax will cost as much to administer as it will bring in returns is not an argument for doing away with the tax.
    - For you must have some regard to make the tax equitable.
    - And to stop demoralization by evasion. [Double line alongside from "That a tax" to "by evasion."]
  - Homeowners ought to be assessed, in the income tax, by same such means.
    - Such Federal assessment of real property might cause a great improvement in the local assessment.
  - What are we to do then with real property within the house – pianos, furniture, etc.
  - The error involved in just neglecting them would vary directly with the durability of the goods.
    - In the case of clothes, the income is yielded almost at once.
    - A tax should be laid on consumption most near to the point where it takes place.
    - Where draw the line between gain and loss?
  - An ideal personal tax would tax people according to their capacity for using income, i.e. ability to consume intensely [Double vertical lines alongside this point].
  - But without an objective measuring rod of that ability, we must stick to an income tax on objective measureables.
  - i.e. we must draw a line somewhere beyond which gain and loss will not hold.
    - What of vacuum cleaners that don't clean, books that misinform, etc.
  - One solution is to add some multiple of the total assessed value to the tax for home use, to cover rental value of house furnishings.
    - As 1% of the total value.

- On the assumption that the rental value of furnishing will be in a fairly direct ratio to the rental value of the house [Double vertical lines alongside this point].
- Especially art collections, automobiles, yachts, libraries, etc.

2– Income in kind as compensation for services

- As when homes are provided to ministers, army and navy officers, etc.
- It is difficult to have a graduated income tax according to the pleasureableness of people's income.
- All receipts ought to be included in the calculation of taxable income.
  - Some separation might be made between compensations in kind for services rendered, and compensation in kind as gratuitous receipts.
  - How value them?
  - Many writers (even Schanz [Georg von Schanz] and Haigue [Robert Murray Haig]) say that gifts and inheritance should not be taxed by an income tax, on the ground that they are already reached by other taxes.
    - Another case of the treacherous "offset" reasoning – doesn't make sense.
    - Also it represents a shift back to the source theory of income.
  - The taxation of gifts shortly before death is no outlet.
    - God knows what the courts will decide in judging the "presumption of death."
  - First one in 1924 – lasted a few years: $50,000 exemption per year – rates severe only over one million.
  - $50,000 exemption – once per person per all time.
    - 1932– Gift tax – paid by donor.
    - For putting any future gifts in the appropriate brackets, the whole sum of gifts given is added and the tax applicable to the total is applied to the increment in any year.
    [In margin: 1932 – 50,000, no tax
    1933 – 40,000, 1% tax
    1934 – 60,000, 2%]
- During any man's life, the theory was, the tax will be the same, on his total gifts, whether he gives it all at once, or in small increments per year – contemplation of death is not involved.

- If this were carried over to inheritance, it would be an impossible tax:–
- The only logical tax is to assess gifts and inheritance as income in the year they are received.

- Estate taxes and inheritance taxes nowhere amount to anything, except in England.
- Estate taxes and inheritance taxes were introduced into Federal taxation in U.S. during Spanish-American and again during World War.
- In spite of all the exemptions, the Federal revenue amounted, in the first two years, to more than all the State revenues from the same source since their beginning.
- Great agitation by trust companies, tax lawyers, etc.
- The result was the credit exemption.
  - The Federal tax is figured up, and then all State taxes paid are deducted from it, up to 80% of the total.
  - This brought in the greatest mass of assessment and administration in the U.S. fiscal system.
  - The States all put up their taxes to get as much as they could.
  - The result was a move toward breaking down the separation of Federal and State governments – a most significant one.
    - Opportunities for a thorough coordination of State and Federal fiscal systems, especially if the credit were put at 50%.
      - The Federal government actually administers the tax, but the States, by getting their laws in line with the Federal, get a share of it.
      - For example, the Federal government could administer and assess an income tax of 8 or 10 billion dollars a year – and, by a uniform system of double taxation by State and Federal governments – hand 50% or so over to the States – who would get the revenue they want, while there would be, actually, only one tax.

Joint Congressional Committee on Internal Revenue Taxation
- "Federal and State Death Taxes," 1933 – (L. H. Parker).
- Recent, comprehensive, good.

- The Income Tax of 1894 (which was declared unconstitutional)
  - Provided for the inclusion of gifts and inheritances in income.
- The Income Tax of the Civil War – provided for rental value as a part of income.

- The U.S. tax – at present – has one great advantage over any other.
  - It taxes capital gains (investment asset held for more than two years) and excludes capital losses.
  - In 1921, it was added that capital gain should not raise the tax by more than 12 1/2% of the total tax without the capital gain – and only on incomes over the 12 1/2% tax limit.

– Regular income $2,000,000.
– Capital gain on land speculation $1,000,000.
– Ordinary tax would be:
    400,000, on the first million.
    600,000, on the second millon.
    <u>600,000</u> on the third (capital gain) million.
– But, this law said that the tax on capital gains could not be more than 12 1/2% of the total tax, otherwise: thus
    400,000, on first million.
    600,000, on second million.
    125,000, on third million.
    is the result.
– However at first there was no provision for capital losses.
  – A capital loss of one million would have reduced the tax by a whole 600,000.
  – This was remedied in 1924 – three years time – by the addition of the 12 1/2% clause to the capital loss side.
– The whole thing was absurd.
– Then came the agitation for removing the whole taxation and exemption for capital gains and losses.
  – They were not income!
  – Other countries did not have them!
  – It kept down circulation of stocks – aggravating both bull and bear markets.
  – Implication is that big holdings of common stock should come "into the market" and thus be sold to stenographers, etc.
  – Why <u>should</u> this tax be any restraint on Mellon's sale of shares in 1929.
    – The death duty is based on a different concept – (why?).
    – <u>Gifts</u> are taxed as to capital gain – original value.
    – Inheritances are <u>not</u> taxed as to capital gain – appraised as to market value.
  – Thus a rational motive for holding off sale until you die, when the tax liability [on capital gains] will evaporate.
  – <u>Necessity</u>, at least, of taxation of every <u>estate</u> <u>as</u> <u>though</u> every item had been <u>realized</u> at the day of death.
    – In addition to the inheritance tax.
    – In order to <u>tax the capital gains</u> (or losses) of the estate as of the day of death.
    – As well as the inheritance tax on the total present value of the estate.

- This assessment is made in any case for the calculation of the present value for the inheritance tax.
- There is every incentive for holders to go into the market to sell securities, in order to realize loss and get reduction.
  (Wash sales – conversion into other stock)
- The agitation was not – as it might have been – for the improvement of the tax, but for the abolition of the tax.
  - Internal Revenue study:
    - Assumption that one had no income, by which they mean no income from capital.
    - Then, a capital gain might be taxed all in one year – and held for several; whereas, if taxed in each year, it would be much less, due to the progressive rates.
    - But, they forgot that most holders of large capital gains were already holders of large incomes, so that the rate applicable in any year would be as large as [if] applicable [to either] at once [sic; Ostrander: I think I blundered here].
- The essential thing is not to separate one factor of fluctuation in income, but to attack the whole problem of fluctuation of income.
  - Possibility of five-year check on realization criterion.

- One ideal system would be to tax the algebraic sum of capital changes (gain or loss) and dividends – levied on the individual.
- Another system would be to treat the corporation as a partnership – for the purposes of tax collection – then tax every individual shareholder according to his share of the total earnings (dividends and surplus).
- Ideally – corporations chartered by Federal government – and limitation of capital issue to half-dozen or dozen types.
  - But no single corporation could use more than two or three of those types.
  - Sharp distinction enforced between owners and creditors.
    - Periodic liquidation enforced by owners whenever their equity falls.
  - It would be very easy then to treat corporation as a partnership for tax collection.
- But with the present system of corporations,
  - A corporate income tax is foolish – not personal; just another way of getting revenue for government.
- But, as long as corporations remain as they are, there is a very good case for a progressive tax on undivided income – progressive according to the proportion of total earnings not distributed.

   – Would be a good tax to the extent that it was a poor revenue producer.
   – It might be possible to pass a law that all corporations distribute 100%
     of earnings as dividends.
   – Finance all new capital out of new issues.
   – Tax-exempt securities (cf. Hardy [Charles O. Hardy]) (look out for
     poor statistical work).
       – Hardy makes the most decisive point against exemption –
         unwittingly.
       – Only case in which there is <u>consumer's surplus</u>.
           – Conversion of income progressively into tax exempt securities.
             [Diagram in form of rectangle with line from northwest corner
             (at point labeled 10,000,000) to point above southeast corner (at
             point labeled 30,000)] [Ostrander: This diagram has no heading.
             What does it mean[?] I must have missed something].
           – At first it would be very profitable to convert, and save tax.
           – But after about 30,000 mark, it would not be profitable any more
             to convert, because the tax would not be severe enough.
           – But the utility of income would be the same, in fact greater at the
             end – a consumer's surplus – by the amount of income tax saved.
       – Tax-exemption, with a progressive income tax, gives about 15%
         bonus to the large holders of them.
           – Government loses.
           – But the government gains if there is a large number of small
             investors and trust companies with an <u>in</u>elastic demand, who tax
             up [Ostrander: I don't understand this. Did I then?] the securities,
             but without appreciable change in their tax paid.
 – The whole thing goes back to the foolishness that has been read into the
   McCauellagh vs Maryland case [M'Culloch v. Maryland, 17 U.S. 316
   (1819)] – it is no impairment of a state or government powers, to tax its
   income.
 – The whole practice must be cut out of the system before it becomes decent
   one.
 – The present tax prohibits deduction of interest paid with which to buy
   tax-exempt securities.
       – But the only way they can catch this is if the tax-exempt securities are
         used as collateral, i.e. if no other collateral is available.
       – Otherwise – it is profitable to borrow at 5% – give stock collateral, and
         buy tax-exempt, 4 1/2% bonds – because it reduces income tax by so
         much.

COURSES FROM MELCHIOR PALYI

MELCHIOR PALYI: INTRODUCTION AND BIOGRAPHY

Edited by Warren J. Samuels

INTRODUCTION

F. Taylor Ostrander had three courses from Melchior Palyi at the University of Chicago during the 1933–1934 academic year: Economics 332 on monetary theory; 333 on business cycle theory; and 334 on the European banking system. Fellow students included Albert G. Hart, later a leading monetary economist, and Rose Director, sister of Aaron Director and later wife of Milton Friedman.

Given the period, namely, the lowest point of the Great Depression, these were apposite topics and Palyi was an appropriate lecturer – at least within the ambit of then-traditional ideas.

Melchior Palyi was born in Budapest, Hungary on March 14, 1892. When he died in Chicago on July 28, 1970 he had had three careers: business and banking, university lecturer, and columnist.

Palyi received his Master's in Law from the University of Budapest and the doctorate in economics from the University of Munich in 1915. He worked at the Austro-Hungarian National Bank and the Hungarian Ministry of Agriculture between 1915 and 1918. During 1921–1923 he taught at the Universities of Göttingen and Kiel and at the Handelshochschule in Berlin (where Roy F. Harrod attended his lectures). During 1926–1928 he was a visiting professor at Oxford, the University of California at Los Angeles, and Chicago. He served as chief economist of the Deutsche Bank during 1928–1933 and advisor to the Reichbank

Documents from F. Taylor Ostrander
Research in the History of Economic Thought and Methodology, Volume 23-B, 305–306
Copyright © 2005 by Elsevier Ltd.
All rights of reproduction in any form reserved
ISSN: 0743-4154/doi:10.1016/S0743-4154(05)23108-5

and managing director of its Institute for Monetary Research during 1931–1933. Upon the ascendancy of the Nazis to power in 1933, he emigrated first to the United Kingdom – serving as guest economist at Midland Bank and lecturer at University College, Oxford – and then to the United States, serving again as a visiting professor at the University of Chicago between 1933–1937 and, after 1940, as lecturer at Northwestern University. Palyi was a columnist for the Chicago Tribune during 1961–1968 and for the Commercial and Financial Chronicle during 1968–1970. In the 1956 *Handbook of the American Economic Association* he self-identified his fields as international economics, and money, credit and banking. His books included *Principles of Mortgage Banking Regulation in Europe* (1934), *The Chicago Credit Market: Organization and Institutional Structure* (1937), *Creeping Paralysis of Europe* (1947), *Compulsory Medical Care and the Welfare State* (1950), *The Dollar Dilemma: Perpetual Aid to Europe?* (1954), *Managed Money at the Crossroads* (1958), *An Inflation Primer* (1961), and *The Twilight of Gold, 1914–1936: Myths and Realities* (1972). Palyi was a contributor to *Adam Smith, 1776–1926* (1928), having earlier been co-author of *Hauptprobleme der Soziologie: Erinnerungsgabe für Max Weber in Gemeinschaft* (1923) and co-author and compiler of *Lujo Brentano: Eine Bio-bibliographie* (1924). A search on JSTOR indicates the following record: journal articles, 7 (2 AER, 1 JPE, 3 JBUC, 1 QJE); panel discussions, 4 (all AER); book chapter, 1; book reviews, 36 (27 JPE, 7 JBUC, 1 APSR, 1 JFIN); reviews of his books, 7; citations to him or his work, 14; his reply, 1.

Palyi was a committed supporter of the gold standard and an opponent of central-bank monetary management (including G. F. Knapp's "state theory of money"), especially of John Maynard Keynes, and of any institution that he perceived to be socialist in nature.

ACKNOWLEDGMENTS

I am obligated to Daniele Besomi, Don Moggridge and F. Taylor Ostrander for help in preparing this biographical sketch of Palyi.

NOTES AND OTHER MATERIALS FROM MELCHIOR PALYI'S COURSE, MONETARY THEORY, ECONOMICS 332, UNIVERSITY OF CHICAGO, 1933–1934

Edited by Warren J. Samuels

INTRODUCTION

Ostrander's notes reveal Palyi's course to have had, in effect, four parts, the first two being principal ones. The first part is a review and interpretation of selected aspects of monetary theory, especially the Banking versus Currency Schools of monetary policy. The second part is an interpretation of recent European history, centering on the rise to power of Adolph Hitler. The third deals with Hayek, apparently through a report by another student, Albert G. Hart. The fourth deals with Keynes, apparently through a (second) report by Rose Director. These are taken up in the same sequence in this introduction.

Monetary Theory and Policy: Banking versus Currency Schools

The central topic here is the relation of central bank and commercial banks in the issuance of notes by the former and of demand deposits and notes by the latter, i.e. the operation of a banking system and a monetary system in the process of being worked out alongside each other. Two major topics are added to the

Documents from F. Taylor Ostrander
Research in the History of Economic Thought and Methodology, Volume 23-B, 307–330
Copyright © 2005 by Elsevier Ltd.
ISSN: 0743-4154/doi:10.1016/S0743-4154(05)23109-7

mix: One is the conceptual framework of that process, namely, the quantity theory, supplemented by the theory of the specie-flow mechanism. The other is the evolutionary systemic context of that process, namely, the transformation of a system of landed property dominated by the landed aristocracy to one of non-landed, as well as landed, property increasingly becoming dominated by the business class. The reformulation of the banking and monetary system was part of that systemic transformation; it was dominated by businessmen and the new monetary and banking system helped bring about and protect the new political and economic system. This is a matter of interpretation – Palyi's interpretation.

One facet of the lectures is Palyi's general attitude toward the quantity theory, indeed substantially all monetary theory, as a *theory of control*. The aspect of quantity theory discussion that loomed so large, namely, automaticity, especially after World War Two, when the quantity theory (*properly* applied) was lauded as the non-interventionist alternative to Keynesian fiscal and monetary policy, is subdued, but not altogether absent.

Another is the evident variety of ways in which the quantity theory was operationalized, i.e. how M, V, and T were conceptualized and handled. This also contrasts somewhat with post-War usage, when the policy choices, hence exercise of control, latent in the different versions would have been conspicuous – though eclipsed by the lauded automaticity, even though conservatives like Frank Knight pointed out the inevitable non-automatic, non-rule, elements of administering the quantity theory.

Third, Palyi and others were apparently prone to naturalize his conceptualization. Thus we read of the "Natural or Equilibrium Rate of Interest" and the "natural abundance of money." Naturalization without precise definition has long been a characteristic linguistic usage in economics. Substantively, Palyi's terms beg important questions: Equilibrium is not a normal state. In fact, it is never achieved. Actually it is arguably only a tool of analysis. What is there about "natural" that warrants its use here? The meaning of the second use is also unclear. "Abundant money" seems to suggest low interest rates and high bank reserves, perhaps inflation; is that the "natural" state of affairs – and, if so, what does it mean, and why? The term comes in a discussion of Adam Smith's theories of money – but he uses "natural" with several very different meanings. Later, apropos of Ricardo, we read of a "<u>natural</u>, cost of production long run theory."

Further apropos of Adam Smith, Palyi's view of him, as reported in the notes, was striking: "He was first a Statesman." But it is in connection with Smith that one reads, "Word '<u>overflow</u>' is a dangerous analogy; replaces an economic explanation by a mechanical concept." One can extend the point to cover the

Invisible Hand, perhaps a metaphor or some other figure of speech, perhaps not. Does one use such language because one hesitates to admit to not knowing the actual economic explanation, or for rhetorical flourish using a widespread trope? Shortly thereafter, one reads, "The attitude of the statesman, don't bother about the abstract, or the obvious." What came to this editor's mind while reading this statement is that laissez faire, quite aside from its use by interested private parties, enables the statesman to do nothing – or to seemingly do nothing, for maintenance of the status quo is a policy, a policy of control, no less than doing something else.

Palyi's reported view of Smith's understanding of the economy of "the simple and obvious system of natural liberty" is pure idealism. Ostrander's notes report Palyi to have said,

> –To Smith there could be no uncertainties (in Knight's sense).
> –All was a <u>rational</u> system; every one knows what to do.
> ...
> –All is automatism, runs smoothly using only individual interest and intelligence as its oil.
> –No concept of Central Bank, or of <u>policy</u>.

This is not Smith; surely it was not Jacob Viner's interpretation of Smith a half-decade earlier nor Lionel Robbins's two decades later – unless one includes the intelligence of the statesman and, on that basis, their policies. But this latter – even though Robbins identifies Smith's Invisible Hand as that of the statesman – is hardly what these men's Mont Pelerin Society colleagues have had in mind. Yet neither is it what one will find in Smith's *Lectures on Jurisprudence*. As for Palyi, some lines later, one reads, "Minor mistakes may occur; <u>major</u> mistakes can not occur, unless the government steps in."

Perhaps even more remarkable is the report in the notes that Palyi argued, apropos of investment abroad being one outlet for excess money at home (whatever that might mean), that this was "rather advanced to hope to find in Smith." Yet Smith's mention (it is not much more than that) of the Invisible Hand in the *Wealth of Nations* comes in the midst of a discussion of businessmen preferring, sometimes, to invest at home rather than abroad because of the greater security found there. The Invisible Hand then leads to greater domestic economic and military power. Smith was more advanced than Palyi understood.

The notes deal with "control." This is a more complex topic than may be obvious at first glance. A Central Bank – doing what a Central Bank can do – is *ipso facto* an agent of control. A Central Bank pursuing a policy of stability – say, "keeping everything stable" – is executing a policy of control. A Central Bank pursuing a policy congruent with dominant monetary thought is executing a policy of control. A Central Bank pursuing a policy deemed more congruent with laissez faire, compared with other policies, is executing a policy of control. An institution not part of the government but performing as a Central Bank, is an agent of control, sharing the powers of governance with the government, even if not officially a part of government. Such an agent of control makes decisions between different elements of "stability" and over which means to use in pursuit of its goals. Conventional institutional and conceptual, as well as ideological, distinctions can interfere with understanding "control." And the foregoing is in addition to the laws governing property, corporations in general and for banking and finance, and money. The U.S. Constitution delegates to the Congress the power to coin and issue money but is silent on the effective delegation to commercial banks of money creation in their lending operations.

Palyi was lecturing at a time when economists and others were still working out understandings of such important topics as the relation, as money, of commercial-bank credit creation to currency; the multiple expansion and contraction of credit; how commercial banks were different from other enterprises, financial and other; the nature of the Federal Reserve System as a central bank, in part given the ownership interest in the hands of commercial banks; the nature and function of reserves; the ends of monetary policy; the rules governing Federal Reserve actions; the relevance of monetary policy; the relationships of monetary variables and of the instruments of monetary policy to such economic aggregates as price level, unemployment, output, and income distribution; and, *inter alia*, the choice between non-banking interests. Palyi was neither an Irving Fisher nor an R. G. Hawtrey nor a John Maynard Keynes nor a Denis Robertson, but he was well informed on these matters.

The notes record Palyi, in his discussion of the Bank of England as a Central Bank, saying of those who operated the Bank and their supporters that "Yet they were liberals, business men supreme, state has no business with business." The "Yet they were liberals" refers to the incongruity of people engaging in control who did not believe in social control. The "business men supreme" refers to their belief that the middle class had taken over, or were in the process of taking over, from the landed aristocracy in both economy and polity. In effect, the Bank helped cement and administer the new regime. As Palyi later puts it, "All the men connected with this discussion were business men." The "state has no business with business" is the sentiment that business deployed to help form public opinion and negate policies to which they were opposed.

Palyi was incisively well informed about these matters; he was, in his time, one of them. That he could distinguish between the sentiments of non-interventionism and automaticity, on the one hand, and systemic organization and control, on the other, is suggested by the following:

–A fundamental was always the attempt to avoid all control but <u>automatic</u> control, but to <u>find</u> that automatic control, and <u>set it up</u>.

Immediately following the foregoing deconstructed sentence ("Yet they were liberals . . ."), Palyi elaborated as follows:

> –Thus, banking is banking, note issue is note issue.
> –Banking business must go on independent of note issue.
> –Note issue must be separate from banking – notes can create rising prices.

The problem was that the Bank of England had a schizophrenic existence: it was both commercial bank and Central Bank.

Immediately preceding that sentence, the notes have Palyi explaining the Bank's monopoly of note issue thus: "But only in order to make more efficient the <u>automatism</u>." The inexorable necessity of choice, in operating the Bank, as to whose interests should count in decision-making, and the actual non-automaticity of the system, is thus finessed and obfuscated by such language. Efficiency is a function of whose interests count, and Bank policy helped to determine that.

Apropos of the Currency versus Banking Schools controversy, the reader should note the following: (1) The former emphasized the creation of money by the issuer of notes, on their own motion, as it were, and the latter emphasized the creation of deposits through the making of loans (and vice versa), i.e. the former considered money creation an independent, exogenous variable, the latter considered money creation a dependent, endogenous variable; (2) The former emphasized monetary (read: quantity theory) factors, and the latter emphasized fiscal or expenditure factors in the determination of the level of economic activity; and (3) The former emphasized stable output and employment, increases in money supply affecting only the price level, and the latter emphasized the variability of output and employment due to changes in spending.

Ostrander's notes include the following:

> –The Currency School answered that if the money <u>spent</u> regulated prices, the money <u>issued</u> regulated them.
> –They both talked about the same thing, at bottom – the quantity theory of prices.
> –The <u>difference</u> came down to the question: Does the amount <u>spent</u> vary in the <u>same</u> proportion as the amount of money issued?
> –Currency School always oversimplified – they also neglected the velocity of money.
> –And they always assumed a <u>rational</u> attitude, thus being led to neglect <u>irrational</u> factors in the situation, such as hoarding.

Yes, they both talked about the quantity theory of money; in a sense, either there was no other monetary theory to talk about or the quantity theory provided the mode of discourse for all discussion. But the question concerning the proportion between the amount spent and the amount of money issued does not get at the causal relations, i.e. the difference in causal relations stressed by the two Schools. Palyi's point about oversimplification is well taken, as is that about rationality. But hoarding – soon to be called liquidity preference – is hardly a matter of irrationality. If people hold more money and hold it longer, one may disagree because they have different expectations and/or are unhappy with the consequences; but to call their motives or reasoning "irrational" is quite another matter.

Notwithstanding his own ideological predilections, Palyi demystifies the concept of "laissez faire" in comparing the Banking and Currency Schools. Each wanted non-intervention in everything but what it thought fundamental to its system and approach. Thus we read,

–Both schools were advocating laissez faire.
  –But Currency School advocated control of the Bank of England in order to keep it from controlling prices.
    –Laissez faire always advocates interference to stop interference; to "restore" "competition."

This is as incisive and insightful as one can get. The problem of control is not whether or not to control but for which/whose purpose. Palyi makes the point yet again:

–A monetary system is a managed system – either conscious or unconscious.
  –This policy of management influences total output, total demand, total value.

Interestingly, Say's law is not a reported topic of discussion.

Recent European History

Palyi's biography indicates his personal predicament in the context of the general predicament of Europe, the latter perceived eventually by many as a conflict between capitalism and fascism and by some as a conflict between capitalism and socialism. Palyi presented in these lectures his own interpretation of recent history. The predicament, in his view, was driven by the force of "an anti-progressive, no-change outlook." The result was a "Return to romanticism, mysticism – all aimed at stopping any change in reality." The institutional transformation began, ironically, as opposition to "changes hurting major groups," brought about, for example, by price changes. The result was a "managed economy" and "autocracy," and the policy scheme "degenerate[d] into isolation, self-sufficiency" and nationalism. In particular, according to Palyi, technological progress was everywhere squelched, despised because of the suddenness and immensity of the changes it brought about. Palyi brilliantly understood what others have called cumulative causation and over-determination: He is reported in these notes to have argued that "Intellectual and economic policies are not one cause [and the] other result – but mutually condition each other, develop side by side. Terribly fateful it was that "Intellectual life turns to the irrational, anti-rational, mystical, romantic."

The foregoing is certainly an unusual view of the mentality of the middle class (Ostrander feels that prior to the Nazis, Palyi was very upper class). So is Palyi's view of labor, given in one line: "Labor carries the ideals of <u>progress</u> – this is why <u>labor</u> seems to be turned against, all over."

Whatever one thinks of individual aspects of the foregoing, and his ideas are certainly open to criticism, Palyi's interpretation suggests one fundamental insight. Every society has the problem of managing not only the conflict between continuity and change but the differential incidences of benefits and costs, i.e. the distribution of sacrifice in society. As some would put it, religion, ideology, economic and political theory . . . must set minds at rest, otherwise (or in conjunction) reliance must be on force; as Vilfredo Pareto (not alone) saw it, a combination of fraud and force. Stability of power structure may be juxtaposed to instability in other respects, and principles of continuity may (selectively) legitimize change and disarm sacrifice.

The connection, therefore, between the two parts of Palyi's course was the importance of the system of control. The system of control may be lauded for its non-interventionism or for its automaticity or for its intervention against intervention. But every economic system – the monetary and banking system and the general economic system – is, in his view, a managed system. It is a system "set up." The management of the system may be built into it and operate relatively invisibly and insensibly, but it is a managed system, managed consciously or unconsciously or, more likely, in some mixture of the two.

The preceding two paragraphs are the interpretation by the editor. In any event, the first part of the course comprised Palyi's personal view of monetary theory. In the second part of the course he seems to have regaled his students with the insights he learned from his personal experiences.

Palyi is reported to have concentrated on the "intellectuals" and their opposition to change, going so far as to say, "Intellectuals agree with the fundamental idea of stabilization *à la* Hitler." This brings to mind the arguments of Ludwig von Mises, Friedrich von Hayek, Joseph A. Schumpeter and others that placed responsibility for "socialism" on the antipathy of the "intellectuals" toward capitalism. Given the status of these men as intellectuals themselves, these arguments of all of them are ironic.

Hayek

Here we find the report on a fascinating dialogue between Palyi and Albert G. Hart, student presenter and subsequently long-time professor of economics at Columbia University. The subject is the Austrian – Hayek's – theory of the business cycle. Among the features of the discussion are questions of: (1) causation between phases of the cycle, rather than causation of the cycle as a whole (as it became with Keynes); (2) causation between variables ("And did interest rate or contraction cause crisis?"); (3) changes in theoretical formulation by Hayek; and (4) changes between considerations of necessity and those of probability and between actual developments and those of logical necessity. Such is either a gold mine or a morass for a theoretician. Clearly demonstrated are the multiplicity of possibilities and attendant difficulties of considering questions of causation and necessity within a model of cumulative causation. Also evident is the practice of introducing new variables into the process of deductive reasoning at one or another stage.

Keynes

The student presenter is now Rose Director, later Mrs. Rose Friedman; the reported dialogue is again between Palyi and Hart. The problem is the complexities and ambiguities of John Maynard Keynes's Treatise on Money (1930). *Inter alia*, the stabilization of employment enters discussion, including the possibility of unemployment in equilibrium, but the question of the relation of saving and investment does not.

NOTES ON MELCHIOR PALYI'S COURSE ON MONETARY THEORY, ECONOMICS 332, UNIVERSITY OF CHICAGO, 1933–1934

1 –Transfer analysis and approaches
   –Nominalistic approach – Quantity Approach

2 – Anti–quantity schools
   –Quantity and velocity functions of something else or constants
   –What is underlying them?
   –The demand for money.
   –The "Banking School" of Thought – but underlies the "Currency School" too
   – the difference between them is on another line.

–The <u>velocity</u> of circulation is a <u>passive</u> factor, or a non-changing factor.

–Cost of production theory of value of metals, and of money generally.

–In case of paper money, it substitutes some <u>psychological</u> factor for quantity – or considers quantitative changes a result of psychology.

–Policy of this approach is "sound banking based on commercial paper" – "automatic control."

–This approach is more developed by businessmen than by scientists.

–Men of <u>not-systematic</u> methods, bankers.

<u>Tooke</u> – descriptive, not abstract.

<u>Adolph Wagner</u> (Germany)

<u>Laughlin</u> (U.S.) – never <u>tried</u> to be systematic ["!" to left of name].

– This approach became that of the 19th century up to the War.

–In spite of Marshall and others.

–Bankers and Central Bankers wouldn't listen to any others.

–Keynes (Indian Monetary Policy – 1912)

–Robertson (Industrial Fluctuation, 1915) [Bracket connects the two lines, Keynes and Robertson, with arrow pointing to next line].

–Both, at this early date, had tendencies more to the anti-quantitative than to quantitative approach.

–Mill – could approach the transfer problem from an entirely different point of view from his approach to bank credit – foolish.

3 – <u>Control</u> (or Quantity) Approach

–Regards the balance of payments as secondary. Sees it in the light of domestic control.

–Whether or not Central Banks could control the country's <u>welfare</u>, in any way.

–Starts out with paper money.

–Bullion discussion. Silberling, H., in Review of Economics and Statistics, discussion in Angell [Norman J. Silberling, "British Financial Experience," *Review of Economics and Statistics*, vol. 1 (October 1919), pp. 282–297, 321–323].

–Thornton, Ricardo, etc. Banking School.

–Money a matter of quantity, which can be regulated by control of its quantity by issue.

–By affecting demand for money by:

–Discount rate

–Open market operations

–Public works (governments).

–Currency School – could <u>Central</u> Banks control business cycle?

–Different schools were a matter of bias as to the function of a Central Bank.

–Both had the same underlying theories – disagreed on a problem of policy, which they regarded as fundamental.

–Discussion of control died down, or seemed to, in last half of 19th century.

    –Some silver discussion.

    –Austrian-Italian discussions of advantages of paper.

–What really happened was that the 20th century began with a new approach to control.

4 – Neo-Nominalist

–Mostly influenced by Currency School – but not knowing those authors.

–Threw over all attempts to study the value of money.

–Threw over "automatic control," "normal adjustment" (Bagehot).

–Studied Central Bank and government administration.

    –Not interested in theory, yet having great theoretical implications.

    –Implied that there were no limits to Central Bank control.

    –Clapp [Georg Friedrich Knapp?] (German) denies that there is any.

      –Value of money – what exists is the monetary policies of governments.

      –But this is a theory of value of money, and a theory of administration.

    –Ansiaux – "The ruling principle of exchange." [Maurice Ansiaux, *Principes de la politique régulatrice des changes*, Bruxelles: Misch & Thron, 1910].

      –Exchanges and gold flows are a result of central bank policies.

–Value of money is determined by government policy.

–Interested in amount of control and the results of its [indecipherable word] interested in stable or unstable exchanges.

5 – Quantity Theory

–Has been used by other approaches.

–A form of approach to supply such as set forth by Bodin and Davenant.

–Value of money not a function of demand, but of factors such as velocity, interest rate, or, if ruled by demand, then demand is ruled by something else.

–Control {need, must} not be automatic.

    –Automatic [?], demanding.

–Marshall, Fisher [indecipherable words]

    –Renewed the old control approach, and united it with the Neo-Nominalist approach.

    –Then came in Keynes, Robertson, Pigou, Fisher.

–Velocity stressed – (l'enfant terrible of previous monetary theory) becomes center of interest.

    –Reformulation of quantity theory in light of Velocity.

      –Dozens of reformulations due to differ concepts of velocity.

  –Changes in it, measurement, causes.
 –Does velocity have a life of its own – or is it a function of other things, or a constant.
 –Most difficult to approach from statistical, descriptive or theoretical points of view.
 –Choice of Index Numbers; what price levels to study, or incomes or employment.
 –Stabilization – toward what shall it be directed?
   –Towards which of the above – Price Levels, Incomes, Employment, etc. (cf. Walsh, "Concept of the Value of Money") [Possibly Correa Moylan Walsh, "Professor Edgeworth's Views on Index Numbers," Quarterly Journal of Economics, vol. 38 (May, 1924), pp. 500–5119].
   –What are the index–number and welfare implications of alternative forms of stabilization? – and price level aspect?

6 –Interest Rate Analysis
 –The new approach of the 20th century.
 –Its reactions on business cycles, prices, etc.
 –Is tied up almost inevitably with Control – we abstract from this.
 –What are the reactions of interest rate on the other aspects of economic system?
 –Problem of the Natural or Equilibrium Rate of Interest.
   –Wicksell – in long run analysis, brings in an equilibrium concept.
    –And studies a non-monetary economy.
   –Visualizing a natural rate, bring the long run savings and investment into equilibrium.
 –Idea of equilibrium, and of interest rate is not new.
 –What is new is the concept of long run and a rate of interest that has nothing to do with the actual rates at any time.
 –The old quantity theory approach looked to money and goods (asked or assumed which is variable which is independent).
 –The new quantity theory looks to the ratio of savings and investment.
   –First appeared in a paper of Jevons, in [18]70s.

Adam Smith's theories of money (Rose Director) [Apparently an oral report by a fellow student].
 –Clearly a cost of production theory of the value of money – taken from Cantillon
 –Also a quantity theory – implied (explicit in Hume).
   –If there us "too much" money in one country, it flows out.
    –Does he imply a rise of prices when there is more than a natural abundance of money? Thus an export of money.

–Or does he think of the channels of circulation being filled, then the excess money being exported, because he couldn't think of any other thing to "do" with it? (Hart) [Possibly a question by Albert G. Hart, another student]

–Palyi: – either Smith didn't like the quantity way of explanation

–or he wanted to write for the businessman and for the government. He was first a Statesman.

–Hume's essay had been read by no one but professors. Smith wanted to avoid that fate.

–Foster, (Ricardo) followed Smith's analysis of gold flow analysis without mention of price changes.

–Most early 19th century writers declined to discuss the mechanism of change; the change of prices.

–Palyi thinks this was taken for granted, as obvious.

Smith states the theory of bank system expansion due to loaning, as distinct from the individual bank expansion (Phillips was not so novel!).

–This is stated, not as something new, but as an accepted fact.

–Later, in early 19th century the issue was obscured by an attempted distinction between bankers' deposits and other kinds of deposits.

–He implies that every paper note coming into circulation replaces an equal amount of silver.

–Implies (by not discussing it) a constant elasticity of demand for money. cf. Smith on Joint Products – for an almost explicit statement of elasticity concept.

–Assumes the replaced silver then flows abroad.

–Word "overflow" is a dangerous analogy; replaces an economic explanation by a mechanical concept.

–Might mean – (1) higher prices at home (Hume), gold flows out.

(2) Investing capital abroad.

(3) Changing velocity of money.

–The attitude of the statesman, don't bother about the abstract, or the obvious [Double vertical line alongside in margin].

–The Banking School accepted all of Smith except this idea of an outflow of money.

–Concept of separation of circulation and transactions into that used for consumers, and that used for dealers [Double vertical line alongside in margin].

–Leads to concept of the Velocity of Money; number of times money changes hands.

–Also the concept of varying velocities of different kinds of money.

–But did not discuss the idea of velocity of credit [Double vertical line alongside in margin].

–To Smith there could be no uncertainties (in Knight's sense).

–All was a <u>rational</u> system; every one knows what to do.

–If too much money, some flows abroad.

(a) – Assuming higher prices at home.

(b) – Investing abroad (rather advanced to hope to find in Smith)

(c) – Changed velocity.

(d) – Money returns to banks.

–All is automatism, runs smoothly using only individual interest and intelligence as its oil.

–No concept of Central Bank, or of <u>policy</u>.

–Convertibility of gold into paper is always assumed by Smith.

–Given this, an excess of credit is likely to return to the issuing bank.

–Ordinary credit can not influence prices.

–Over-issue is only temporary – doesn't influence prices.

–The banker is a human, non-automatic, element – but he also is intelligent – i.e. liquid.

–Smith said credit should be adapted to the needs of trade – the concept of, if not the word, elasticity.

–Minor mistakes may occur; <u>major</u> mistakes can not occur, unless the government steps in.

–The individual country can never have a very different value of money from the whole world that is on the same standard – for very long, or more than cost.

–In the single country, value of money is based on interaction of supply of and demand for money.

–In world, it is based on labor costs of production – the labor which gold commands is the result of the labor that goes into production of gold.

a – Is labor more valuable because it takes more of it to produce the gold metal, and it commands more – cost of production theory of value.

b – <u>Or</u>, is labor less valuable because more of it entering into production of gold, it exchanges for less in way of gold.

–A demand, productivity concept of value.

–Smith is not explicit – ambiguous.

–<u>Say</u> starts from one side, von Thünen from the other.

–a, is the big point in Smith's theory; in a certain period, if labor produces one-half as much silver, each unit of labor brings one-half as much silver; but one-half as much silver brings prices down by one-half, so that labor still commands as many commodities.

This made Smith feel he had a complete system; and could base it on the constant labor value.

–But, b, Smith changes at times to the concept of a shifting labor value.

–It is easy to have a labor theory, so long as we think of physical units of goods.

–b, brings up a possible quantity theory.

–Ricardo had labor value theory, and a quantity theory of value of money.

–Labor cost is the quantity of money.

–Or, cost of production labor value theory of specie money, and a quantity theory of inconvertible paper.

–And, a natural, cost of production longer run theory, with a market, quantity, short run theory.

–Bullion Controversy begins in 1799.

Bullion Report:

–Palyi says – it implies that control is able to keep even an inconvertible paper up to "full value."

–It implies that the way to do this is via the discount rate.

–Said that it didn't matter what the country banks did, it only matters what the Central Bank does.

–Assumed a constant relation of country issues to Bank of England issues.

–Currency School claimed, control only Bank of England. Banking School claimed that commercial bank deposits had a life of their own – and must be regulated as well.

–Bullion Report implied that control of quantity of paper was all that was necessary to keep paper up to full value.

–Although they urged immediate return to gold.

–Because they had no faith in the Bank of England, whom they blamed for everything – depreciation, over-issue, higher prices, too low interest rate.

–Bullion Report was not very clear on the subject of the proportion of Bank of England notes to country bank notes.

–Implying both that there was a constant proportion, and that there was a decrease in that proportion.

–So that both Currency and Banking Schools could look to the report for a basis (though they dealt with deposits, and the report with notes).

Banking School:

–Bullion Report had implied that the Bank of England was not behaving itself like a good central bank. – Idea of the functions of a Central Bank first set out by Torrance [sic: Torrens], 1808.

–Had been 100% liberal, yet it implied the necessity of some control by a central bank.

–This idea made it very unpopular, in such an age of laissez-faire and government "hands off."

–The Currency School solved this problem by creating a control, yet one in which [neither] the government, nor any individual, had no [sic: a] part.

  –Setting up an "<u>automatic</u>" control.

–In 1810 – the Bank of England argued that they were just as any other bank.

–By 1832 – the Bank of England agrees that they have responsibility for the whole system.

  –Germs of the idea of open market operation, at this <u>time</u>.

  –Bank kept circulation stable, but allowed earning assets (securities), cash, ratio of cash to liabilities, to vary.

    –This brought a too liquid position in depression, a too pinched position in boom.

–The Bullion Committee had urged <u>control</u> of <u>paper</u> currency.

  –The Bank of England, by the 1920s, had accepted the principle of control of a <u>Gold</u> standard currency.

  –But the Bank sought to evade the problem of deciding a <u>policy</u> of control, by keeping everything <u>stable</u>.

    –Once, in 1819, they had attempted a policy of credit rationing – this had not worked.

–1832 – Announcement of policy by Bank of England – based on <u>Currency School</u>.

  –For every pound of paper, a pound of gold goes out of the country.

  –Wanted a strict limitation of paper issue (£14million), and above that pound for pound, paper for gold.

  –Wanted to maintain the <u>mechanism</u> of the Gold Standard – but wanted to <u>shorten</u> the length of time needed for the working out of that mechanism.

    –In order to stop inflation and end the trade cycle.

    –The Bank of England might over-issue, but would be brought to account immediately.

    –For Bank of England had continued, in past, issuing paper money long after exchange had fallen, gold was out-flowing, and trade balance turning unfavorable.

    –Forcing the Bank of England to watch the exchange rate and change its policy accordingly.

    –The Bank of England should have a monopoly of note issue.

    –But only in order to make more efficient the <u>automatism</u>.

      –Yet they were liberals, business men supreme, state has no business with business.

      –Thus, banking is banking, note issue is note issue.

–Banking business must go on independent of note issue.

–Note issue must be separate from banking – notes can create rising prices.

–Currency School held that there was a constant proportion between total of Bank of England notes, and the total of country bank notes: – control Bank of England notes and the total circulation would be controlled.

–First commercial banks in London, 1826 – by 1932 the amount of deposits was negligible.

–Though the Currency School did not deny that deposits could increase and perhaps affect prices, but their amount was so small, they neglected their effects.

–Also, thought that bills increased because prices were higher, because times were booming.

–Keep prices under control, why bother about control of bills.

–Note issue, is simply an extension of gold mining as a source of currency.

–A fundamental was always the attempt to avoid all control but automatic control, but to find that automatic control, and set it up.

–All the men connected with this discussion were business men.

–The Bank of England was responsible for the cycle.

–The Currency School involves a definite explanation of the business cycle.

–Never stated that the Bank of England started the cycle, did not answer that question.

–The Bullion Report was the first action taken by a parliamentary committee against a central bank.

–Banking School (Tooke, McCloud [most likely Henry Dunning Macleod], Fullerton [most likely John Fullarton (1780–1849)], Gilbert [most likely J. W. Gilbart (1794–1863), first manager of the London and Westminster Bank, author], Nelson [Simon Newcomb, Nelson Aldrich?], and Stigler).

–Blamed fluctuations in trade for the business cycle.

–Not money issued, but that money used for expenditure, with its turnover, was the regulator of prices. – The spent-margin, consumers outlay – the flow of money from income, not the issue of money.

–The Currency School answered that if the money spent regulated prices, the money issued regulated them.

–They both talked about the same thing, at bottom – the quantity theory of prices.

–The difference came down to the question: Does the amount spent vary in the same proportion as the amount of money issued?

–Currency School always oversimplified – they also neglected the velocity of money.

–And they always assumed a rational attitude, thus being led to neglect irrational factors in the situation, such as hoarding.

–HeSlferich (German, 1838) [likely Karl Helfferich, but 1838 date is puzzling].

Two types of Quantity Theory:

(1) Mere functional relationship; algebraic

–A formal expression for the demand for money (Pigou).

–On one side is money; on the other side is the physical aspect – no causal explanation [Single vertical line alongside in margin from (1) to here].

–Banking School – there can not be an excess or deficiency of money. Price level is influenced by physical side only.

(2) An explanation of the cause of exchange.

–Currency School-Money is coin and paper – thus both influence prices. Money is caused by humans – i.e. the Bank of England or balance of trade.

–Credit is money, only in the sense of function, is not in the senses of causal meaning; it is derived money.

–Tooke's income approach is the same as the Quantity Theory – there is still the functional relation of the two sides.

–Money increases, more paid out in salaries, wages, more spent, thus higher prices.

–Leaves out the time necessary.

–Most elementary form of Quantity Theory:

–A pile of money units, a pile of goods – each exchanges for same unit of the other.

–Then comes idea of a flow of money against goods.

–Currency School thought in terms of a fixed volume – annual.

–Total product (cost of production = wages fund = capital) is moved by money.

–Fixed standard of life for labor, fixed capital.

–Banking School – As long as the process goes on nationally, there can be no disturbance; even the creation of new money does not disturb.

–Currency School – the creation of new money by Bank of England caused a disturbance – temporary.

–Bank of England should restrict credit according to the output of new gold.

–Banking School – the attitude of the Bank of England is not important – any disturbance comes from the goods side; it is the Bank's business to provide money (even when it goes to speculation).

–If it does not, other banks will.

–Fundamentally, there is not a great difference between the two schools.

   –Both are practical men, little theoretical interest.

   –Rather it is a difference of interpretation of certain facts – and of the policy.

–Both schools were advocating laissez faire.

   –But Currency School advocated control of the Bank of England in order to keep it from controlling prices.

      –Laissez faire always advocates interference to stop interference; to "restore" "competition."

      –Malthus-Say correspondence.

–Both schools agreed on the matter of necessity of liquidity. – Only one business for a Commercial Bank – commercial loans.

–But Banking School analyzed and built up the theory of liquidity – Currency School was not interested.

   –One type of practical man was impressed by the power of the Bank of England over credit and crises – short run.

–The other type of man was interested in the long-run aspects.

–The Currency School agreed with the Banking School as to long run point of view – but could not agree in short run, mostly because Banking School would not discuss the short run [Single vertical line in margin alongside these three lines].

–The "City" must have been at that time tremendously nervous, speculative – liable to panic.

   –The Banking School said, what of it, if we are liquid, everything will be all right (plus panic policy of accommodation).

   –The Currency School tried to keep the panic from occurring.

–The whole discussion was almost superfluous – a classic example of improper definition of the problem.

Changes in prices coming from side of changes in value of money do not influence all prices by the same amount.

–Individual prices have a tendency to move at least in the same direction.

–Total output and total demands are different from individual output and individual demand.

   –Are what we are really interested in is prices.

–Mathematical approach cuts free from the influence of monetary factors. – Money changes only the constants.

–Monetary theory takes output as a whole and relates it to money.

–A monetary system is a managed system – either conscious or unconscious.

   –This policy of management influences total output, total demand, total value.

In Europe, after the war a shift from discussion of socialism <u>vs</u>. capitalism, to discussion of monetary policy in relation to either.

Ansiaux
   French policy of bank rate since 1860 was to keep it stable.
      –The Bank of France does not <u>make, elle souffre</u> changes in interest rate.
      –Gold flows disturb equilibrium.
      –The Bank was thought to be a place where the interest rate was always <u>stable</u>
      – where the businessman <u>could</u> always get money at 4%.
   –Stabilization of interest rate was a national policy for nearly a century – for all parties.
      –[Pierre Joseph] Proudhon wanted stable rate.
      –Blanc [unlikely but possibly Louis Blanc; possibly Jerome-Adolphe Blanqui] wanted stable rate.
   –Stability of gold reserve also wanted (had highest per capita gold stock of any country before war.
   –Narrow gold points.
   –Had silver 5 franc pieces that were legal tender – were paid out to those who wanted to redeem <u>too much</u> gold.
   Ansiaux returns to <u>Mercantilist</u> viewpoint re gold.
      –Classical theory said a country never can lose its whole gold stock.
      –Aimed at policy of stability – of interest rate and of employment, of prices.
      –Since there was this high gold reserve, flowing out of gold meant flowing out of <u>the currency</u>.
      –He fought against the French version of the <u>Banking School</u> – pure automatism.
      Palyi says Ansiaux wanted mainly a <u>stable</u> interest rate. The classical theory, even where it worked, brought a quickly fluctuating rate.

Protection –
   a – <u>Progress</u>-protection – like 17th century Mercantilism.
      –Aimed at developing the State, industrializing it.
      –Russia, Turkey, Persia, Japan.
   b – <u>Stability</u>-protection – aimed at <u>freezing</u> everything <u>as is</u>.
      –Very different from 19th Century protection.
      –Protection of <u>every vested interest</u> (rigidity)
      –And assignment of <u>permanent status</u> in society.
         –Italian corporations, Nazi corporations, land laws.
      –Forbidding of <u>new</u> technological devices; i.e. no part of system can be allowed to <u>change</u> – as that breaks down rigidity.

–This rules Germany, Italy, Austria, and indirectly, less developed, in England, France and Gold countries.
  –Gold and paper are equally managed.
  –Prices are to remain stable – where they are – neither rising nor falling.
–Stabilization of every aspect of economic structure.
–Corollary: National self-sufficiency.
–Even inflation is aimed only to keep things rigid and stable.

Intellectual Background:
  –All of intellectual background is imbued with an anti-progressive, no-change outlook.
–Return to romanticism, mysticism – all aimed at stopping any change in reality.
–This is the Farmer's attitude, and that of the shaky banks.
    –In economic aspects it was begun to aid these two groups.
    –Not true of countries where farmer is interested in exports.
  –Western Europe – already industrialized, wants to end progress.
Eastern Europe – the scene of progress towards industrialization (is this an inevitable partner of a certain degree of industrialism? – what degree? – in other ages? ["F.T.O."])
Expens[ive] – reduces productivity, increases cost of administration (wherever unemployment increases – bureaucracy increases – as an attempt to sit on the cover even more).
  –Is weighted neatly on urban rich, rural rich escape.
Before War – for economists, there were only two alternatives – Capitalism or Socialism.
  –Now, there is no difference – Proudhon, the only writer to attempt to bring these together, has won.
  –Not even in Russia is the Capitalism vs. Socialism struggle carried on.
  –Everyone is socialist now. Dominant groups believe in neither Capitalism [n]or Marxism (progressive socialism).
  –They believe in socialism, nonetheless, of the Proudhon-small middle-class-idea of socialism – i.e. stability, anti-progress (Give credit to those groups who would not be able to get it under free capitalistic system).
  –Intellectuals are against changes hurting major groups (as by price changes; thus: managed economy).
    –Are in favor of autocracy.
    –And degenerate into isolation, self-sufficiency, nationalism.
    –Propaganda agencies in all countries.
    –In most countries, the intellectuals have given up their liberal ideas – are not liberal capitalists or liberal Marxists (i.e. they believe in free progress).

–Technological progress is squelched everywhere – it is so sudden and immense in its changes.

–Intellectual and economic policies are not one cause [or] the other result – but mutually condition each other, develop side by side.

–Intellectual life turns to the irrational, anti-rational, mystical, romantic.

   –But there are <u>arguments</u> for the irrational approach.

   –Literature turns to the irrational, non–discussional.

   –Philosophy also turns – and brings up the medieval idea of states (St. Thomas Aquinas, <u>minor</u> guilds) – Medieval idea never worked for a long period – except in <u>Germany</u>, Middle Italy – (France).

Intellectuals agree with the fundamental idea of stabilization *à la* Hitler.

   –Liberal philosophy and politics develops <u>emergency</u> plans.

   –Stabilization philosophy and politics develops <u>permanent</u>, 1,000 years, plans.

   –Labor carries the ideals of <u>progress</u> – this is why <u>labor</u> seems to be turned against, all over.

   –Criticism is a <u>discursive</u>, rational thing – thus squelched.

   –The persecuted groups are usually [the rest of the page is blank].

<u>Hayek</u> – [oral report] by [Albert G.] Hart

1 – Wants <u>equilibrium</u> as defined by mathematical economists – it <u>does</u> not come about, Hayek wants to know why and what to do about it.

2 – Equilibrium of neutral rate of interest ["natural" written in above "neutral"].

3 – Equilibrium in the supply (constant) of circulating mediums.

   –Is Hayek interested in changes in velocity?

–Hart admits that Hayek never speaks of equilibrium in connection with 2 and 3, only in connection with 1 [above].

–Equilibrium always refers to <u>price</u> (Given supplies and demands <u>at those prices</u>).

– full employment of <u>labor and capital – at that price</u>.

–Palyi: As Hayek never <u>says</u> anything, Hart is correct in trying to interpret.

–<u>Hart</u> says, <u>Hayek regards these three as identical</u> (that is one reason for his never discussing equilibrium in terms of 2 or 3). <u>If they are not identical, Hayek' s theory collapses</u>. – although Hart and Palyi do not think these three are the same.

–Hayek assumes a quantity theory of money, and a transactions-velocity constant.

–Increased production = falling prices = falling wages.

   –Roundabout production, more intermediate goods, same quantity of money, falling prices.

i.e. equilibrium <u>is</u> not disturbed by technological change, if value of money remains constant.

–This is what an explicit, rational, Hayek would have said, if he existed.

–What Hayek said was that a change of volume of money consequent on technological change disturbed equilibrium.

Palyi: Hayek has <u>no theory of depression</u>, thinks there is no theory of <u>cycles</u>. Only, there is a <u>theory of boom</u> – and depression is an inevitable consequence.

–But boom is not an inevitable consequence of depression.

–In a static, non-monetary equilibrium, all maladjustments would move themselves out easily.

–It is the introduction of <u>money</u> that prevents such automatic readjustment.

–<u>Horizontal disproportionality</u> – between one group of consumers' goods overproduced, another group underproduced [In margin at top of page just above the preceding: "Between goods taking the same length of time"]. – many will cause no <u>further</u> maladjustment.

–<u>Vertical disequilibrium</u> – is self-adjusting only if the rate of interest remains the same, i.e. only if money – length of production increased or shortened – mostly, a maladjustment <u>within</u> producer's goods.

–Marshall, says Palyi, knew that interest rate and capital were intertwined, but knew also that <u>other</u> elements were also important – such as, especially to Palyi, wage rates.

–Hart says that most English textbooks, especially Henderson, say this same thing, that the rate of interest has an effect on the vertical distribution of resources.

–Palyi will not let Hart say that all textbooks say the same thing as Hayek, i.e. that interest rate <u>rules</u>, is the only factor in, the vertical distribution of resources.

–Palyi says – Hayek does not compare <u>the</u> interest rate with other costs, except the <u>market</u> rate of interest.

–Hart – says Hayek protects himself by ruling out technological changes.

–Palyi says – does this mean that every producer has taken <u>full</u> advantage of the <u>existing</u> technology?

–Hart says <u>Yes</u> – each producer produces at the <u>most favorable</u> use of existing technological factors.

–Palyi qualifies this: most favorable, <u>at the</u> existing price level, but any change <u>in</u> prices will make <u>other</u> use of that technology profitable.

–Hart: the will of the consumers is the all-regulating God to the Austrians.

–Banks expand loans; increase circulating medium, causes vertical disequilibrium. Nobody has <u>willed</u> the saving which follows (forced).

–Quantity theory re nominal purchasing power (prices).

–If this did not happen, there would be no business cycle [Arrow in margin from "Banks expand loans" to this line].

–As it does happen – there must, still, be some end to the creation of credit, which brings on the crisis.

–A rise in rate brings on contraction, or stoppage of expansion, this brings on crisis.

–Did interest rate or contraction cause crisis?

–Hart says, the theory of origin of the cycle is different in Geldtheorie and in Prices and Production.

–Palyi asks: is there anything in Hayek that shows why a stoppage of expansion is necessary? It may be likely to end, or common sense may show that it must end, but what is there to make the end logically necessary?

–All that Hayek says is that the reserve ratio is impinged upon – this is his necessary cause of the contraction.

–Palyi says there are three points of controversy here.

–If there were no reserve ratio – would there still be a necessity of stoppage?

–Or, if there were a reserve ratio – what would of necessity turn the stoppage into a contraction?

–Hayek: when expansion stops, a shift in the proportion between the demand for producers' goods and the demand for consumers' goods occurs, which is a crisis.

Keynes – [oral report by] Rose Director
   –Labor Standard and Purchasing Power Standard
   –Consumption goods (price level ["standard" written in above preceding two words] – composed of a composite unit of consumers' goods).
      –A cost of living standard.
      –Not interested in stabilizing prices far away from consumers.
      –But is interested in stabilizing this standard of consumption goods.
   –Currency Standards:
      –Cash Balances Standard: Marshallian, Keynes (early) and Hawtrey.
      –Cash Transactions Standard: Fisher.

–An index number must be considered, as at a fixed point of time and with fixed commodities–thus it must be changed with change in time and commodities.
–Keynes drops his critical point of view of Book II, in Book III [*A Treatise on Money*, 1930].

–Having chosen the price level he wants to stabilize (consumption goods), and the index he wants to use, he forms his equations.

–Equilibrium implies full employment of labor – and factors of production.

–Stabilization may mean stabilization of a price level, or of employment of labor (efficiency earnings).

–Palyi thinks Keynes chooses stabilization of employment but in Vol. II he comes back to a wholesale index number – as more easy.

–Hart says it is possible to have the equation in equilibrium, still have unemployment.

–Equations: –Amount spent on consumption goods is equal to amount received for value of consumption goods sold – no problem of time – a mere transfer problem.

–But he says also, that this is equal to cost of production of consumption goods – this raises a problem.

–Efficiency earnings = cost per unit of output.

–If second term of equation = zero, then price (per unit) = cost (per unit).

–Hart – What of costs? Are Keynes's costs really costs?

–If so the common sense approach is adequate.

NOTES FROM MELCHIOR PALYI'S COURSE, BUSINESS CYCLE THEORY, ECONOMICS 333, UNIVERSITY OF CHICAGO, 1933–1934

Edited by Warren J. Samuels

In his opening methodological lecture, Palyi contrasts correlation, first with statistical analysis, and second with causal analysis and explanation. The students are cautioned that the "Correlation method creates the presumption of an accurateness which it does not in reality possess" and that "Business cycle theories are <u>not</u> generally *a priori* – not if worth anything, but are <u>induction plus inspiration</u>." One wonders what his answer might have been if queried whether *a priori* theories *of any subject* are "worth anything" and whether "induction plus inspiration" has or creates the presumption of accuracy greater than correlation.

Palyi stipulates that "No theory can have logical flaws; nor contradict the same author's price theory; it can not violate the facts; must explain and cover <u>all</u> relevant facts." In retrospect, these requirements are difficult to achieve: Logical flaws can be due to errors of logic per se or to putative conflicts within the economics or to the Duhem-Quine problem. Not contradicting the relevant price theory both privileges price theory and can involve a conclusion denying the possibility of business cycles. A theory not violating the facts depends on what theory the facts are constructed on. One doubts if any theory can "explain and cover all relevant facts" (quite aside from the statement being a tautology). Neither

Documents from F. Taylor Ostrander
Research in the History of Economic Thought and Methodology, Volume 23-B, 331–346
Copyright © 2005 by Elsevier Ltd.
All rights of reproduction in any form reserved
ISSN: 0743-4154/doi:10.1016/S0743-4154(05)23110-3

internal consistency nor coherence, as tests of theory, apply themselves. Palyi himself cautions, "Most theories discuss realms in which verification from facts is impossible." Finally, the assumption of one-way answers to a problem, while still conventional, is limiting, especially in view of developments in terms of cumulative causation, general interdependence, general equilibrium, recursive modeling, and over-determination. This particular assumption is odd given that Palyi incisively if not brilliantly distinguishes between theories providing explanation in terms of causation and those in terms of mechanism, the latter "dismissing an <u>ultimate</u> cause." It is also inapposite to conventional price theory, in which changes in demand can lead to changes in price and changes in price (due, say, to changes in supply) can lead to changes in the quantity demanded, i.e. not a simple one-way relation but the price *mechanism*. It is not too much to say that eventually models will increasingly comprise not causal relations but adjustment mechanisms (for example, Keynes's model comprises the following mechanisms: multiplier, multiplier-accelerator, liquidity preference, marginal efficiency of capital, and so on) – with due recognition of the difficulties escaping the language of causation.

To the historian of economic thought, Palyi presents a two-part argument of enormous interest: "Most theories are created for the <u>purpose of expounding some policy</u> or another – those that do not do so are for the most part poor theories." The first is a candid statement of genesis – one taking the externalist or relativist position – and the second a subjective view of theory choice. The statement is no fluke; on the next page, one reads, "Individuals who study things do it with the aim of understanding . . . and aim of deciding policy." Here we have, briefly stated, an exposition of the social role of the man of knowledge – including the claim to be a player in the game of power.

As for preferences for different methodologies, Palyi suggests "Different methods are mainly a matter of different personalities." He "Regards the theoretical [approach to] analysis as a phenomenon to be explained" and "Also approaches the social, political, etc. backgrounds of the <u>theorist</u>." In these matters, Palyi is anticipating several developments in the study of science and in the philosophy of science in the latter third or so of the twentieth century. That includes the variety of factors used in explaining one's preference for theory etc. and the preferred approach to policy.

When Palyi moves on to discuss business cycles, he seems, in part, to echo Thorstein Veblen's emphasis on a pecuniary economy and, thereby Wesley C. Mitchell's business-cycle research program in which "changes in market valuation are of supreme importance."

Words carry definitions, and definitions in economics often tend to embody and give effect to theories (some, of course, do not); thus are models instantiated. Palyi is not unusual in this respect and in not grounding the theory underlying the definition: "The concept of cycle," he is recorded in these notes as saying, "must be restricted to the short-term <u>shift</u> from one employment to another." Subsequent discussion, however, indicates Palyi's awareness of the problem of defining business cycles in terms of one of its many facets. He also notes, "There is no precise, clear definition of the cycle." There were, of course, many definitions, each more or less precise, each more or less clear. But phenomena do not label and define themselves; both label and definition are a function of non-deliberative and deliberative endeavors. Another interesting aspect of all this is that those who accepted the "reality" of the business cycle, each tended to define and mean different things by it.

John Maynard Keynes first arises in the context of a discussion of the classification of business cycle theories into those generated by extra-economic and by intra-economic forces. The latter are self-generating, and their readjustment is self-correcting. Palyi says of Keynes, probably already the world's foremost advocate of active monetary management of the economy, and Palyi's <u>bete noire</u>, that he "accepts the self-correcting theory, but neglects it and demands artificial correction."

The discussion is structured, by six dichotomies: (1) that between intra-economic and extra-economic, driven by some undefined specification of the economy; (2) that between natural and artificial, also driven by some more or less ambiguous notion of what is normal, i.e. natural; (3) that between discussions of one cyclic stage leading to another (initially a narrower division) and discussions of forces driving the system (the stages being only stages, not solely causal in themselves); (4) that between depression as the name given to the downward stage of a cycle, eventually called a recession or downswing, and depression as a phenomenon of a vaster order of magnitude – amplitude and duration – than a recession; (5) that between theories of mechanism and of causation, the former said to constitute an economic approach and the latter a sociological one – and that "A combination of the two gives the cycle-theory problem;" and (6) that between initiating factors and (adjustment) mechanisms.

The relationship between cause and adjustment mechanism is not always clear. A cause in one theory or model can be an adjustment mechanism (even if not so labeled) in another. Both can co-exist in the same theory or model, e.g. the mechanism being (part of) the process through which the change wrought by causes works its way through the system. The difference between cause and mechanism may be minimal in the context of cumulative causation, in which a consequence in one period becomes a cause in the next. One might read Palyi's reported discussion on psychology, for example, in such a context; also Keynes's later model in which income is a function of spending and spending, in part, is a function of income (and, in part, of psychology).

Further in these notes, after a review of various business-cycle theories, Palyi notes the important distinction and the important limit: "These theories show a part of the mechanism through which a change works itself out – do not explain the change."

Palyi is recorded as saying that "Most psychological theories assume changes in the psychology of the entrepreneur or capitalist," but "There are no psychological theories of changes in labor psychology, or of changes in consumer psychology" and that the latter, changes in consumer psychology, "is especially important for hoarding." Three points: (1) The focus on the entrepreneur or capitalist is because of their central role in investment and, following Keynes, the central role of expectations in investment decision making (through the marginal efficiency of capital); (2) Selig Perlman, for one, did have a theory of labor, and capitalist, psychology (the former job conscious and the latter opportunity conscious), though not of changes in their psychology; and (3) Changes in consumer psychology are not unimportant for hoarding – liquidity preference – but are more important for spending in the generation of income. Hoarding is not the crucial alternative to spending on new goods; it is one leakage of several from the spending flow in which portfolio investment *vis-à-vis* real investment looms larger.

Palyi noted that "High costs either make for depression because [of] low profits or they make for boom, because [of] high wages." If he had translated those insights into a tendency model of spending, he might have produced a Keynesian *General Theory*-like model before Keynes. Why he did not is suggested in part by the language of causation he uses in discussing Jevons:

– More saving in the upward trend – i.e. more production of capital goods.
– Less saving in downward trend – more production of consumption goods.

In the first statement, Palyi envisions saving financing investment; he thereby neglects: (1) investment financed by new bank credit; and (2) that more investment leads to greater saving, not the other way around. In the second statement, he sees saving and investment as alternatives, whereas less saving is due to less income, which also means less consumption, thence less investment, further lowering income, et seq. He soon thereafter says, in connection with Spiethoff, that, "People invest more – for some reason or other – not merely save, but invest in capital goods." This seems to avoid the false saving-or-investment dichotomy and also make room for portfolio investment as an alternative to real investment (the model used to make the foregoing points is more complicated; hence *ceteris paribus* attaches).

Perhaps most remarkable about the notes is the total absence of the word – and concept – "unemployment." The only time it arises in the pages relating to

Economics 333 is in the title of Arthur Cecil Pigou's *The Theory of Unemployment* (1933) in the Bibliography. If the term was used, Ostrander did not record it; certainly, if used, it did not rise to the level of importance to lead Ostrander to record it. One major shift in economic discourse suggested by these notes, therefore, is that which makes unemployment (more or less negatively correlative with income) a central concept of modeling, analysis and policy – a result of the Great Depression and either a part or a result of the Keynesian revolution.

Some readers might infer from Palyi's lectures that monetary and banking institutions are created on the basis of pragmatic political and economic decisions and ideas (theories), then theories are created to explain those institutions and how they work. The theorists treat the existing institutions as natural (and even "automatic") and, sometimes, their revision or replacement as unnatural, though these latter come about in much the same way.

ACKNOWLEDGMENT

I am indebted to Daniele Besomi for remarkable assistance in identifying several of Palyi's citations.

APPENDIX A
NOTES ON MELCHIOR PALYI'S COURSE ON BUSINESS CYCLE THEORY, ECONOMICS 333, UNIVERSITY OF CHICAGO, 1933–1934

(1)

Economics 333 – Business Cycle Theory   Dr. Palyi

BIBLIOGRAPHY

Aftalion, A. (1913). *Les crises économiques et financières*. Paris.

Bouniatian, M. (1930). *Les crises économiques*, 2nd edition, Paris.

Cassel, G. (1924). *The Theory of Social Economy* – 2 Vol. New York.

Cox, G. V. and Hardy, C. O. (1927). *Forecasting Business Conditions*. New York.

Hansen, A. H. (1927). *Business Cycle Theory*. Boston and New York.

Hawtrey, R. G. (1929). *Currency and Credit*.

Hayek, Fr. (1933). *Monetary Theory and the Trade Cycle*. New York.

Juglar, Cl. (1869). *Des Crises Commerciales et leur Retour Périodique*. Paris.

Keynes, J. M. (1930). *A Treatise on Money*. London.

Kuznets, S. (1930). *Equilibrium Economics and Business-Cycle Theory*. Quart. J. of Econ., 44.

Lavington, F. (1922). *The Trade Cycle*. London.

Lescure, J. (1910). *Des crises générales et périodiques de surproduction*. Paris.

Marx, Karl (1890). *Capital*, Vol. III. New York.

Mitchell, W. C. (1927). *Business Cycles, the Problem and its Setting*. New York.

Pigou, A. C. (1927–1929). *Industrial Fluctuations*. London.

Pigou, A. C. (1933). *The Theory of Unemployment*. London.

Robertson, D. H. (1913 [1915]). *A Study of Industrial Fluctuation*. London.

Robertson, D. H. (1926). *Banking Policy and the Price Level*. London.

Schumpeter, J. (1912, 1929). *Theorie der wirtschaftlichen Entwicklung*. Leipzig and Munich.

Spiethoff, A. *"Krisen"* in *Handwoerterbuch der Staatswissenschaft*, 4th edition.

Stucken, R. (1926). *Theorie der Konjunkturschwankungen*. Jena. [New edition: *Die Konjunkturen im Wirtschaftsleben*, Jena, 1932.]

Tugan-Baranovskii, M. (1901). *Studien zur Theorie und Geschichte der Handelskrisen in England* – Jena. [*Les crises industrielles in Angleterre*, Paris, 1913.]

Veblen, Th. (1904). *The Theory of Business Enterprise*. New York.

Wicksell, K. (1908). *Geldzins und Güterpreise*. Jena.

Wicksell, K. (1922). *Vorlesungen über Nationalökonomie*, Vol. II. Jena.

Zimmermann, P. (1927). *Das Krisenproblem in der neueren nationalökonomischen Theorie* – Halberstadt.

(2)

I-<u>Methodology</u>
Ups and downs, show themselves in <u>more</u> or <u>less</u> typical or rhythmical cycles.
1. <u>Quantitative Approach</u>
    – <u>Correlation</u> (+1 to –1).
    – 1. correlation: a always met by $-b$
    ± 0. correlation: a never occurs with b.
    – H. M. Moore
    – Became the typical <u>American</u> approach.
    – Statistical analysis does not imply correlation – rough consideration of data
    – pre-war.
    – Only correlations above about + 0.8 mean anything – + 0.5 means only a
    probability.
    – Correlation method creates the presumption of an accurateness, which it
    does not in reality possess – the old method not so far wrong.
    – Measurement of <u>time-lags</u>
    – Correlation applied not between items of the same type but between types.
        – Correlation coefficient might be very different according to the time-lag
        chosen.
            – Comparison of changes per period of time and per segment of cycle
            with other changes per time-period and cycle-segments.
2. <u>Theoretical Approach</u>
    – Does not <u>compare</u> data, figures.
    – But <u>analyses</u> data and figures.
        – Brings figures into a logical system.
        – Establishes <u>causal relations</u> and <u>explanation</u>.
        – Statistical approach does not lead to causal or explanatory study.
            – Merely correlates a few series.
        – Theory implies <u>systematic</u> approach – and has as much validity as systems
        usually have.
        – Business cycle theories are <u>not</u> generally *a priori* – not if worth anything.
            – But are <u>induction plus inspiration</u>.
    – We shall classify and analyze business cycle theories.
        – No theory can have logical flaws; nor contradict the same author's
        price theory; it can not violate the facts; must explain and cover <u>all</u>
        relevant facts.
        – Most theories discuss realms in which verification from facts is
        impossible.
            – Data not available.
            – Where one theory tries to explain causal sequence and another tries
            to explain consequences.

        – A <u>theory</u> must assume a one-way answer to the problem, does the
cause lead from <u>a to b</u> or from <u>b to a</u>.

        – Business Cycle theories – may mean:

         (a) An explanation in terms of <u>cause</u>.

         (b) Or in terms of <u>mechanism</u> – dismissing an <u>ultimate</u> cause.

        – Most theories are created for the <u>purpose of expounding some policy</u>
or another – those that do not do so are for the most part poor theories
[Single vertical line alongside].

3. <u>Historical Approach</u>

    – Statistical is historical, and historical uses statistics (might use correlation
too).

    – Historical goes back further than material for correlation is available.

    – Not theoretical, is based on facts.

    – Arrives at some generalizations.

    – Different methods are mainly a matter of different personalities – historian
is more interested in differences in cycles than in similarities.

4. <u>Sociological Approach</u>

    – Regards the theoretical analysis as a phenomenon to be explained.

     – In terms of institutional, historical, sociological, psychological etc. points.

     – Also approaches the social, political, etc. backgrounds of the <u>theorist</u>.

    – Individuals who study things do it with the aim of <u>understanding</u> – 1.0 seek
a causal theory – and aim of deciding policy.

    – We concentrate on historical and theoretical aspects – mainly latter.

5. <u>Policy Approach</u>

    – Statistical approach aims at business man's policy, and social policy.

    – Theoretical and historical approach[es] aim at policy, or criticism of policies.

    – Historical approach studies the different <u>policies</u> used in treating cycles in
the past.

II. What do we mean by business cycles?

    – It <u>seems</u> to be a specifically <u>capitalistic</u> phenomenon.

      – We do not know what have been the conditions in feudal and pre-feudal, or
would be in a <u>socialist</u>, society.

      – They were not without <u>fluctuations</u> and <u>changes</u>.

      – What would be their place in a socialist society with no market?

    – What <u>is</u> a <u>modern</u> cycle, what is not?

    a – Cycle = <u>regular</u>.

      – Konjunktur – does not imply so much of regularity.

      – <u>Time</u>-regularity – old concept – ten years.

       – Begun by Marx, on European 19th century experience.

– <u>Pattern</u>-regularity – rhythmic progress of certain phenomena.
– No movement of time not under cyclical conditions – i.e. <u>no</u> static interval
  – more recent view – continuous.
  – Although it may not yet be apparent which way the cycle is tending –
    an <u>appearance</u> of equilibrium – old view was that cycles were bubbles
    on a straight line – sporadic.
b – <u>Palyi says</u> the <u>modern</u> cycle is different from older change in business
  activity.
  – Old: good crop always was a boon, war always a depression.
  – New: good crop may bring depression, war may mean a boon.
  i.e. paradoxical aspect, not of the depression (abundance yet poverty), but
  of the cycle (abundance yet poverty, <u>and</u> riches yet scarcity).
  – Previous societies were not based on <u>valuation concepts</u> – few markets,
    little barter, little money.
  – Modern society – changes in <u>market valuation</u> are of supreme importance
    [Double vertical lines alongside this point].
– How define what is the <u>up</u> and what is the <u>down</u> of a cycle?
  – Often we think we are sure but find it hard to <u>define</u> our standards.
  – At other times we are mistaken – Germany, in 1930, thought herself at
    bottom of depression; after crisis of 1931, found she had been having a
    small boom.
a – Usually <u>employment of fixed capital and of labor</u> – is the criterion [Single
  vertical line alongside this point].
  – A good crop does not mean a higher employment of factors.
  – A war does not mean a lower employment of factors.
  – In earlier times, the employment of factors was <u>quite stable</u>, no labor
    market, no capital market – thus crops and wars are the only fluctuations.
b – Or, sometimes, employment of capital in the form of money – utilization
  of <u>monetary</u> capital.
c – Labor employment seems to be the infallible test.
  – 1930: – Less labor employed per unit of capital than in 1900.
    – The relative proportion changes, but the <u>average</u> did not change, and
      it's generally agreed to that the <u>average</u> employment is what counts.
d – The concept of cycle must be restricted to the short-term <u>shift</u> from one
  employment to another.
  – Hawtrey, cash reserves fall = boom, [or] vice versa.
e – Capitalization and earnings.
  – Capital values rise in upswing – more profits, also lower income.
  – Define business cycle in terms of increased capitalization (?).
  – But interest rate.

- Employment aspect is disregarded by the valuation concept.
- A rising tendency is the characteristic of the cycle – not a large stability.
- There is no precise, clear definition of the cycle.
- A cyclical situation is either a rising or a falling phenomenon.
- Eng[lish]-Cambridge school is unbalanced: employment index leaves out cycles, or provides them ahead of time.
- Valuation must be the same, all over, impossible to take account of money change.
- i.e. the cycle remains a matter of judgment – the closer we get to it, vaguer it is.
- Fixed capital and labor averages may go in opposite directions – these two can not be averaged.
- Seasonal fluctuations are only really cyclical ones – connected with physical facts – do not involve unknowns – might be forecasted.
- Cyclical changes are irregular in time – they do reoccur, but at irregular intervals – of major swing.
- Cycle = changes from boom to depression – is obvious, is glaring.

Theories of the Business Cycle
- Classified, according to the points considered by their proponents as the relevant ones – not ultimate classification.
- Changes in volume of employment and in capital values.

Bergman – Geschichte des Krisentheorie, 1890s – not using word Konjunktur! – contains over 40 theories, then incomplete [Eugen von Bergman, *Die Wirtschaftskrisen. Geschichte der nationalökonomischen Krisentheorien*, 1895] [Single vertical line in margin alongside this reference].
- Interest in cycles is itself a cyclical matter.

Herkner, 3rd ed., Handworterbuch der Staatswissenschaften (Krisen) [Entry, Heinrich Herkner, "Krisen," in Conrad's *Handwörterbuch der Staatswissenschaften*, band IV, 1st ed., 1892, pp. 891ff, 2nd ed., 1900, pp. 413–33].

Spiethoff, 4th ed. – after war [Likely "Business Cycles," *International Economic Papers*, vol. 3, pp. 75–171, 1953, partial translation of "Krisen," in *Handwörterbuch der Staatswissenschaften*, 4th ed., 1923].

Hansen – Most recent classification [Alvin H. Hansen, *Business Cycle Theory: Its Development and Present Status*, 1927 and *Economic Stabilization in an Unbalanced World*, 1932].

Mitchell – Business Cycles [Wesley C. Mitchell, *Business Cycles*, 1913 and *Business Cycles, The Problem and Its setting*, 1927]

Clark, J. M. – Study of the Business Cycle (method of approach to the cycle) [John Maurice Clark, *Strategic Factors in Business Cycles*, 1934] [Single vertical line in margin alongside Herkner through Clark references].
– Are all out of date.
– Hansen is best – but discusses "monetary" and [blank space: "cyclical?"] theories as distinct.
  – Modern theory lumps them together.
  – Is an empirical classification – what has been theorized.
– Classification should be in terms of all possible fields of explanation.
  a – Extra-economic forces
    – Harvests, sunspots, wars not of economic origin, psychology, change of policy or institutions or technology, gold production,
      – Accidental, sharp breaks.
      – Normal would persist unless outside disturbance.
      – Both boom and depression are accidents.
  b – Intra-economic forces
    – Changes occurring out of normal course of economic process – almost necessarily arising or tending to arise.
    – Waves – normal, is rhythmic abnormality.
  – Some theorists are to be found in both camps, e.g. monetary theorists believing in accidental origin, or in inherent economic origin.
  – Marx is under b) – a perfect example, the crisis arises out of the profit system.
  – Most post-War theorists are in between these two approaches – try to make room for everything.
  –b) [supra] implies concept of equilibrium, the cycle is disequilibrium.
  a) [supra] believes in a normal.
  – A difference of generations – men growing up in 1870–1890, believe in depression, men growing up 1900–1925, believe in boom.
    a – accidental; b – self-generation.
      – Cycle does not need such a theory.
    – If (a) is involved, the maladjustment might continue indefinitely.
    – If (b) – then the readjustment is self-correcting.
      – Keynes accepts the self-correcting theory, but neglects it and demands artificial correction.
  – Subdivisions of (b) intra-economic forces:
    (i) Mechanism applies only to one cycle, a boom generates a depression.
    (ii) Mechanism is complete.
      – Depression generates a boom, which generates a depression, boom, etc.

– Why does every one begin with the boom – a universal "arbitrariness"?
  – It is impossible to have a depression without having had a boom.
  – It seems, empirically, that no depression <u>has</u> to precede a boom.
  – Palyi calls this common sense underlying theoretical approach.
  – a – Accidental theories are as numerous as possible accidents, cf. Robertson, Industrial Fluctuations.
  – To what extent can this be <u>called</u> a theory?
  – A <u>pure</u> theory is a theory of mechanism, not of causation – <u>economic approach</u>.
  – The study of causation underlying is a sociological approach.
  – A combination of the two gives the cycle-theory problem.
– Theories of mechanism:
  – Quantity of money – <u>monetary</u> theories – includes hoarding, which is initiative as well as mechanism – supply and demand of money.
    – Non-monetary:
      – Equalization of production and consumption (not initiative) – supply and demand of goods.
      – Psychological – not accepted by Palyi – except as an <u>addition</u> to the monetary and non-monetary mechanisms.

I. <u>Extra-Economic Theories</u> (<u>change</u>, is fortuitous, not inherent)
  1. <u>Crop theories</u>, etc. <u>physical</u> theories
  2. <u>Underconsumption</u> theory (purchasing powers) – forgets cost side
  3. <u>Cost-Profit</u> theory; <u>entrepreneurial</u> attitude – forgets sales side – 2 and 3 mutually exclusive
  4. <u>Psychological</u> theory
    – External psychological forces creating cyclical changes.
      – Mob or mass psychology.
      – The entrepreneurs are the mob, in the first instance.
    – Something in human nature causes such changes in psychology, or they are inexplicable.
    – Ups and downs of optimism and pessimism.
    – Why does everybody have the <u>same</u> sort of cycle of psychology?
      – Individual psychology does not explain this.
      – <u>Social psychology</u> may explain it – or an outside, economic, force.
  <u>Taste</u> – leaders get optimistic for some reason or other – others follow by imitation – what vision?
  <u>Propaganda theory</u> – it causes booms – but, more propaganda in depression that in boom.
  Among economists, there are two main lines of approach.

      a – Psychology not a cause, only a <u>medium through which econ[omic]</u> <u>change works</u>.
        – <u>Schumpeter</u> – entrepreneur psychology goes up and down – but, in the ultimate, it is the banker's readiness or unwillingness to lend that causes this entrepreneur cycle.
– Even this is not ultimate explanation.
      a₁ – Psychology not a cause <u>or</u> a medium, only a <u>mirror</u> of what happens in the economic realm.
      b – Psychology is a <u>result</u> of a cyclical force, and <u>cumulates</u> the original cycle.
        – Is not a causal theory.
          – The cycle occurs, psychology carries it further.
        – It is cumulative, and should carry the cycle always in one direction, never creates the break.
          – Perhaps at a slower rate.
          – If true, <u>other assumptions</u> are necessary.
            – e.g. that some people will take fright from the ever-increasing optimism.
Lavington, <u>tries</u> to have a theory of psychological cycles, but always falls back on to material bases in the ultimate.
Pigou was more skeptical of the theory, put it as an addition.
Robertson hinted at the theory in Industrial Fluctuations.
Keynes uses the Lavington kind of assumption in <u>Treatise</u> – bullishness growing into bearishness by the very fact of bullishness.
        – Palyi says this is not necessary to the equations ([James E.] <u>Meade</u> to the contrary).
        – i.e. the psychological theory does not <u>explain</u> the cycle.
          – True, everything goes <u>through</u> the human mind.
          – Most psychological theories assume changes in the psychology of the entrepreneur or capitalist.
            – There are no psychological theories of changes in <u>labor psychology</u>, or of changes in consumer <u>psychology</u>.
          – Latter is especially important for hoarding.
  5. <u>Confidence Theories</u>
    – Never written up, but held by business men, bankers, etc.
    – Is usually a subdivision of monetary theories, because it is usually confidence in money and banking institutions that is concerned.
    – Rational, or irrational, theories of confidence.
      – One or the other must be made.
    – Sometimes it is connected with cost-profits discrepancy theory.

– Or with this <u>and</u> monetary theory.

Most of these theories can be combined with others.

  – Except for <u>purchasing power</u> and <u>cost-profits</u>. –High costs either make for depression because low profits,

  – Or they make for boom, because high wages.

6. <u>Saving-Investment Theory</u>

  – <u>Tooke, Juglar</u> (<u>not</u> an over-production theory – for "surproduction" [overproduction], for the French, is a technical term meaning crises).

    – Capital goods production increases faster than other production, in boom, falls off more than other production in depression.

      – At first was not a theory – but seemed to be an empirical fact.

  – <u>Jevons</u> – gave up sun spot theory; did not give this up.

    – More saving in the upward trend – i.e. more production of capital goods.

    – Less saving in downward trend – more production of consumption goods.

    – Not a theory – again an apparent observation.

  – <u>Keynes</u> expanded this, but in connection with monetary theory.

  – <u>Spiethoff</u> – tried to explain cyclical changes in terms of more or less production in capital goods.

    – Too descriptive.

    – People <u>invest</u> more – for some reason or other – not merely <u>save</u>, but <u>invest</u> in capital goods.

      – Why? Why do people put money in <u>short-term</u> or <u>long-term</u> goods.

        – F. W. Herrmann – a German concept, as opposed to English concept of <u>productive vs.</u> unproductive goods. Also Knies, Schmoller.

        – Because capital goods production rises, it is the upturn of the cycle.

          – <u>Why</u> does long-term goods production rise?

            – It might bring lowered costs, and thus make sure a boom.

        – <u>Investment</u> theory and <u>Underconsumption</u> theory are contradictory.

          – By first, so long as investment continues, boom continues.

          – By second, if more is spent on investment than on consumption, a depression occurs.

– These theories show a part of the <u>mechanism</u> through which a <u>change</u> works itself out – do not explain the change.

  – Two <u>explanations of change</u> in saving and investment.

    – <u>Overproduction theory</u> – though this is older, historically, than the savings-investment theory.

      – A tendency to produce more than can be sold at a given price level.

        – Therefore <u>changes in price levels</u>, which cause <u>maladjustment</u> [Double vertical lines alongside this point].

          – i.e. overproduction theory is really a <u>price level theory</u>.

– Underlying idea is of a balance of consumption and production – Adam Smith, [also] Sismondi and Malthus.
  – Say answered this – a general overproduction is impossible, goods buy goods – physically, production and consumption are equal [Single vertical line alongside this point].
    – But even Say agreed that some price level changes would be necessary, though he did not see their effects.
    – But mobility and competition are not perfect, thus overproduction occurs.
  – Why should production go further than ability to sell at expected price?
    – The question is left open.
    – Really, it says, if a change occurs, or a cycle starts, overproduction will result.
7. Monetary Theories
  – Shift arising as an accident.
  – Bullion and "mercantilist" theories.
  – Banking School – were opponents of Currency School; did not have much theory of their own.
  – Said country banks' notes had same effect on prices as central bank notes. Fullarton had gone so far as to say that all changes in money were results of business changes.
  – Banking "School" was not a school. Tooke, Wilson, McCloud [Macleod], Bagehot. Tooke had a moderate restricted monetary theory of society.
    – Pure gold movement wouldn't do much to cause cycle.
    – Credit caused the cycle.
      – Currency theorists said every change of credit influenced cycle.
      – Banking School said only large changes of credit influenced cycle.
    – Tooke: credit expansion per se is not so bad as wrong credit expansion.
      – i.e. not in response to demand for credit.
    – Credit goes the wrong way because the Bank of England lowers its rate too low.
      – Market is tight because of speculation – bank should raise rate.
    – Banking School had no theory of business cycles.
      – Causation was hardly discussed.
    – Emphasized (religion) of [sentence not completed].
    – Currency School – a largely price theory.
  – After 1850 the Currency-Banking opposition spread and became national.
    – England: survival of Banking School.

– France: survival of Currency School (cf. Banque Enquets 1860 – Bank of France unable to control by discounting theory).
  – This [was] emphasized.
Juglar – post war – first hist[orical] statistical, theory of cycles.
  – Credit expansion (Loveley) [Pierre Émile Levasseur or Adolphe Landry?].
  – Interest rate policy of Mill succeeded.
Bimetallists
Contemporary Cycle Theory – monetary
  – International Gold Supply Theory – long run, especially.
    – Chevalier (after Australian gold supply found)
      – Always enough gold, now, but in the past not always enough.
    – Cassel – Gold shortage theories. (Person [Warren M. Persons])
    – Does not describe the cycle.
      – Disregards rest of Quantity Theory.
      – Disregards the rate of production [?].
    – Liquidation of gold exchange standard took place in 1931 and after – until then it had expanded, then became stable – this contraction of credit occurred only after the crisis.
  – Discount Policy – historical and institutional approach turned away from the automatic structures and towards the conscious human control.
    – Interest rate controls the market – Marshall, Hawtrey, Hahn.
    – Hawtrey – higher interest rate – causes more capital to flow into investment, at the same time causes the demand for investment to fall off.
      – Consumer has growing income, growing outlay – diminishing hoards.
      – i.e. Monetary approach affects consumer.
      – Credit supply becomes a function of credit demand – and is a permanent feature of society – and can go on for good.
        – Until someone fails to play his part and hoards – for any reason.
      – Monetary, discount, interest theory.
      – Trader has to sell stocks, when interest rate rises; buys stocks when rate low – thus the interest rate affects prices.
      – Is a mercantile theory – dating from 19th century England (Ricardo).

NOTES AND OTHER MATERIALS FROM MELCHIOR PALYI'S COURSE, EUROPEAN BANKING SYSTEM, ECONOMICS 334, UNIVERSITY OF CHICAGO, 1934

Edited by Warren J. Samuels

That Melchior Palyi taught a course at the University of Chicago on the European banking system is unsurprising, given his background provided in the biographical sketch presented above.

Palyi clearly knew the material on which he lectured. A number of topics are perhaps of particular interest.

One is his attention to liquidity, a topic on whose expansion and partial reorientation John Maynard Keynes was then working.

Another is Palyi's careful attention to the difference between markets in theory, or in the abstract, and in practice:

> –The jobber's risk is helped out by a certain amount of <u>inside</u> support – especially when new capital issues are offered – secret procedure.
>> –Jobber becomes a sort of commission merchant for the issuing house.
>> –To what extent this procedure is used ordinarily, with old securities, is not known.
>> –This would be denied, but everyone knows it exists – nor could the market function in its present organization, without this.

Documents from F. Taylor Ostrander
Research in the History of Economic Thought and Methodology, Volume 23-B, 347–400
Copyright © 2005 by Elsevier Ltd.
ISSN: 0743-4154/doi:10.1016/S0743-4154(05)23111-5

(The foregoing brings to mind U.S. history of private practice and of government regulation with regard to combining security underwriting and stock brokerage – not least the revelations of market manipulations by several firms). A careful reading of Ostrander's notes indicates that Palyi did not take markets as givens. Instead, he devotes considerable attention to the actions of firms – driven by risk management and the quest for profit – constituting contributions to the continuous formation of markets. That there is no unique, singular money or capital market is an important implicit theme recorded in the notes. Nor are markets beyond government control. At one point the notes record, "the market could be manipulated."

The foregoing notwithstanding, Palyi is recorded interpreting the British capital market in terms simultaneously of control and of absence of control – in part also illustrating, inter alia, the role of selective perception in defining reality:

- At present is more completely ruled by government than any other market.
  - Through cooperation of insurance companies, banks, and the Bank of England, i.e. Treasury.
  - No issue comes out with[out] the approval and cooperative action of these four agencies.
- In spite of its lack of organized <u>organs</u> – no rules or laws – it is the best organized market in the world today [1934].
  - Fear of financial punishment, patriotism.
- Capital flows into the channels desired by the leaders of the market – mainly into government bonds.

Thus does Palyi present money and banking, and monetary policy, as matters of *power*. Policy will be made but by whom and which policy? The notes read:

- It became a matter of <u>power</u> – who had the political and legal right to control?

One larger story implicitly told by Ostrander's notes from Palyi's lectures in Economics 334 concerns the muddling-through process through which the money supply, or a very large proportion of it, was tied up with private banking and the money market with the capital market, i.e. commercial combined with investment banking. It would have been too much to expect a monetary and banking system, or systems, to produce a money supply in such a way that neither introduced instability on its own nor reinforced instability arising elsewhere in the economy, perhaps especially but surely not only during the period of finance capitalism. As indicated in comments pertaining to other archival materials in this series, among the vast bodies of ignorance and misconception were such topics as the nature of money, the nature of a central bank, and the role of reserves. Banking, in particular, was tied up with the economic and political power structure of society and its accompanying set of perceptions of morality; "sound money," a concept perhaps more honored by malfeasance and violation – on the canon's own terms – than strict adherence, was the pretentious and presumptuous language of the domain of public economic morality. Even in the more recent, presumably more enlightened, age, the language, now that of "fiscal responsibility," is subordinate to political machination; all is pragmatism, as it must be, though the combination of pretense and belief in "absolutes" is irritating.

Palyi dissolves any notion that the evolution of monetary institutions and instruments – types of banks, notes, accounts, and so on, is driven by anything but comparative ease of doing business in a world of uncertainty and instability.

Palyi's other larger account covers the development of banking and bank credit creation. Banks were dependent on deposits of savings (actually holding money in one liquid form) and on the vicissitudes of their borrowers' abilities to repay loans. The array of institutions forming money and capital markets and conducting business was highly variable, as were the financial arrangements constructed to cope with risk and uncertainty, most if not all of which were seriously adversely affected by the inflation of the 1920s.

Highly suggestive, even fascinating, is a recorded discussion of the putative realities of money and banking:

- Thus the balance sheets were mysteries.
    - Huge concealed reserves, window dressing.
    - Development of a journalism of analysis of such statements – it became a widespread and popular <u>sport</u>.
    - Stockholders meetings were the battleground for political and personal feuds – canopy-prestige played large part.
    - Banks had habit of naming directors and presidents.
    - Politics entered the sphere – banks, big men, industries lined up on one side or another of political struggle as an outgrowth of the struggle over <u>corporation management</u>.
        - Hitler government ended some moves towards reform. Now public is entirely out of influence. Management supreme – but large profits not safe.
    - Reserves built up – it was considered a <u>robbing</u> of the bank, to pay out most profits in dividends – paid small dividends – all the deposits decreased.
    - Some signs of reform.
- Present Nazi line of policy is to return to liquidity under the <u>old</u> banking firms – not to change anything – at least this is the Schacht idea, and he rules the banking system – for the time.

This seems to be the way things are, not an idealized world of purely conceptual markets. At another point, Palyi may well combine both approaches:
- A market is an interrelated whole, functional.

Apropos of the fundamental conceptions of economic theory, one may consider the following from the notes:

> French money market is the most competitive in Europe – the five groups (including Bank of France) compete, and the banks within these groups compete.

This is a remarkable statement, remarkable chiefly for what it does not say, which severely compromises what it does say or seems to say. The meaning of "competition" is left unspecified; an honorific, it will mean different things to different people. But nothing is said; only a primitive term is voiced. Nor is anything said about the role of structure, to wit, a market formed by five groups. And nothing is said about how both competition and results are driven by structure.

And apropos of the fundamental conception of politics, Palyi is recorded as delivering a sophisticated version of a common view:

> In financial matters, French psychology thinks of the government as an institution into which you put money, from which you get interest. In reality it is an [word indecipherable] and corrupt group of men.

Contemporary Public Choice theory, or one version thereof, identifies the first view – government as a benevolent society – as wishful but widely held thinking and includes the second view – its definition of reality – among its indictments. A different version of Public Choice theory interprets political markets along the lines of economic markets, even to the point of achieving one or another notion of optimal results.

Perhaps the most remarkable part of the notes records Palyi on the deepest topic usually either finessed, trivialized or excluded by concentrating on money and banking theory, namely, social control:

Control Aspect – social control aspect of credit structure.

- Partly connected with money, and partly with capital distribution aspect.
- Social-political aspects of credit.
- Not control of money, or control of capital distribution – these are tools, of 1st and 2nd aspect.

This is only a hint and providing a hint does not make Palyi a political sociologist, but it does differentiate him from conventional U.S. – U.K. economic analysts.

At one point the notes record Palyi as saying,

> At present, France has no long term debts, except government debts (and a few semi-governmental utilities).
> This is main explanation of adherence to gold standard.

This is strange: The usual explanation is that the gold standard will – is thought to – protect long-term debts. Few such debts would explain the opposite of what Palyi claims.

Shortly thereafter Palyi declaims that

> Safety first is the absolute principle of French savers.
> Only exception was during and just after the War – his fingers were burned then.
> Even another war could not lead to acceptance of paper money.

This is in error. France has had paper currency for many decades now.

Palyi repeatedly tells of European bank failures and their subsequent reorganization by government. This history has three significances. One is that banking is systemically too important – with regard to money supply, money market, and capital market – to be treated like any ordinary non-banking business. Second, such reorganization is akin to the working of the bankruptcy laws in the U.S. – where corporate bailouts are not unheard of also. Third, the foregoing points are true but modified by arguably dysfunctional behavioral responses, e.g. less due diligence – though what constitutes diligence varies with institutions and circumstances.

Palyi is reported as saying, "<u>Austria</u> – 1931, breakdown of Credit Anstalt started the world financial crisis." This may or may not make sense, depending on the meaning of causation and of "financial crisis." The failure of Credit Anstalt could have been either causal, itself a result of the total depression situation, or one historical development. As for "financial crisis," it could have "commenced" in 1929, 1931 or 1933. Each issue also depends on the domain contemplated, e.g. North Atlantic economy, southern Europe, etc.

The notes also record something of a history of monetary policies, a history that leaves much to be desired. A conspicuous source of difficulty is myopic adherence to and application of erroneous or only partly meaningful theories. It is, of course, easy to be critical in retrospect; the principal story is that of learning through trial and error. Still, the adverse consequences of many policies seem to have been neither unintended nor unforeseen (especially the latter) but disregarded.

NOTES ON MELCHIOR PALYI'S COURSE ON THE EUROPEAN BANKING SYSTEM, ECONOMICS 334, UNIVERSITY OF CHICAGO, 1934

General

Banks: – a) Take credit – unless they run entirely on their own capital – giro banks.
        b) Transfer credit

    c) Create credit

    d) Is a broker of credit

– In making a definition the wish is always the father to the thought – each school of banking defines the "proper" function of banks as what they think it should be.

– Our definition: "public or private institutions acting in the credit market."

    – The giving of credit, or the taking of credit, are not, separately, the credit market – they are the supply and demand of that market.

    – The bank is the agent between supply and demand.

    – Historically – a clear distinction between: (a) short-term; and (b) long-term market – 18th, 1st half 19th [centuries] [In margin: a – money market, b – capital market].

      – The bill, and the mortgage.

      – Formulated almost at the end of its period of truthfulness.

    – Then a break down of the concept – the long term paper is the best short-term investment.

      – Where is the distinction?

      – Last half of 19th century – asks what is the difference?

    – 20th century – denies the distinction.

    – The old liquidity concept is now viewed only for the central banks – though in Anglo-Saxon countries even central banks deal in "capital" market.

    – The only valid distinction today is:

      – Easily liquidated paper

      – Ultimately unliquid paper.

    – This defines the extremes. Between them is the great majority of paper – which it is nearly impossible to define.

      – Consuls may be bought for short term.

      – Day and hour market is again different from week or month market.

        – Here – changes in valuation do not easily occur over hour periods.

– "Money market" comes to mean – market dealing in fixed capital value.

– $100 is always $100 for it.

– "Capital market" becomes the market in which capital value can change – up or down – $100 may grow or shrink [Single vertical line alongside this and preceding point].

– The fixed value concept can only be found in the short-term money market – where money is exchanged against money – of the same value.

– It is one of the usual problems of economics – one can not make a definition without including in it so much that the whole story is told in the definition [Single vertical line alongside this point].

– Our only theoretical concept – is an understanding of what is meant about a function by those who work with it.

– General Banking Literature
  – <u>Laughlin</u> – <u>History of Money</u> – perhaps best for history – but <u>very</u> poor, at that.

– What are the problems of banking? – especially for Europe.
  – General description of <u>individual markets</u>.
  – Policies – central, commercial, and capital-market banking.
  – Practice and theory of international banking practice and policy.

1. <u>The British Money Market</u> – oldest:
<u>Short Term Market</u>
  A – Supply
  1 – Its supply comes through channels of the <u>Big Five</u>.
    – They are deposit banks, but also largest savings banks – rivaled only by Postal Savings.
    – Three Scottish Banks, few provincial banks, Irish Banks, merchant houses taking deposits – are only of negligible importance.
    – Change from past: (1) supply is concentrated; and (2) supply consists of savings deposits, to a large extent. – which formerly went into long time securities, growth of savings relative to other deposits – new concentrated banks, rocks of stability, attract savings that would not have gone into institutions that were less "<u>public</u> agencies."

  2 – Also important supply of foreign funds – from all over the world.
    a – Funds deposited in London – the <u>safety place</u> of the world.
      – Funds sent to London to take advantage of interest rate changes – from Holland and France, especially.
    b – Acceptance credit – bills with signature of London bankers or acceptance houses, were sold all over the world – <u>i.e. credit was.</u> <u>imported into London</u>.
      – Acceptance houses were extremely liberal in accepting up to five times their resources.
      – By giving credit, London got credit.
      – The acceptance house gives the credit, but the person to whom it sells the bill is the ultimate giver of credit.
      – A source of huge turnover, not so huge supply of funds.

  3 – Surplus funds of the capital market:
    – Was it a constant sum? Always new funds, always losing old funds, i.e. a constant level? <u>No</u>.

– But <u>more regular</u> than in U.S. and France. Why?
  – Close cooperation between <u>individuals</u> and corporations in the market.
– Central bank does not have many direct credit relations with other banks.
– Only acceptance houses, and some unstable industries (in recent years, and very much a secret), get direct credit from it.
– Some of the Big Five have done acceptance business, get credit to that extent.
– It took the War to get cooperation (and peace) between the Big Five and the Bank of England.
– The Central Bank buys and sells paper.
  – Treasury bills, foreign exchange, government bonds, acceptances.
  – Through the brokers, it sells and buys short term paper.
– The Bill Brokers, or Discount Merchants, buy bills on the market, sell then to the Bank or vice versa.
  – Thus the Bank has no <u>direct</u> contact with Big Five.
  – The Big Five first <u>clear</u> among themselves. They sell the surplus or borrow the deposit from the brokers.
  – He [the broker] is the <u>center</u> of the money market – his needs or conditions <u>reflect</u> the needs or conditions of the market.
    – He deals only with the surplus of funds or surplus of demand.
    – There is almost <u>immediate</u> adjustment.
  – The banker has no worry about liquidity, the broker takes care of that worry.
  – Thus the market is the most automatic in the world.
  – The main problem of bankers – surplus or deficiency of immediate funds – is taken care of by the brokers [Double vertical lines in margin alongside preceding six lines].
– The charge is <u>very</u> small; the rate is practically fixed, at least known in advance – small margin for bargaining.
– The Bank changes its rate only on Thursdays.
  – It may charge <u>less</u> than that rate, but not more than it, <u>during</u> the week.
  – [i.e.] another problem made nearly automatic.

B – <u>Demand in London Money Market</u>
  – <u>Big Five</u> – keep 11 or 12% <u>reserve</u> – but this only at the end of the month, between it is below that – reserve mostly a deposit at Bank of England and cash.
  1 – Commercial bills have disappeared from their assets.
    – Their place was taken by acceptances and Treasury bills.
    – Since the war and depression Treasury bills are predominant.

2 – Acceptances had signature of a bank, of two reputable exporters, or importers [In margin: "reimbursement bill"].
– Exceedingly safe – documents attached.
– New form of acceptance after War: bankers acceptances – no documents – signed by two banks [In margin: "quasi-reimbursement bill"].
– During War – large volume of short term Treasury bills put out – this could not be easily changed into long term debt because the long term market was full of government debt.
– Thus the "short-term" bill is not short term, but lies on the market.
Cheapest rate – first class commercial (bank) bill – small in volume.
Next cheapest – Acceptances – small in volume, especially in depression when international trade falls off – some replaced by "Standstill bills".
Next cheapest – Treasury bill.

3 – Triumphant march of long term (government) bonds into the assets of banks.
– Formerly about 20% of assets was in this field.
– At present over 30%.
– Midland Bank loans to Manchester textile firms,
– Expected to be repaid out of the boom, or by selling securities.
– Lloyds Bank went into shipbuilding credits.
– Barclay's Bank (and Lloyds) went into foreign hands [?] (now mostly closed down at a loss).
– These advances have been frozen ever since the crisis of 1921, which lasted till 1934.
– Some still remain as "advances," about one-half has been written off as a loss.
– New advances since then – perhaps one-half of total really <u>are</u> short term.
(1) Commercial bills fallen off largely.
(2) Acceptances fallen off, or made standstill bills – long term, but a market still exists for them.
(3) Growth of long term credit assets, and of: (a) frozen advances (no market); and (b) other frozen assets as acceptances (some market).

– British money market used to be highly liquid and sensitive.
– Financed international trade in staple goods with world wide market.
– Acted in the world's central capital market.
– London was central <u>share</u> market, for international securities of greatest esteem.
– Paris was central bond market for world.
– Conservative tradition of banking, high liquidity, high confidence.

– Post-war money market is relatively liquid and sensitive.
  – Still deals in securities, but also in government bonds.
  – Runs danger of short-term capital outflow.
(British Consuls fell in value 20% in 10 years before the War – Lloyds lost
more than their capital, in this).
– Very fine division of labor.
  – Is the broker an outworn tradition? Or is it a delicate and important
    machine to keep liquidity[?]
  – Errors of judgment are more frequent outside London than there.

The British Capital Market
  – Stock Exchange – a very closed market – few members.
    – Jobbers – decide which securities are to be handled, specialize in certain
      securities – about twelve parts to the market.
      – Each jobber is expected to be ready to buy and sell any amount on
        demand and to state his price – he is the market, and takes a huge risk.

    – Broker – has connection with the public – through "outside brokers" –
      thousands of outsiders interested in this sort of thing.
    – The jobber's risk is helped out by a certain amount of inside support –
      especially when new capital issues are offered – secret procedure.
      – Jobber becomes a sort of commission merchant for the issuing house.
      – To what extent this procedure is used ordinarily, with old securities, is
        not known.
      – This would be denied, but everyone knows it exists – nor could the
        market function in its present organization, without this.
    – Two sources help to finance the jobber:
      a. Issuing houses.
      b. Banks – acceptance houses.
      – Give short (broker's loans) daily loans to help him balance his accounts
        and carry over.

Promoters – bring new enterprises onto the market.
  – A boom phenomenon.
  – Played small part after War – disappeared with the coming of
    depression [FTO asks:] (where?).

Underwriters – technically the issue house performs this function.
  – A new issue brought out by a small issuing house is underwritten by
    several underwriters – promising to take up any part of the issue not
    taken by the public.

- Actually, they do not ever take up any of the issue – merely loan their <u>names</u> for a good sum.
- For commercial, utility, foreign issues.
- Are supposed to be <u>good</u> names.
- The better the issue, the lower the underwriter's fee.
  - This fee is published – and the amount of it is watched closely as an indication of risk.
  - Even when no risk has been assumed this fee has some meaning – indicating the underwriter's fear for his name.
- Issue Houses specialize on certain issues.
- Speculative securities are issued by acceptance houses.
- Two kinds of issues.
  - a – Sold <u>abroad</u>.
    - Insurance companies and investment trusts buy up the whole issue (Big savers).
  - b – Subscription sales.
    - Very speculative – <u>public application</u>.
    - Issuing house can assign issues to appliers <u>just as they</u> please – denying or giving heavy allotments.
    - Appliers –
      - <u>Investors</u> – provincial solicitors, trustees, chartered accountants, investment trusts.
      - Brokers, bankers, merchants – "<u>Stags</u>" are speculators – <u>subscribe</u> more than they want – assume subscription price [is] too low.
    - Stock exchange marketing starts <u>after</u> subscription is closed.
      - May start at a <u>higher</u> price.

- i.e. issuing is done by appealing to savers, and to "<u>Stags</u>" – who hold issues only a short time, till savers take them (presumably at a higher price) – but this latter is speculation.
- Stag subscribes for unknown, unsure <u>future</u> buyers.
  - Also speculates by subscribing to more shares than he wants, even for speculation, in order to get a large amount by the final allotment – in order to get enough.
- Stags lost heavily by speculative errors in 1927, 1928, – and, of course, in Depression.
- Capital market was predominantly foreign until recently – pre-war London the center of <u>international securities</u>, and <u>international commercial equities</u> – and of governmental bonds of British Empire.
- After War – the center moved to New York.

– Recently – more <u>governmental</u> bonds (more were being issued – by Socialist governments, public utilities being taken over).
– Financing British industry.
  – <u>Leading big firms</u> get credit from London Stock Exchange.
  – Other firms get advances from Big Five, which freeze into capital (long run) advances.
  – London is not not interested in small issues – neither brokers, issuing houses are interested.
  – Provincial stock exchanges take over some of this financing.
  – Major trade in England is carried on by <u>small firms</u> – cutlery, <u>coal</u>, leather, shipbuilding, tramp shipping, steel.
  – Real estate credit is not carried by London.
  – How is it financed then?
    – <u>By the unorganized capital markets</u>.
      – Financing by <u>local</u> supply meeting <u>local</u> demand.
        – After chance intermediary of bank branches.
      – Personal connections, relatives.
      – Thousands of intermediaries bringing together the minute demand and minute supply of capital.
      (The Financing of British Industry – von Wieser – in German)
        – Breweries are financial houses.
    – Same is true of agricultural and real estate credit markets.
      – But bankers enter here.
      – Cooperative firms finance house building.

Summary:

Money market and capital market are separate; connection between them made by:
  a. Carryover of stock exchange brokers [and] jobbers by banks and discount houses.
  b.
  c. Acceptance market and issuing market are in the <u>same hands</u>, though they are vitally opposite functions – are intermediaries in both markets.
  d. <u>Close</u> correspondence of short and long term rates.
    – Discrepaancies in unnormal [sic] times are less than anywhere else (but Dutch market).
Source of long time savings
  – Savings deposits are not pure savings but are intermediary between savings and demand accounts.

– Savings are formed [formal, found?] in investment trusts, trusts and insurance companies.
   – Play safe, buy those securities which <u>have</u> gone up – an attitude typical of <u>amateur</u> speculators.
   – To maintain an issue, or get it started, requires close cooperation of brokers.

– On whole the <u>British capital market</u> is not very sensitive – is ruled by traditions of decency, and lack of government interference (example of Russia borrowing from London during Crimean War – with knowledge of Foreign Office – war is war – business is business).
   – At <u>present</u> is more completely ruled by government than any other market.
      – Through cooperation of insurance companies, banks, and the Bank of England, i.e. Treasury.
      – No issue comes out with [out] the approval and cooperative action of these four agencies.
   – In spite of its lack of organized <u>organs</u> – no rules or laws – it is the best organized market in the world today.
      – Fear of financial punishment, patriotism.
   – Capital flows into the channels desired by the leaders of the market – mainly into government bonds.
   – Large inflow of funds from foreign sources.
      – Less because of interest rates, than because of less risk, speculation in exchange.
      – Of huge amounts.
   – Appearance of great stability, of strong and wise leadership, of conservation.
   – Is founded, in part, on a great weakness – this large element of funds which is at the control of political (mainly) factors.
   [Brace in margin encompassing preceding four lines]
   – The London market is predominantly international.

2. The French Money Market
   – The Bank of France
   – The commercial banks
   – The savings banks.
– Bank of France – conservative, high reserve backing, and discrimination in choice of securities.
   – Not allowed to hold government bonds – i.e. no chance for open-market operations for government policy.
   – Its assets mainly commercial bills and bankers bills.
   – Average of 80% gold reserve for <u>all</u> its liabilities – theoretical possibility of great credit expansion, actually, little danger.

– Reorganization in 1928 – before that it had no legal gold reserve, but
Parliament decided from time to time how many notes beyond gold reserve
could be issued – usually this amounted to 50% backing.
   – Since 1928 – the Bank of France [has been] organized on a Kemmerer-
   Dawes plan – gold reserve behind notes, behind deposits – about <u>35</u>%.

– Position of Bank of France on money market.
   a – It is a <u>competing</u> bank with the other commercial banks – and regarded
      as such.
     – A hangover of the 100 year old tradition.
     – It is a commercial bank, with some privileges which other banks do not
       have. – and pays for its monopoly – by taxes, by sacrificing potential
       earnings in the form of high reserve.
     – It is owned by private people, and governed by them. By now [no] law
       can the government <u>force</u> the bank to do anything.
     – It is the leading commercial bank in France.
     – It competes in the market – not for deposits – pays no interest on them
       – but for <u>bills</u>.
     – It has 200 branches throughout the country, and <u>tries</u> to make good its
       lack of earning power in other lines.
     – It charges a higher rate than other banks, but takes bills other banks
       wouldn't take – <u>small amounts</u>, and unknown, small names.
   b – There is close cooperation between the central bank and other banks –
      and the other banks <u>need</u> this.
     – The Bank of France must have three names on its bills, thus it can only
       get <u>them</u> from other banks, who provide one of the signatures.
     – The other banks usually sell their notes to the Bank of France – just
       before they come due, and the Bank of France has to collect them.
     – Close central bank control of the market.
     – Banks which do not have commercial bills get <u>no</u> support from the
       Bank of France.
       – Thus the banks have to keep sufficient sums on hand in the form of
        commercial banks [bonds?].
     – Yet the Bank of France has no <u>obligation</u> to help other banks.
  – Bank of France is extremely liquid – was so even before the War (Bank of
   England was <u>smaller</u> than Bank of France – had much more government
   paper among its assets).
     – Mainly commercial paper.
     – Some (commercial) paper of government type.
     – High gold liquidity.

– Bank of France has a strong position on the market – because it competes with the other banks (Bank of England has a strong position on the market, because it does not compete with other banks).

– The Commercial Banks
  – Crédit Lyonnais
  – Societé Generale
  – Comptoir d'Escompte
  – About twelve other, smaller banks (one of the larger was formerly an association of provincial banks – combined into <u>one</u>, after War).

  – The French banking system is the most secretive in the world – those who know about it, don't write – nothing is to be found in the literature about the inner workings.
– Current Accounts – are quite liquid – due to extreme conservatism of loaning.
  – Banks are very courteous to people bringing money – but immediately suspicious of anyone trying to take <u>out</u> money.
  – Fairly smaller volume of loans due to this conservatism.
    – Though France is the center of several minor world markets, its volume of commercial dealings is small.
1 – Commercial credit.
2 – Credit to Stock Exchange.
3 – Acceptances.
4 – Short term treasury bills.
  – Some long term government bill.
  – Some railroad bills (subsidized by governments).
  – French monetary policy makes clear separation of subsidized from other businesses – except for tariff.

Growth of advances in boom – decrease of them in depression.
  – <u>Direct</u> advances by credit accounts.
  – But these are so liquid that they can be diminished by 50% in bad times.
– Growth of <u>paper</u> in depression – falling off in boom <u>portefeuille</u> [portfolio]
    – No distinctions made in statements.
    – Everything is most secretive, yet every one talks about the secrets.
  – At first a large proportion of treasury bills, 75% of portfolio in early years of depression was <u>government</u> paper.
  – But now 50% is government paper.
  – Ordinarily even less of portfolio is government paper.

– Very little saving in commercial banks.
– Velocity of circulation – French statistics are almost useless (P. des Essars)
– Checks used less than in any other major country.
– French deposit banks are primarily for deposits.
– There is less connection of the money and capital markets than in England –
  for savings to not flow into the commercial banks. English banks – though they
  do no admit it – have changed substantially – taking up long term deposits,
  and longer assets.
– But, the French banks are issue houses – issue bonds, are investment bankers
  (commission business) – thus connected with capital market.
   – They sell bonds through their hundreds of branches, to their depositors.
      – Sell mostly bonds – especially government bonds.
3 – Savings Banks
  [Three inches of space]

4 – Cooperatives – large number, but have no money, thus unimportant
      (Frenchman wouldn't lend to his neighbor, rather to a savings bank).
   – Financed by government subsidies.

5 – Banques d'Affaires – industrial banks
   – Work as real issue houses (as acceptance houses in England) – issue the
      exotic paper.
      – Banque de Paris et des Pays-Bas – Hungarian President – always
         successful.
         – Does everything a financier can do – within law and decency
            (sometimes those are in doubt).

   – Loan money – sometimes on commercial bills.
      – Also direct to industries (few major modern capitalistic industries are
         very strongly and concentratedly organized – textiles, chemicals, iron
         and steel, perfumes, autos, shipbuilding and shipping).
      – Close connection between Banque d'Affaires and Affaires [business] –
         personal.

   – Commercial banks enormously reduced their loans to industy in the
      depression – showing that they were really short term loans.
   – Banques d'Affaires did not reduce their loans to industry – i.e. they are
      supplying the real working capital of industry.
      [In margin at top of page: "Banques d'Affaires are holding companies –
      connected with building up trusts, and participating."]
   – Banque d'Affaires is the field of banking failures.

- Which shocks the Britain, but not the Frenchman, who knows exactly what these banks are, and expects such; – Anyone who puts his money in one knows what to expect (Oesterreich scandal in 1931 – loans to Italian artificial silk trust [Austrian Kredit Anstalt failure in June 1931, precipitating world financial collapse]).
- Are speculative, private and high interest rates.
- Capital market rates rule, due to risk.
- But deal in both capital and money markets, and hold securities in both – hold very speculative securities in capital market.
- It is one of the usual aspects of a French depression that no commercial banks fail (not one since Crédit Mobilier) while some Banques d'Affaires always fail.

6 – Private Banks
- Rothschild – a private trust fund – not very active, and no business to do.
  - Are brokers, foreign exchange speculators.
    - Paris became center of speculation in unbalanced exchange during Europe's post-war epoch.

- Rothschild houses are not very close any more.
  - Vienna house was most active, failed with Kredit Anstalt failure.
  - German house failed. English house is most active acceptance house in London.
  - Lazard [Frères] (New York City and Paris houses, not very close).
  - Before War – commercial banks had five times as many deposits as Banques d'Affaires.
  - After inflation – commercial banks failed to keep up to expansion – suffered from flight of capital – while Banques d'Affaires grew greatly.
- Savings banks

7 – One semi-governmental acceptance house – to try to replace London as an acceptance center – but needs brains and bankers as well as a house.

French money market is the most competitive in Europe – the five groups (including Bank of France) compete, and the banks within these groups compete.
  - High standards of liquidity, and, on the other hand, occasional flights into wild, exotic speculation.

The Bourse
Parquet – central part of building – open only to members – handles first class securities.

<u>Cambios</u> – more numbers, opens earlier, closes later, competes with parquet, handles second class securities.
– This division corresponds to the usual French separation of speculation from good security.
– Before War Paris was the cheapest money-market.
  – Most bond issues did not remain on the market, but disappeared, governments did not worry about this effect on future markets.
  – French rentier the truest <u>investor</u> in the world – never parts with his holding.
  – France had advantage: (1) cheapness; (2) prevalence of gilt edge securities; and (3) power of observation.
  – Only paper officially listed on Bourse could be disposed of. Bourse was controlled by government,
    – So it could kill an issue easily, with help of newspapers and banks (cf. [Herbert] <u>Feis</u>, "Europe, the World's Banker, 1870–1914" [New Haven, Connecticut: Yale University Press, 1930])
    – Speculative element of non-foreign issue.
    – Corruption.

  – In financial matters, French psychology thinks of the government as an institution into which you put money, from which you get interest. In reality it is an unreliable and corrupt group of men.

Real Estate Financing
  Crédit Foncier (German origin)
    – Organized for getting funds for municipalities.
    – Gives credit on mortgages, issues bonds on the mortgages.
    – At first (1850) had a legal monopoly; later lost this, but little competition has grown up.
    – Its securities are not gilt edge – but about 80 (conservative valuation).
    – In general, real estate market is unorganized (including financing <u>new</u> houses).

<u>Farm Finance</u> – largely financed by private sources, by short term; also by banks on a purely commercial basis.
  – In 19th century France developed the best theory and best practice of farm credit.
    (1): Crédit Fonctionnaire, long term, slow amortization, if at all; (2) short term working credit; and (3) intermediate.
<u>Colonial Bonds</u> – growing volume on market but total still small.

– At present, France has no long term debts, except government debt (and a few semi-governmental utilities).
– This is main explanation of adherence to gold standard.

Supply of Savings:
– Regularity of their flow results from conservative habits and customs of people.
– Slow industrialization.
– Safety first is the absolute principle of French savers.
  – Only exception was during and just after the [First World] War – his fingers were burned then.
  – Even another war could not lead to acceptance of paper money.

Liquidity of French Market:
– Based on highest (per capita) gold reserve of any country.
– The wealthier rentiers escaped the effects of inflation by selling out early – poorer ones last.
  – During inflation, most French reserves went abroad, returned afterward.

German Money Market
– Its peculiarities
  – French and English – assets grew by a growth of savings in early 19th century.
    – Then, in later half of 19th century – liabilities grew – people got richer, put money in banks, which means that bank liabilities grew first.
    – England, based on colonial action, on textile exporting.

  – In Germany the course was quite different.
  – 18th– early 19th century – no wealthy people, no deposits based on growth of richness. – Little banking.
  – Right up to present – the German banker had to begin by extending credit – to create rich men – whose deposits would pay back credit.
  – This is the Crédit Mobilier type of banking begun in France – carried out in Germany.
    – Credit given to people whose wealth consists of mobilized securities, or whose production is based on securities.
    – Two brothers, [names unknown], pupils of St. Simon translated his ideas into action by this bank – but tried to build a deposit bank on the basis of long term investment.
    – Credit Mobilier failed – but the Darmstadter Bank began, on the same lines, in Germany.

- The Germans <u>realized that they did not have a deposit bank</u>.
- They made a distinction between deposits on the basis of their <u>source</u>.
    - Not a theoretical distinction based on <u>creation</u> of deposit, but a <u>personal</u> distinction based on the relations of bank to customer.
    - <u>Deposit Account</u> – one-sided; the customer comes to the bank with money, puts it in; does not withdraw it all, uses it as a savings or checking account.
    - <u>Current Account</u> – industrialist comes to bank, asks for credit, bank gives it on short terms but it is two-sided – and the deposit tends to <u>grow</u>.
- Germany did not have a thriving industry and rich men – but it had promoters who were interested in making money out of banking, and at the same time were interested in promoting industry.
- The German was willing to lend to industry – small total of government bonds which were mainly railroad and thus in a way industrials.
- Savings were urged into industrials creating a market for them.
    - It took an actual industrial development to form a stable background.
    - Crédit Mobilier based on such an assumption but it was too early for that to occur in France.
    - Some continuation of these ideals was to be found in the Banques d'Affaires – until after the post-war inflation.
- German banks get a basis for long-term expansion on a basis of short-term deposits (French and English system based on deposits coming into the bank, German system based on credit creation going out of the bank).
- Darmstadter Bank – takes deposits, introduces giro activities, loans to industry on the assumption of providing working capital.
- Importance of the <u>promoter</u> – who went into banking and as a banker-promoter created industry – the promoter aspect has died out (just as the promoter aspect died out in England).
- Most provincial banks either failed, or moved to Berlin or larger centers, or were merged.
    - Tendency for banking to concentrate in Berlin – especially after branch banking was started in the 70s – especially after stock exchange came, and due to political centralization there.
- i.e. this sort of bank can only exist on a <u>national</u> area – it tends to fail in provincial, industrial [sentence incomplete].

- Three major banks growing out of many mergers – are national with
  <u>many branches</u> throughout.
  - Deutsche Bank and Disconto, Gesellschaft Dresdner Bank
    (Darmstadter und National Bank) 1931, Kommerz und Privat-
    Bank
- Two smaller banks – no branches – but quite important.
  1. Berliner Handelsgesellschaft.
  2. <u>Reichsk</u>redit Gesellschaft
     - Founded in 1918 by government – its function is to be the
       government's <u>teller</u>.
     - Takes deposits from private people.
     - Has assets of long-term government securities, and some
       industrials.
- A few provincial joint-stock banks – as:
  - Allgemeine Deutsche Kredit Anstalt (ADCA) – Leipzig.
  - With close connections with some British banks.

German Commercial Banks
- Crédit Mobilier type – deposit and current accounts, short term financing
  of industry.
- After War – banks drew on each other – used acceptances as means of
  getting credit from Reichsbank.
  - On assumption that acceptance is a <u>suris</u> [?] of the amount loaned to
    customers – the bank <u>substitutes</u> its name for the customer's name.

1 – 1st class liquid assets – by old rule 50–60%.
  a – cash – 3% (i.e. "secondary liquidity") – actually fell below that.
    - Because of close cooperation of commercial and central banks.
    - No merchant exhausts his credit limit by his loaning through commercial
      bills (Central bank has a list of <u>every</u> merchant – and his credit limits. If
      two names or three names are on a bill, it scarcely can happen that all
      three credit limits can be exhausted).

  b – balance of the central bank.

  c – commercial bills – were disappearing before war.
  d – acceptances – each bank has a <u>limit</u> at the central bank – for acceptances
    bearing its name.

  e – treasury bills – growing slowly.

Commercial bills disappearing – to cartelization of industry – growth of "common-interests" – "Interessengemeinschaft" – in private firms, or interlocking directorates.

- Thus there is not any commercial transaction between two firms thus allied. It is impossible to draw a bill – one pocket drawing on the other pocket.
- Structure of credit given moved from a short-time type to a different type.
- Theory that it was not good to have a circulating signature – really the merchant did not want to promise to pay in three months – renewing – when he <u>knew</u> it was a three year credit.
- Also a disappearance of the wholesale merchant.
  - Especially in coal, also in other heavy industries.
  - He was the kind who <u>could</u> sign a three month bill.

<u>2</u> – Investments – small in amount.

a – Loans on securities – stock exchange loans – small in amount, play role of barometer.

b – Government securities – play a growing role.

c – <u>Direct</u> investment in branches and mergers.

d – Private investments

  - Investment is not a regular business of German banks – they do it only temporarily – [such] as to support a new issue's quotation, play this item down – hide it – generally undervalued.

3 – Current advances – more than 60% of assets, today.

- <u>Circulating capital</u> – merchant's credit, to move goods; cash and banker's deposits commercial credit – three months
- <u>Working capital</u> – to add to fixed capital the amount necessary to carry goods through production and storage from origin to sale – a short-time industrial capital – industrialist's [?] credit.
- Intermediary credit – a "<u>short</u>" fixed credit.
  - Especially for <u>new</u> industry.
  - The theory [is] a rationalization.

(– German, Italian, Eastern European banks were issue houses)

  - Getting the supply of fixed capital from the public and advancing it to industry.
  - But in actuality, the issue is made to the firm first, then, as it becomes possible, the securities are unloaded to the public.
  - Pre-war <u>practice:</u> – Banks tried to get new customers on short-term credit, were willing to give longer credit to old customers – and were willing and anxious to <u>speculate</u> in new fields of major importance.

- As "prince-consortium," gas modernization.
- In each crash in Germany, 1900, 1907, some banks had over- financed some <u>too</u> speedy modernization.
- With war – came inflation.

- Post-war <u>practice</u>: – Inflation 1921– 1923 changed standards of valuation, and personnel.
    - Banks emphasized <u>material</u> collateral, but gave credit in old ways (accepted collateral of same firm's securities, to whom they were loaning on real estate).
    - Tended to put more and more into long term holdings of assets – and the security collateral became less and less, so that the crash left them with ten times as much holding of real estate as their accounts had showed.

<u>Private Banks</u> (Mendelsohn's Bleichrader)
- Foreign exchange business – dealers, arbitrage agents.
- Issue houses – brokerage – speculative.
- A few substantial firms – the rest very speculative, on the margin.
    (Provizialbank – Vereingsbank in Munich is owned by the Mendelsohns) – who also own mortgage houses, etc.
    (Warburg firm in Hamburg) (New York connections)

<u>Specialized Banks</u>
- <u>State banks</u> – owned by every State – no connection with Federal government.
    - Are <u>tellers</u> for the Stage governments – but deposit bankers too – i.e. minor joint-stock banks.
- <u>Railroad Bank</u>
- <u>Post Office Bank</u> – important for <u>giro</u> transactions, <u>large</u> number of deposits – thus [having] a place in the money market.

<u>The Central Bank</u> – before War – a strong central bank – which bought commercial bills on the market – two names, on French pattern – buying bills 8– 12 days before due, to <u>collect</u> them.
- After inflation – a new development.
    - Three name bills, didn't buy so many (commercial bills dying out).
    - Market needing support – (Deutsche Bank only <u>two</u> times in 30 years was indebted to the Reichsbank – since the inflation it has been indebted every day).
    - Central bank wanted to <u>create</u> an acceptance market – bought acceptances, but no market was created – they weren't then sold.

- Country came out of the inflation without savings or cash – most new saving was "forced."
  - Central bank created small coins, and legal means of payment.
  - Commercial banks have to create credit and get it into circulation, this takes time.
  - Central bank got foreign exchange by borrowing under Dawes & Young plan.
  - Central bank forbidden to have more than a minimum of government bonds.
    - 35% reserve for note issue.
    - 10% might be in foreign exchange reserve.
    - Allowed to use "Lombard credit" – i.e. collateral but gilt edge collateral; signature of government paper and private debtor – interest 1% higher than discount rate – wide margin.
      - Used by bankers who had no bills.
  - Central Bank – privately owned, best stock in country – pays dividends in depression.
  - Pre-war – Governor appointed by government, and he chose ten other directors, with advice of stock holders.
  - Post-war – under Dawes Plan – a committee of leading bankers – nominated by Hindenburg for life – who choose the Reichsbank chairman.
    - This board abolished by Hitler – Governor appointed by government – and the bank allowed to make deposits on basis of government securities.
      - This became a main means of keeping up the market for government operations – i.e. buys, but does not sell government bonds.
        - This is the channel through which inflation would develop if it came in Germany.
        - Central bank buying government securities with printed paper money – advocated by rabid partisans; opposed by Bank.

Savings Banks – municipal savings banks; everywhere.
  - Before War were the equal of commercial banks – 20 billion as against 22 billion in deposits [of commercial banks].
  - Commercial banks repudiated all deposits after inflation.
  - Savings banks gave back 10–50% of pre-inflation deposits (large number of assets were in mortgages – these were revalued 10–18%) – it was

largely a bookkeeping account – the depositors did not withdraw this deposit.

– Also – a great hatred arose against the commercial banks and the people turned to the savings banks. $13\frac{1}{2}$ billion deposits in savings banks – 12 billion deposits in commercial banks.

– Even though commercial banks hold foreign short-time funds, their deposits were less than savings banks' deposits.

– They carry on financial operations for the cities, who own them – tellers.

– [Arrow in margin, indicating this sentence should precede last sentence]. Social changes helped this move savings by lower classes – more people who saved in lower classes.

– Saving deposits.

– Giro deposits – for exchange within the savings banks system (expens[ive] if outside system) [Used] first on urban land.

– Government regulation – 25% in mortgages, 50% in government and municipal gilt edge securities, 25% for money market.

– Each savings bank has an independent manager – small, bureaucratic official, without banking experience.

– Gave credit to municipalities, and to local <u>small</u>, handicraft merchants.

– Ten provincial giro organizations of savings banks – with one Federal organization at the head – to collect the surplus funds of local banks – and of provincial funds – put it into better use.

  – But these also invested in municipalities – the same ones.

  – Thus a frozen system – but unfrozen by Reichsbank help and change to government financing of cities.

  – Carry on giro clearings.

Cooperative Banking

– Savings of members – and credit of the combination.

– A group of artisans, none of whom can get credit, alone, can get much more than would ever be given to them – in total – when they act as a cooperative.

1 – Urban

– Now cover all Germany, and have two central organizations.

  – <u>Schultze-Delitsch</u> firm, Dresdner Bank, cooperatives section.

  – Pre-War – some minor subsidies from government, not very great success. The minority which were successful became closed shops, closed to new members. As long as they were unsuccessful, they sought new members.

– Did not solve the problem for the <u>city</u> <u>middle class</u> – but helped that
class over a difficult transition period.

– Capital destroyed in inflation – have not recovered since, government
subsidy.

2 – Country Cooperative – Raiffeisen
– <u>All</u> over the country – under the leadership of the local teacher or priest
– they collected the savings of the peasants and villagers (there were
no village savings banks to compete, as in the cities – even members
there <u>put</u> their money in municipal savings banks, <u>asked</u> for credit from
cooperatives).
– There was little or no competitive spirit among peasantry – but
similarity of action.
– They were highly successful.
– Were <u>copied</u> very successful in Denmark, Sweden, Holland – were not
so successful in Latin or Eastern countries.
[In top margin with arrow to "Denmark" in preceding sentence: "– Very
successful, turned to long term, mortgage credit – owned by governments,
subsidized."]

– They give "short-term agricultural" credit – three years – for buying
livestock, for getting means of production into the village.
– Buys and sells goods – is the scientific bureau of the village.
– Were successful in purpose – but often failed through lack of financial
ability.
– Thus a spread of branch banking.
– Federal Central Cooperative Bank (formerly Prussian).
– Branches lost their capital in inflation – now the credit is not collected
by the central bank, but it is the agency for distributing the government
subsidy to the branches.
– Operates in Berlin Money Market.

<u>German Capital Market</u>
– Very complex.
– After 1900, capital outflow, on balance.
– After War – capital inflow the main element.
– Mortgage Bank Institute (Pfandbrief).
– <u>Landschaft</u> – 1780 – combines the large landowners of a section – procures
credit for them, on the basis of combined action.
– It developed that the Landschaft gave credit and took credit, with mutual
responsibility.

– This <u>mobilized</u> the credit of the farmer into long-time securities (When this was transferred to France – it became joint-stock). Later this idea was brought back to Germany as:

– <u>Mortgage Banks</u> – strictly regulated –
  – Minimum capital – no bond can last longer than the (three year) mortgage.
  – Could only loan <u>conservatively</u> – but government couldn't regulate what that meant.
  – These were very successful – never in trouble – reduced interest – but Landschaften were failures <u>in every country</u> and every crisis – government subsidy had to be brought in.
    – Loan largely to city real estate.
  – Most European countries have highly developed mortgage banks.
  – Their bonds were <u>very</u> popular – no made gilt-edge, but regarded as such.
  – The public still believes in <u>substantial</u> things, such as real estate – like mortgage bonds better than government bonds.
  – Lost capital in inflation – but were rehabilitated after inflation, 18% payment made – sold new bonds all over the world, got much foreign capital.
  – In a depression, their interest falls off only by 15% – the statistics of outflow of interest are the best indications of soundness.
  – Pure mortgage banks and mixed mortgage banks – latter do not <u>mix</u> the kinds of business, but carry on other activities, commercial, etc. – can not use mortgage funds in commercial activity – but no legislation as to what they do with commercial funds.
    – Ay bank can give real estate credit.
    – Only these banks can issue bonds based on that credit.

<u>Commercial Banks, Private Banks</u>
  – The <u>banks</u> led the capital market – and were main <u>issue houses</u>, as well as distribution, or <u>marketing, houses</u> (this system copied over Europe).
    – Problem of competition between two organizations selling securities.

<u>Rentenbank Kredit Anstalt</u>
  – When its monetary role was ended, in 1925 –
  – Became a government mortgage bank – to get two large loans from U.S. to give credit to large farmers (Junkers) – whose credit standing and credit had disappeared.
  – Experiments made with <u>Industrieschaften</u> – giving mortgages on industries in local communities – failed.
    – Industries in a locality, not analogous to <u>land</u> in a locality.

– With few minor exceptions, the few big banks have the <u>issue</u> business in
their hands (or rather in their <u>branches</u>!).
  – Stock exchange loses importance.
  – Banks can place issues: (a) through exchange; and (b) through bank
  branches.
    – Open market loses importance.
    – <u>Bank-customer</u> role assumes importance.
      – Dresdner Bank alone placed 12% of new issues post-war
    – We can infer that this movement would go further.
    – The issue is quoted on the stock exchange, and subsequent placement
    is made by banks to their customers – who <u>can</u> operate on the market.
    – Bank charges a commission as if it got the shares on the market.
    – Though possibility of placing anything outside the banks becomes
    small.
    – Stock exchange loses importance, less turnover, draws less short-term
    capital.
    – Thus Berlin did not become as large or as important a capital market
    as Paris or London.
      – When it had all the advantages except a rapid turnover of money.
      – Overpowering influence of big banks.
    – The stock exchange quotation became relatively easy to manipulate.
    – There are less speculators, and less bank financing of new issues on
    exchange.
    – Another cause turning the financial system away from short term
    credits (speculative) towards more direct, long run, productive credit.
– Present crisis has destroyed all matters of refinement – pre-war discussion
  of exchange was whether it was 5% better or worse – when the question
  is one of 50% better or worse – refinements don't much matter.
Houses built by speculators – all over continent (except in France) – mainly
small apartments.
    – Who put up some capital – get rest "Bankkredit" – short term credit
    from banks –
    – After War government established two Bankkredit institutions.
    – After one-quarter of house is up, bank gives second batch of credit,
    then third and fourth installments (two months each).
    – All installments mature when house is ready.
    – Then mortgage bank steps in, out of its credit and <u>sale</u> price of house,
    the Bankkredit is paid.
    – But speculative risk on the presence of a buyer – business cycle and
    overproduction.

- Buyers are men who <u>believe</u> in real estate, want a Rentenhaus for old age.
- In case the credit goes wrong, no buyer, the mortgage bank may step in – if it and the bank are friendly – and if house comes up to <u>all</u> requirements.
- Bank has security – it does not want to be a housekeeper.
- Bank then sells mortgage bonds of mortgage house.
- Builders borrow from contractors, who do their own speculating, borrowing from banks.
- In case of pinch, it is the small artisans whose credit is cancelled.
- After war, when real long time pinch came, it was solved by <u>government subsidy</u>,
  - True all over Europe – if things go wrong, and the building is up to government regulations, then a government subsidy is forthcoming.

<u>Organized and Unorganized Capital Market</u>
- Foreign deposits in commercial banks were more than half of total deposits
  - at peak of boom.
  - Were like bankers balance.
- Before 1931, the good bank had a substantial part of its credit line unused.
- Some amount of liquidation must be carried out by the central bank in a crisis, in order that confidence in the whole <u>banking structure</u> be trusted.
- The tradition of the capital market (Oriental development) is the exclusion of the (small) shareholder. The arrangement is dictation.
- Thus the balance sheets were mysteries.
  - Huge concealed reserves, window dressing.
  - Development of a journalism of analysis of such statements – it became a widespread and popular <u>sport</u>.
  - Stockholders meetings were the battleground for political and personal feuds – canopy-prestige played large part.
  - Banks had habit of naming directors and presidents.
  - Politics entered the sphere – banks, big men, industries lined up on one side or another of political struggle as an outgrowth of the struggle over <u>corporation management</u>.
    - Hitler government ended some moves towards reform. Now public is entirely out of influence. Management supreme – but large profits not safe.
  - Reserves built up – it was considered a <u>robbing</u> of the bank, to pay out most profits in dividends – paid small dividends – all the depositor deserved.

– Some signs of reform.
– Present Nazi line of policy is to return to liquidity under the old banking forms
– not to change anything – at least this is the Schacht idea, and he rules the
banking system – for the time.

Dutch System
– Most conservative, most liquid.
– Five big banks – one failed, but supported by other four – had done some
speculation.
  – Liquidity – especially in secondary assets.
    – No bonds in assets – only a few handled as a brokerage business.
  – Most reserves are international bills.
  – Most illiquid credits are those to Germany by Holland (not so large
  credits as were given by London) (longer than given by Paris and
  Brussels).
    (– The long term, private credits have a somewhat favored position in
    Germany).
  – These are standstill credits.
– Dutch have lowest interest rates – short-term.
  – Loaned to London.
– Narrow gold points – six hours by airplane to London and Berlin and
Paris. – Much arbitrage.
– Many funds have flowed in.
– Wealth of Holland is a commercial wealth – its banks true commercial
banks.
– A country of small, intensive farming.
  – Cooperative organization.
  – Farmer's short term credits financed by commercial banks – on sale of
  product – products are standardized and sold on world market.
  – Savings – built on savings of a few rich people and many small savings.
    – Large artificial silk industry – low taxation attracted it from
    surrounding countries.
  – A country of the large rentier – more millionaires per family in Holland
  than anywhere else.
  – A new development – many private bankers.
    – Especially after and during inflation.
    – And again especially after Hitler.
    – A few large ones, many small ones (Mannheim-Mendelsohn)
      – Came to rescue of French government in 1933.
    – Bring life and new ideas into a highly conservative market.

Scandinavia
- Long term credit for farmers – on model of French – selling bonds abroad, now mostly at home.
    - Seem to be quite sound, especially in Norway and Sweden.
- Denmark – a weak banking system. Two main banks, one failed in 1928 had to be restored with government aid. – A German type, mortgage-deposit bank. It was the deposit side which failed.
    - Denmark has a long history of failures of importance.
- Sweden: four main banks – one failed with Kruger, 1932, was reorganized by government.
    - Concentrated all its funds into the financing of one industry – the match trust – Skandenovisk Credit Aktienbologet.
        - Tried to combine deposit banking and financing.
    - The other big bank, Swenska Handelsbank, very conservative.
- Norway: three main banks – mixture of English and German ideas.

Italian Banking
- 1900, banking really began in Italy (aside from note issuing houses and private banks) by the Banco Commerciale di Milano – a branch of the Deutsche Bank.
- Two other Milanese banks, recently merged.
- Banco di Roma – of German type.
- Three main ones now, with many private bankers who are mainly brokers, not important.
- After War – separated from German influence, but continued along the same paths – what had been 60 years experience in Germany was compressed into 20 years.
    - Even more financing of industry than in Germany – and its industry was not highly competitive at home, and was not important on world markets, and it had no international capital market, for its securities.
    - In order to have an industrial revival or growth – the banks had to step in, and after 1925, Mussolini had them step in even more.
- As a consequence, the banks became, in the crisis, entirely illiquid.
    - Government stepped in – credit to industries, limits on withdrawals, etc.
    - Finally, 1933 the government had to step in to change and save the whole banking system.
        - As one industry after another failed – they were reorganized along Fascist lines into Corporations – cotton, silk, steel, shipbuilding – each having its own bank, guaranteed by the government.
    [In] January 1934

- The three big banks have to hand over to these government
  (corporation) banks <u>all</u> their claims against these industries – plus
  collateral.
- Getting in return a claim against these government banks – i.e. credit
  from the government.
  - They are now "liquid" again, and separate from industrial financing
    and operating as English type <u>commercial banking</u>.
- The whole of Italian industry is now owned and controlled by the
  government.
  - Bolsheviks want to do this, Italians don't want to – only difference.
  - Except for the few lines which have not failed – motor cars, etc.
  - Nationalistic government can't allow industries to fail and have
    unemployment.
- As to <u>minor</u> banking institutions, no one seems to want to write an
  article about them, even the Italian branch of Palyi's bank would not
  describe them.
- Government banks for industries – are to give new credits.
- In each case, it is a government bank connected with the Corporatzione
  to take care of the industries within that Corporatzione.
- New money went into shipbuilding, utility plants, and public works –
  i.e. one way of distributing government money.

<u>Italian Capital Market</u>
- Home investment in securities was always small.
- Depression and Fascism bring out an interesting slant on the Italian capital
  market.
- No wealthy middle class – small savers.
- <u>Postal Savings</u> Banks have existed for years [in margin: $4\frac{1}{2}\%$] – and other
  cooperative savings banks compete to get small savings which always put
  their money into <u>these</u> channels.
  - No political revolution can change this.
- All these savings, ultimately flow into the <u>Treasury</u> – not as revenue, but as
  credit – whence it is spent on <u>public works</u> – military.
  - Benefiting mainly the <u>Italian workers</u> – who put the money back into the
    Postal Savings – whence it is spent again – the Italian workman lives
    moderately and doesn't raise his standard of living by getting money – it
    has seemed to fall slightly.
  - A flow – but the practical effect is a government <u>deficit</u> – for the income
    from such public works is <u>nil</u>, while $4\frac{1}{2}\%$ is paid in interest on Postal
    Savings.

- Some inflation – yet a (controlled) stable exchange [rate].
- Highly successful, large international connections.
- Attacks speculative money – this element growing in Amsterdam.

The Eastern European Banking System and Money Market
- "Banker" is a word for business man.
  - In order to get more capital, he founds a bank, and uses his depositors' money.
  - Has direct control over industry – the entrepreneur is a banker.
  - Countries of small capital, with a small class of intelligent, enterprising entrepreneurs, growing population, government protection, inflow of foreign capital.
  - Hungarian cereal market is composed of the big banks, who also build elevators.
  - Bankers provide mortgage credit to small farmers and large "magnates," landowners.
    - They have so many grips on a product like wheat, it is practically in their hands.
  - If they do not fail, they acquire great popularity.
  - Having capital power, they get political power – all these countries having more or less corrupt governments, mostly more.
- Tremendous diversity of function;
- Absolutely speculative, unconservative.
  - Look at Western bankers as old ladies, not understanding real business.
  - Lose more money than Western bankers, but it is usually other people's money.
- Gambling, aggressive, entrepreneurial element – Jewish, Armenian, quick, unreliable.
  - The system and the intelligence are on a par – which causes which.
- Great developments of this kind of "banker's brain" – overpopulated now.
- Used to carry on large amount of arbitrage in securities and foreign exchange.
  - This is ended by the Depression, especially by central bank action and control over securities.
  - Requires great knowledge of many foreign markets, how they will go, and how to make money by ten minute deals.
  - Before the War, all the big arbitrageurs came from Vienna, Budapest, or Prague – where they learned the game in its most developed environment.
- Close connection with agriculture.
  - Most banks are in both industrial and agricultural fields.

- The larger proportion of either depending on the character of the country.
- The theory grew out of this practice that "the best security is land."
- Dictatorship by the central bank.
  - The power of the commercial banks to resist the central bank depends only on their political power.
    - Jewish private bank in Petrograd, refused to loan to government – was systematically and efficiently bankrupted in short time by the central bank – it was found to have 50% more liquid assets than liabilities.
  - Due to lack of open market.
  - Lack of that kind of confidence known in New England or England.
    - People put money in bank, but don't trust it – quasi-confidence, in a characteristic state of all Eastern system.
  - Given the kind of entrepreneurial spirit there is, the banks are often in trouble, usually illiquid.
  - The government can and does interfere for its ends.
  - The first thing a new country did after the War, was to create a central bank.

- Austro-Hungarian central bank was a part of the old Dual Monarchy.

Austria – 1931, breakdown of Kredit Anstalt started the world financial crisis.
  - Tremendous number of banks – after four big banks, four middle-sized banks, a great number of small banks.
    - Half of the small ones failed after the inflation, another large number failed after the stabilization of the Franc. Vienna had speculated largely against the franc.
    - Ziegbort – his bank was the personal trustee of the Hapsburg fortunes; it was the monarchy's bank, the bank of the big industries, especially those connected with the government (steel, munitions).
      - It, of course, failed after war, everything on which it was based had disappeared.
    - Kredit Anstalt – the Vienna Rothschilds, with large connections with international banking, and close connections with the Paris, London, Berlin, and Amsterdam Rothschild houses [In margin: "– failed in 1931"].
    - Wiener Bannuverein – owned by a Belgian and Swiss syndicate, after war.
    - Escompte Gesellschaft.
  - The major part of Austria's standstill debt is now made long term – got a very favorable agreement with its foreign creditors.

- But the affairs became worse.
- April, 1934, Escompte liquidated, Wienerbanuverein and Kredit Anstalt merged – only one big bank left.
- Government bank; municipal bank – socialist, a savings bank, the bank of the people of Vienna – more liquid than any other bank – goes on now as a Fascist bank, equally liquid and well managed.
- The entire banking structure of the Danubian, Balkan and Polish countries was closely connected with Vienna. But the nationalization of these banks has gone on rapidly, and nears completion.
- Postal Savings – "Savings" bank only a name.

Switzerland
- A few big, strong banks.
- Many medium sized banks.
- Loaned on short term to hotels, small industries.
    - Middle class control of country.
- The outgrowth of this was short-term bonds issued <u>by the banks</u> – promises to pay, in large sums, sold publicly, one or two years or one-half year or three-quarter year.
    - A unique system – made up to order.
- Germany copies this immediately (Professor Landeman) after inflation – sold $25,000,000 bank bonds in New York.
- R. F. C. [Reconstruction Finance Corporation] in U.S. – capital bonds sold to R. F. C. by banks in return for funds is slightly similar.
- Are issued by commercial banks – who go into the capital market and sell bonds of large amount – non-personal; banker doesn't know who owns it.
    - Freely negotiable, rate 1% lower than central bank rate.
- Explanation:
    (1) rapid growth of a banking system, but in a country were the instinct to save, to put money away in the tea-pot, is highly developed.
        - The Swiss, no matter how wealthy, lives on the lowest possible standard, <u>saves</u> his money – little corruption of old ideals.
        - In order to get hold of this kind of money, the banks used the short term bond.
        - Since depression, they are not so popular as the rate is low.
            - And not so important, because of the growth of savings.
    (2) The French idea of the danger of loaning on short term to industry permeated society.

- It was dangerous to loan outright, so they sold securities on short- long-term and loaned <u>this</u> money to industry on short-loan-terms (i.e. use of long term deposits for short term industrial credit).
- Consequence was a comparative short-term liquidity.
- Difference of Swiss from German and Italian banks came on gradually, they did not loan so much to industry, at least not to Swiss industry.
- But loaned a lot to Germany, especially just before the War.
  - As well as a few big, liquid banks, there were many medium-sized banks, that were highly speculative – very sensitive to depressions. At least one failed in every depression.
    - Then was usually reorganized with fresh capital, fresh assets, fresh speculation.
    - The German side of Switzerland is less speculative in this way than the French side – almost all the banks of this type in Geneva have now failed.
  - During War they built up a new banking system for taking care of foreign funds. This is now made use of again.
    - At least 2/3 of their 4 billion [Swiss frans] or so deposits are foreign funds (to say nothing of hoarding).
    - These made possible the saving of the banking system – from the losses (in speculation) during inflation, from large losses on foreign industrial credit in [the] depression.
    - Their function as safekeeper of Europe's capital made it possible for them to keep going, write off losses, become the most liquid banking system in Europe.
      - What do they do with the funds?
        - They pay no interest, but charge a fee for "expenses."
        - Put funds in their railroads (best system in world), in electric power plant (supply much of Europe), in electric charges, etc.
      - Switzerland is a refuge for wealthy people.
        - Low taxation (arbitrage in taxes, by moving about within 24 cantons) – due to political situation (capitalist, well-to-do, Liberal party rules, with Agrarian party).
          - Not interested in collecting high taxes, on inheritance or property.
    - Swiss <u>central banks</u> gave credit to German industry, mortgages to Germany, Hungary, etc.
      - Is now frozen, and in standstill.

- These were a danger spot, but were saved by inflow of funds, of which a large share has gone into Cantonal banks.
- Mortgages were always a problem in Swiss capital market but no mortgage banks were formed until introduced by Professor Landmann very recently, are not yet very prevalent [Julius Landmann, author of *The Swiss Banking Law*, Washington, DC: Government Printing Office, 1910].
  - The scandals connected with mortgages in a country with no organized market for them – has created a prejudice against them which prevents such a market from growing.
- National Bank
  - Highest gold reserve today – has always been high – originally to protect a highly speculative banking system.
- The country of greatest secrecy in its banking statements – never allowing government interference with private accounts.
- German reserves were leading during War and inflation, Austro-Hungary next.
- Now, probably, the French are leading.

Descriptions of Scientific workings of European capital and money markets
- Pre-war:
  - Jaffe
  - H. Withers [Hartley Withers, author of *The Meaning of Money*, 5th ed., New York: E. P. Dutton, 1930; and, with others, *The English Banking System*, Washington, DC: Government Printing Office, 1910.
- Post-war:
  - W. Leaf – only money market [Walter Leaf, chairman, Westminster Bank, Banking 1927]
  - Lavington – best for money and capital markets – theory overwhelming [Frederick Lavington, author of *The English Capital Market*, London: Methuen, 1921].
    - Many statistical surveys but no functional analysis.
- Western centers – gold countries – no descriptions.
  - e.g. Holland – a good study of colonial banks and a study of bank balance sheets.
  - France – never has been a good study – "Die Französischen Grossbanken" – pre-war.
- A market is an interrelated whole, functional.
  - Great need of study of money markets of foreign lands – a good doctoral analysis – getting at the financial heart of a foreign land.

Banking Policy
- – 1 – <u>Monetary Aspect</u> – Banking is a source of money supply, through notes,
deposits, velocity of circulation, demand for money, gold movements.
    - – Reserves, liquidity, instability or stability.
    - – Banks are a part of the monetary system – recognized as such by the end
of 18th century, but periodically forgotten, especially in depressions.
        - – A long 40 or so year move of literature away from and towards
monetary aspects of banking.
        - – <u>Currency</u> School – only note banks have a monetary aspect.
        - – <u>Banking</u> School – commercial banks have <u>as much</u> monetary influence
as note banks – but neither has any influence to speak of.
        (cf. Banques Inquiètes 1860–1864 – especially Thiers, Say).
        - – Under stable conditions, a tendency to disregard the monetary aspects
of banking.
    - – Up till now there has been great neglect of the monetary aspects of the
capital market – cf. David Hume.
        - – We study the differences, variations away from the usual, general
theories, assuming theoretical knowledge.

- – 2 – <u>Social Functions of Banks (Credit Institutions) in the Distribution of
Capital</u> → (available funds)
    - – between production and consumption
    - – between productions of different types
    - – between production for present and for future uses.
- – Tendency among theorists at Cambridge and Oxford is to challenge the
existence of this aspect unless as an offshoot of the monetary.
    - – It is sometimes true that the working of this function affects the money
supply.
        - – Which is to say that they influence each other – this is different.
        - – But they remain two separate aspects of the social organism.
    - – Price system, government, credit institutions.
- – Affect of credit institutions on the flow of saving – on the supply of savings
– on the psychology of the saver. On the flow of savings towards the banks
or towards other spots.

- – 3 – Control Aspect – <u>social control</u> aspect of credit structure.
    - – Partly connected with money, and partly with capital distributive aspect.
    - – Social-political aspects of credit.
    - – Not control of money, or control of capital distribution – these are <u>tools</u>, of
1st and 2nd aspect.

- Velocity of circulation – <u>bank notes</u> (<u>versus</u> gold and government money) – are now generally used all over Europe.
  - In Russia, Turkey, Balkan States – government money and central bank notes are the same.
  - Switzerland – has large gold circulation.
  - France – some gold circulation.
    - Until recently, French central bank converted only into gold bullion.
    - Recently coin conversion begun – because gold bars were being cut up and circulated.
  - Per capital holding of legal currency – 1919, highest for France – lowest for England.
  - Check circulation seems to rise with increased wealth of country.

<u>Giro</u> – and <u>check</u>.
  - 17th century banks, deposit banks for silver, owner could give an order to the bank having some silver transferred to the account of another depositor.
  - Giro of Italian origin.
  - Check of English origin.
  - Both meet in Germany.
    - Giro – depositor writes to bank, which transfers money, writes other customer – has no clearing system.
    - Check – sent to creditor who sends it to bank – two letters.
  - There has grown up a whole literature on the subject of the relative cheapness of the two systems.
    - We have, in effect, some giro transactions in U.S.
    - Check is a negotiable instrument.
    - Giro creates no such instrument – money kept in the bank.
  - What effect do these have on velocities of circulation?
    - Which system works most quickly?
    - In one place, in one bank – giro quickest, and cheapest.
    - Giro system presupposes two accounts [in the same bank], check system is based only on one.
      - Giro to that extent a high development.
    - Influence of taxes. Most European countries have, or are, taxing checks – no taxes on giro transactions.
  - <u>Commercial bills</u> – circulated in France, Belgium, Holland as a giro without a bank.
    - Are preferred by banks – likely to develop into circulating paper – in countries where small scale units of industry are common.

Volume of Circulation
  – Note issue reserve backing: –
    – English system – fixed except for gold inflow.
    German system – 30% backing.
    French system – no legal ratios. Parliament set limit – varied – anywhere.
  – But no legal restriction on deposit expansion.
    – England: discount charges easy, effective (money market quickly reactive
      to changes of rate, international market), change in rate brought changes
      in bills, easily.
      – Had lowest gold reserve – didn't need more.
    – France had highest reserve, no legal ratio.
    – i.e. legal arrangements had little effect on actual situation.
    [Pre-war]
    – Foreign exchange was pre-war not permitted as reserve for central banks
      anywhere but in Austria-Hungary (1902).
      – Belgium controlled the fluctuations of exchanges within the gold points
        by holding some foreign exchange.
    – Other liabilities than notes were not restricted by legal ratios.
    – Reichsbank and Bank of France took up bankers acceptances instead of
      commercial bills.
    – The commercial banks rather than the central banks were concerned with
      controlling bankers' acceptances and commercial bills.

Post-War –
  – British system remained the same, with a much higher fiduciary issue.
    – i.e. lower gold convertibility for total liabilities – 25–35%
    – No gold circulation in the country, and a change of rate would no longer
      draw in from the nation' s gold reserve.
    – Bank of England changed its discount policy – did not change it with
      small inflows or outflows of gold – tried to keep it low.
    – Also little hope of getting gold from abroad.
  – Reichsbank – reorganized 1924 ([Edwin Walter] Kemmerer).
    – Reserve ratio (35) – 40% of which (5) – 10% could be in foreign exchange.
    – No definition of what foreign exchange meant!
      – Could buy foreign exchange of a paper standard country!
      – No questions were asked as to this definition.
        – No literature.

  – Eastern countries got moving reserve ratios, raising each five year period.
    – Theory that they couldn't yet afford high ratio.
  – Germany – three signatures on commercial bills.

- Implication that third signature should be a bank.
  - Theory that central bank should become a bankers' bank (U.S. system Kemmerer's ideal).
  - Usual European central bank policy is to extend about one-half of credit to banks, rest direct to customers.
  - Restriction on credit to government (Germany about 100,000,000 RM's limit).
- Bank of France had complete freedom of choosing its bills-before War.
  - Is now prohibited from making <u>any</u> advance to government.
  - Which prevents effective open market operations to make policy effective.
  - Gold requirements proved to be of almost no importance.
    - In good times they were not reached.
    - In bad times they were overstepped
      (Reichsbank raised its rate to 20% for a week, kept it at 10% for two months, then gave up the fight).
  - In Britain, pre-war – internal and external gold flows – with Bank of England <u>policy of action</u> to protect its relatively small reserve, by frequent discount changes.
- Bank Act restricted notes to gold holding.
  - Currency School said there must be a discount rate to protect gold reserve.
  - Bank policy oriented by foreign exchange and internal movements.
    - Seasonal flux in exchanges.
    - Psychological flux in internal flow.
    - Last half of century marked by lack of theoretical approach to, and by lack of any discussion of, the workings of the Bank.
      - Gossen and Bagehot described, with admiration.
      - Policy was thought to be perfect, final.
    - Upturn of cycle, gold inflow, lowered rate.
    - Downturn of cycle, gold outflow, raised rate.
    - Gold flows were more numerous than cycle episodes.
      - Numerous changes of rate – sometimes 12 per year.
      - This had a disturbing effect on business cycles.
      - Also the whole business community became <u>financial</u> minded.
      - A high standard of financial intelligence and financial journalism.
        - Coffee merchant had to know his own market, but also had to watch the Bank discount rate.
      - But the journalism centered on the Bank of England and its rate
        - that was all.

- Also, the bank concentrated to so large an extent on the annual and seasonal trends, that it practically forgot the long run cyclical trend.
- And its short run view of gold movements led it to push upswings, and squelch downswings.
  - It was mainly a short-term capital movement policy.
    - Boom meant a boom in foreign securities.
    - Depression meant a drop in foreign securities.
    - No questioning by Parliament or by public or by writers (of policy).
- After War – Bank of England dealt almost entirely in government securities.
- 1931
  - Gold reserve   £120,000,000
  - Earning assets   £300,000,000
                           420,000,000
  - Government paper £400,000,000
    (Foreign and dominion, balances with foreign central banks)
  - Some frozen credit £20,000,000
    420,000,000
- No change of theory – liquidity the aim, but the primary factor in that liquidity was shiftability.
  - Government paper most shiftable.
  - Some question whether long or short run government paper was best.
    - Actually not much difference.
    - But a hangover [?] of policy made.
    - Bankers favor short term.
  - Disappearance of commercial paper.
  - Policy of cheap credits.
    - Low rate – easier forms, i.e. lack of bill, (which can lead to bankruptcy in five ways, and is very public)
    - The same debtor is not inclined to take a credit in bill shape, when he would take a credit in another form.
    - Treasury pressure – to get cheap credit for its huge floating debt.

Open Market Policy – carried on by Bank of England for half a century.
  - Its relation to market is least organized, least bureauocratic – no rules, laws, etc.

- Will not charge more than the announced (Thursday) rate but may, at its discretion, charge less.
- Men in charge of Departments of Bank of England are not as in other countries bureaucrats or lawyers or economists, but business men.
- However a tendency towards more stability of personnel is becoming more apparent.
- Has no rules as to what to buy or sell.
- Buys and sells government paper: (1) as a government agent; and (2) as a controller of the money market.
  - There is no actual or theoretical difference between discount rate and open market operations.
  [Single vertical line in margin alongside preceding two sentences]
  - No rules of eligibility.
  - Equalizaton fund – to be regarded as an <u>additional</u> open market operation.
    - Or a third department – with Banking and Note Departments.
    - No need of "Devisenpolitik" if gold standard works.
    - £175,000,000– then £200,000,000 more.
      - In Treasury bills – given to Bank of England – selling them to get money to buy foreign exchange.
      - Gold Report of League of Nations written very much under influence of Bank of England.
        - Central banks should interfere to make more quick and easy changes (short term) that would occur anyway.
          - But should not touch long term, more permanent changes.
          - How distinguish? – a classical theory lies behind this.
          - Bank of England seems to have applied this policy to foreign exchange.
          - Exhaustion of first 175 million Francs intimates that they have been stopping long run depreciation.
      - Bank of France – Say, Thiers – Banque Enquêtes, late [18]60s.
        - Does not <u>make</u> the bank rate but <u>states</u> its – registers it.
          - Two Banking Schools both opposing Currency School.
            (1) Notes and deposits the same – <u>both</u> should be controlled.
            (2) No difference between central bank and other banks, i.e. no policy at all.
              - Theory of Bank of France – classical
              - Central bank does not <u>influence</u> the market.
            - Bank of France before the War – overwhelmingly competitive – since War, pretty much so.

- Its Department heads – from lower ranks, bureaucratic, or university men (Degree of "Inspecteur de Finance" – after Ph.D.)
  - No business men came in, but bank men go out to business.
  - Lower salaries than commercial banks.
- A gradual change in theory is visible – started by Poincaré (replacing Morot by Moreau).
- Pre-war policy – <u>keep rate constant</u> – did it.
  - A few years with no change – few changes.
  - Claimed this was the State of France – no change of market rate, no change of bank rate.
  - As a matter of fact – the market rate <u>did</u> change, and the Bank of France did not follow the market – in fact, its stable rate kept the market stable rate [sic: market rate stable].
  - Bank of France founded by Napoleon 1804 (?) as a competitor to Bank of England.
    - Wrote on the door – "keep rate stable."
    - Aimed to keep rate low – to provide low charges to small borrowers. – It is not possible to change a low rate much.
  - Not a great deal of competition – any very high rate was suspected of being unsafe – psychology of high grade safety.
  - Internal gold flow, external, seasonal changes.
  - How manage these with stable rate?
    - <u>Extra</u> discount rate which <u>changed</u>.
    - Gold coming in, fairly steadily.
    - Surplus exchange balances tended to be transformed into gold, by Bank, banks, or commercial merchants.
    - During inflation – Frenchmen exported capital (– willing to die for his country, but not to let his money die) – tourist traffic; exports of goods and securities.
    - At end of inflation – a large sum of claims on foreign currencies, were sold to Bank of France which kept them as a reserve.
    [In margin alongside the preceding sentence: Large man saved his money, small man lost his. But there was thus no breakdown at the end of inflation – as in Germany and Belgium].

- In 1930 the Bank of France had 30– 40 billion Francs in gold exchange, as well as the same amount of gold at home. Treasury had more francs [in gold]. – The export of capital helped create the boom in England and the U.S.
- After 1931 – foreign trade and tourist traffic fell off, but the gold reserves held abroad came home – inflow of gold from liquidation of foreign exchange.
- Since October 1933 – until March 1934, they lost gold – no more for exchange reserves, exceedingly unfavorable trade balance (1933 – one billion Franc's foreign exchange) (1933 – 40 billion of gold was added to French balance).
- While gold flowed in, France expanded currency – internal system in bad shape.
- After 1933, October – a new British-like policy.
  - Rate goes up to $2\frac{1}{2}\%$ – considered by Bank as a proof of flexibility – says it will keep rate low.
  - Used 10% of exchange as reserve behind notes – reluctance of credit to expand.
- To get away from discount rate changes, France introduced the gold <u>premium</u> policy.
  - Many 5 Franc silver pieces in circulation. Bank gave these, unless one paid a <u>premium</u> for gold, which widened the gold points.
  - During inflation, 5 Franc pieces disappeared (melted).
  - Gold arbitrage was interfered with as much as possible.

The development of large scale monopolies, cartels, trusts, corporations, etc., means a lessened effect of discount rate.
  - Discount rate efficiency depends on the income structure.
    - Germany, little efficiency.
    - France, little industry, but what there is is highly industrialized – thus little efficiency of rate.
    - Same all over Eastern and Central Europe.
    - Mainly Britain is organized otherwise.

- Cooperation of central bank and other banks is essential for the operation of the discount rate.
    - A conflict would mean little effect of rate.
    - The slight rise of rate usually means little additional cost to banks.
  - Concentration of banks brings out cooperation of central and other banks.

- Big banks have to assume greater public responsibility, and can not stand strong criticism of central banks.
- In England – very curious cooperation.
    - Big Five have accounts at Bank of England but have no representation on Board of Directors of Bank of England – claim to have no formal or informal connections.
    - Yet more cooperation than in other countries, except Holland and Belgium.
- In Germany and Central Europe – banks are largely stock exchange institutions – less interested in money market control.
    - Usually have two groups on Board of Directors – the money market cooperators and the stock exchange-minded men.
    - Germany was working over the problem – how get control over the money market by rate.
    - Foreign exchange policy could be handled by central bank alone.
    - But what of trade cycle policy? – It became a matter of power – who had the political and legal right to control?
        - Reichsbank mobilized Parliament, the press, etc., to back up its claims to control.
        - After War – much greater cooperation – due mainly to personality of Schacht, who was close to banking leaders of money market, and followed their interests.
            - Luther (1930–1933) was not a banker, left things to bureaucracy, less cooperation.
            - Nazi bank is supreme.
        - The more one goes towards the East the more important become non-economic, political and personal factors.

Closed, quasi-gold standards: (a) without communication.
Open gold standard; and (b) without fixed par.
   – a and b are the two essentials of the Gold Standard!
In 1921 crash – Reichsbank acted on usual theory, the bigger the crash, the higher the rate – put its rate of up to 20%
   – As England put its up to 10% and drew in foreign capital.
   – But Germany had foreign exchange restrictions, and standstill arrangements.
   – It took two months to learn that these things violated the assumptions of the usual theory.

The business cycle is a vital part in the efficiency of the discount rate.

- Given a boom – can not the central bank stop it by action on the stock exchange, without affecting the rest of the market.
- Given a credit expansion, and market boom, can it be stopped by diverting credit from the market – a matter of allocation.
- i.e.: separate the credit market into parts and treat them separately.
  - Heretical from 19th century point of view.
- An attempt to eat the cake and have it too.
- To stop a boom without having any disturbing effects.
- It is possible to make some separation of credit markets.
- Especially in Eastern countries – where, if central bank refuses credit, it is impossible to get it elsewhere except at great cost.
- Two experiments, in Belgium and Germany – cf. National City Bank, monthly reports, 1929.
  - Germany – 1926, a boom, due to influence of American credit – unsound.
    - Commercial banks saw this as unsound, as dangerous; put pressure on the Reichsbank to stop it.
      - Also wanted to avoid signs of prosperity in hopes of a quick end to reparations.
    - The Bank felt (correctly) that it could get further by stopping new credit to <u>stock exchange</u>.
      - The boom period was considered all over Europe to be a depression period.
        - Mainly thought it could stop the stock exchange boom alone.
          - High rate – boom kept on.
          - Lowered rate – boom went higher.
          - Raising rate, may create <u>more</u> demand for money on money market.
            - Due to demand and flow of <u>international</u> money.
      - The discount policy failed – open market operations were impossible – by restriction on government bonds.
      - May 12th [year?] – circular sent to customers of all banks – asking that margin be raised from 25 to 50% on stocks – sent out by commercial banks at behest of other banks.
        - And rates raised.
        - This was <u>successful</u>. The market broke the next day – never to recover.
          - Thus the stock market boom was broken while a boom was going on in railroads and industry.
            - It might have been that the boom was not entirely broken, i.e. stocks remained at a boom level, even after break.

- People had too much respect for the stock market boom – which would probably have broken anyway.
    - The method was too rough, aimed at doing <u>something</u> – but they underestimated the sensitiveness of the market.
- Competitive concept – the older bankers did not like the competition of the younger banks.
    - The conservative bankers did not like the new speculative banking.
    - Schacht sided with older bankers – he was a personal enemy of Goldschmidt – the head of the Damstadter Bank.
        - But Goldschmidt was not even slightly phased by this move.
    - One result – the stock exchange lost its international position as a distributive center.
        - And lost its national position.
        - Banking and industry were thrown even closer together – more directly.
        - Stock exchange lost popular respect – which felt that it was <u>manipulated</u> – and, while they were willing to lose money in an open break, they could not stand the thought that the game was managed.
        - It is very unfortunate to have a crash just before a depression.
            - The Post-Inflation crash, 1924.
            - This 1927 crash.
            - The 1929 crash.
            [Brace combining preceding three lines, and alongside in margin: Too many for five years]
        - Strengthened belief in <u>bonds</u> – therefore the debt structure piled up more and more in bonds – which had gold clauses, were rigid.
            - The consequence was an even worse allocation of credit [Double vertical lines alongside in margin]
            - Proved that the market <u>was</u> separated into sectors, which would be attacked separately.
            - But failed in achieving a real reallocation.
            - Was fundamentally wrong, as it was based on an <u>incorrect</u> view of the underlying business cycle.
- Belgian Central Bank did the same thing shortly after.

- without commercial bank cooperation, without advertisement, without affecting margins.
- Was an open market operation – in foreign exchange!
  - Had no large portfolio of government bonds.
  - But had a lot of foreign exchange.
    - Sold that, contracted the whole credit market.
    - But the check worked mainly on the stock exchange boom – working only slightly on the rest of the market.
  - Belgian commercial banks turned away from stock exchange and began to invest in industry – for the higher interest rates there.
    - They had been earning high rates on broker's loans, there.
- Conclusion – it is difficult to generalize as to central bank policy – and it is hard to see the end of supposedly short-run policy.

Commercial banks have no money-market policy (U.S. banks have a little).
- No European banking system has any legal regulation as to reserve requirements – even legal limits for central banks are a post-war development.
- England
  - 20% was regarded as normal reserve pre-war.
  - 10% post-war – but this mostly window dressing for end of month – 6% the average during week.
- On Continent – Gold bloc countries have highest reserves.
  - Germany – pre-war 5–6% on monthly balance sheets.
  - Germany – post-war 3–4% on monthly balance sheets.
    - Both figures were lowered during month.
  - Same for Italy.

Macmillan Report unanimous on the point that window dressing should be abolished – i.e. reserve raised during month.
- Depression has raised cash reserves.
- Discount policy does not depend on cash reserves.
  - During War central banks put pressure on commercial banks to increase their reserves with central bank.
  - To stop their expansion (increasing reserve is deflationary as long as it increases).
  - France – Traditional holding of gold above reserves for arbitrage; in depression the commercial banks have taken to hoarding.
  - Have window dressing of inverse type in order to keep private the large amount of gold holdings.
    - Fearing a loss of prestige if so large a proportion of non-earning deposits were discovered.

– 35% reserves, but reduced to 12%.
– After runs, the French and English reserves were highest.
– After runs, the German and Italian reserves were lowest.
[Brace in margin encompassing preceding two lines]
    – The former met runs by liquidation, the latter by rediscounting with central bank.
    – The former lost 25–30% of debtors.
        – 2 billion Francs liquidation, at least.
        – Debtors were put under pressure, and repaid their debts.
        – Selling out of debts, without new credit extended (a deflation of business).
– French business firms hold a larger cash reserve than English or American firms.
– In Europe everyone watches the central bank reserves, no one watches the commercial bank reserves.
– In U.S. – we talk of expansion possibilities of commercial banks.
    – Banker wants to make money – as much as he can – expansion based on reserves.
– In Europe – banks answer the demands for credit. Expansion creates reserves – due to low reserves.
– In U.S. – there never has been an expansion of credit to the limits of reserves.
– There is not much difference – no better or worse.
– Low reserve banks were sometimes more sound, sometimes less sound than U.S. banks.
– Western European view – cash reserves limit expansion.
– Eastern European view – secondary reserves are what count in liquidity.
[Double vertical lines alongside preceding two sentences].
[In margin at top of page:
    Frankfurter Zeitung-leading financial paper in Germany
      – first financial paper in Germany – 1870]
– A certain minimum ratio of liquid to total assets should be maintained, 60%
– Grew out of empirical observation, became a principle.
– But banks began to question the origin of the principle – why not 50% or 40%.
– As a matter of fact, the 60% turned into a maximum instead of the minimum.
– The theory was, in general, that mortgages were the best credit.
Liquid assets [In margin alongside: Liquid by banker's theory]

– International short term paper and documents (best)
– Regular (internal) commercial paper (worth as much as the banker's opinion of it)
– Treasury bills
  – Germany's pre-war government financing was very primitive – not realizing that the market could be manipulated and that short term borrowing was profitable.
– Finance bills – means of stabilizing seasonal fluctuations – in England, Switzerland, France.
– Banker's acceptances
  – German types – industrialist draws on banker, or banker draws on banker – with a trust or Interessengemeinschaft, it represents two names that are the same, does not represent a commercial transaction.
  – Reimbursement credit – documents attached or awaited, but no one asked what documents they were.
    – Major part of Germany's credit from abroad came in this way.
    – Ultimately, why ask for a bill at all – why not just give the credit?
      – These became a sort of fixed total of credits – bills discounted at London, Paris, etc. – arranged by telephone.
    – Became – quasi-reimbursement – in banker's parlance, not taken up by public.
    – In post-war period the distinction broke down – no scrutiny of bills drawn – commercial accounts – no matter how bad – were turned into commercial bills – for window dressing and profit.
    – These bills could not be discounted at the Reichsbank.
      – Yet the Reichsbank was responsible for the whole business – it gave a credit line on the name of the firm – liberally – and didn't ask about the assets which were not discounted in order to get that credit.
– i.e. a complete change in the liquidity concept – post-war.
  – Germany, 20%, 1929. would be a liberal estimate of liquidity.
  – Less in Italy.
  – Central Europe – the standards of liquidity fell into disrepute.

Banking Policy

Capital allocation is long term and implies a long run monetary policy.
– Pre-war literature analyzed money from the banker's point of view.

– No banks analyzed capital allocation market.

– As a problem <u>outside</u> the banking system.

– Embargoes (external, internal) subsidies, taxation exceptions (<u>Kapitalpolitik</u>-Goldschmidt).

[In margin at top of page: Bresciani – The German Inflation]

  – Capital in form of circulating medium, may have higher velocity of circulation in deposits than in cash.

  – Volume of money in a country with banks is much greater than in a country without banks.

  – A country with <u>rising</u> monetary circulation.

  – Effect on allocation of capital?

    – <u>Forced savings</u> ([D. H.] Robertson, 1924).

      – Have same effect as voluntary saving.

    – Post-war literature does not consider the long run, or international exchange position – but the short run approach.

    – Regulation by balance of payment is either omitted, or considered (or as Jewish).

  – Old theory, commercial credit was self-liquidating.

    – But <u>is</u> it self-liquidating?

    – Bill broker in England.

    – In other countries – European – and U.S. no bill brokers – bureaucratic standards develop.

    – Does commercial credit <u>add</u> to the <u>total of credit</u>, or add to the <u>velocity of circulation</u>.

      – Little influence on prices.

      – Large influence on capital allocation.

        – Velocity rise easily corrected.

        – Total of credit rise corrected only with some strains.

    – If commercial bill form is stuck to and individual scrutiny of their purpose kept to – then the addition to credit may be avoided.

    – And if the <u>temporary</u> addition to capital is represented by a <u>physical, valuable</u> good.

    – Old theory said the value of goods <u>decides</u> the volume of credit – which makes no addition to the capital supply – volume of goods creates a demand.

    – But the psychological knowledge that one <u>might get</u> credit may easily influence the volume of goods.

    – There is a difference between bank financing of marketable goods – commercial credit, and bank financing of industrial expansion.

      – Different assets in bonds of bank.

– The former gives <u>little</u> chance of <u>growth</u> of industrial expansion.
– The latter gives much chance of growth of industrial expansion.
– Banking purists demand restriction of granting of commercial credit to <u>raw materials</u>.
  – In order to prohibit the giving of credit five times over to the <u>various stages</u> of production of a physical good.
  – It would be just the same to say that <u>only finished</u> goods should get credit.
– <u>But</u> the short term character of the credit, and the length of time required for an industrial process combine to make it less serious:
  – this granting of several credits to the same physical good.
"Brokers loans" and the supply of credit.
  – <u>Some</u> capital <u>is</u> tied up in stock exchange – even continental writers could not deny this.
  – Impossibility of accurately estimating the amount of balance
    – incoming or outgoing – which will exist at the end of a day.
    – Thus necessitating reserves for reserves if the bank is sound.
    – Although weekly averages make the problem easier.
– The total value of brokers loans does not make a large proportion of credit in the stock exchange – except in the U.S.
  – But even if the total credit in stock exchange is <u>not</u> a large amount of total supply of credit – it has a tremendous <u>influence</u> on the total allocation of credit.
    – If a crash comes – the banks may not lose their brokers loans, but they lose because the stocks they own fall in value, and because the industries to which they have extended credit are weakened, their credit imperiled.
    – The social importance of broker's loans is not great, by itself, only when taken in conjunction with the other loans by banks.
    – Other short term bank paper is invested in other banks – postponing the problem.
Long term paper of banks.
  – The better developed the capital market (the less "long term") – i.e. the less difference between long and short term paper.
  – What does long term paper holding mean?
    – cf. [Harold] <u>Moulton</u> 1928 and his criticism of the liquidity concept.
      – Government paper alright for deposit backing.
        – Even admitting that a modern government can not lose credit (not true).

– Or that its paper is the <u>last</u> to go (wasn't true in Germany).
  – Even then major losses on government paper are possible.
    – Lloyds Bank's lost as much as its capital between 1900 and 1914.
      – It had put its money in Consuls.
      – Which were raising [sic: rising] in interest rate, losing capital value.
    – No possibility of a margin on government bonds – except appreciation against depreciation.
    – In depression a bank loses on land assets and wins on government assets.
    – But in boom it loses in government assets, wins on industrial assets.
    – Even then [arrow from "depression" two lines up], government may force a reduction of interest rate.
      – But the real question is – to what extent does the banking system loan to industry?
        – With the exception of Eastern Europe where the backs are [incomplete sentence].
        Finis